Agroforestry: Practices and Management

Agroforestry: Practices and Management

Editor: Malcolm Fisher

www.callistoreference.com

Callisto Reference,
118-35 Queens Blvd., Suite 400,
Forest Hills, NY 11375, USA

Visit us on the World Wide Web at:
www.callistoreference.com

ISBN: 978-1-63239-795-9 (Hardback)

The publisher's policy is to use permanent paper from mills that operate a sustainable forestry policy. Furthermore, the publisher ensures that the text paper and cover boards used have met acceptable environmental accreditation standards.

Trademark Notice: Registered trademark of products or corporate names are used only for explanation and identification without intent to infringe.

Printed in the United States of America.

Cataloging-in-publication Data

Agroforestry : practices and management / edited by Malcolm Fisher.
    p. cm.
Includes bibliographical references and index.
ISBN 978-1-63239-795-9
1. Agroforestry. 2. Agriculture. 3. Forests and forestry. I. Fisher, Malcolm.
S494.5.A45 A37 2017
634.99--dc23

Table of Contents

Preface

Agroforestry is a step towards sustainable development. It aims at optimizing productivity and profitability by unifying agricultural shrubs or trees with forestry techniques. This book provides comprehensive insights into the field of agroforestry and outlines its processes and applications in detail. While understanding the long-term perspectives of the topics, the book makes an effort in highlighting their impact as a modern tool for the growth of the discipline. Those in search of information to further their knowledge will be greatly assisted by this book. This book aims to equip students and experts with the advanced topics and upcoming concepts in the area of agroforestry.

The main aim of this book is to educate learners and enhance their research focus by presenting diverse topics covering this vast field. This is an advanced book which compiles significant studies by distinguished experts. This book addresses successive solutions to the challenges arising in the area of application, along with it; the book provides scope for future developments.

It was a great honour to edit this book, though there were challenges, as it involved a lot of communication and networking between me and the editorial team. However, the end result was this all-inclusive book covering diverse themes in the field.

Finally, it is important to acknowledge the efforts of the contributors for their excellent chapters, through which a wide variety of issues have been addressed. I would also like to thank my colleagues for their valuable feedback during the making of this book.

Editor

Dynamics of Two Multi-Stemmed Understory Shrubs in Two Temperate Forests

Xuejiao Bai[1,2], Tania Brenes-Arguedas[3], Ji Ye[2], Xugao Wang[2], Fei Lin[2,4], Zuoqiang Yuan[2], Shuai Shi[2,4], Dingliang Xing[2,4], Zhanqing Hao[2]*

1 College of Forestry, Shenyang Agricultural University, Shenyang, Liaoning Province, China, 2 State Key Laboratory of Forest and Soil Ecology, Institute of Applied Ecology, Chinese Academy of Sciences, Shenyang, Liaoning Province, China, 3 Centro Cientifico Tropical, San Jose, Costa Rica, 4 University of Chinese Academy of Sciences, Beijing, China

Abstract

A multi-stemmed growth form may be an important trait enabling the persistence of individual shrubs in the forest understory. With the aim of evaluating the role of multiple stems, neighbor competition and soil nutrients in shrub performance, we study the dynamics of two temperate multi-stemmed shrub species. We modeled stem growth and survival of *Corylus mandshurica* and *Acer barbinerve* in two temperate forests with differing structure in northeastern China. One forest was an old growth broad-leaved Korean pine (*Pinus koraiensis*) mixed forest; the other was a secondary poplar-birch forest. Growth of the two species and survival of *C. mandshurica* increased with stem number in the old growth forest, but not the secondary forest, suggesting the benefits of a multi-stemmed growth form are facultative. *C. mandshurica* also suffered more from overstory neighbor competition in the old growth forest, which may suggest that this species is less shade-tolerant than *A. barbinerve*. Moreover, the performance of the two species were clearly influenced by understory neighbors and soil variables in the old growth forest relative to the secondary forest, which may be due to different forest structure. We conclude that multiple stems are not always important for the persistence of shrub species. Even within the same species, the multi-stemmed benefits might be facultative, differing among forests and neighborhood compositions.

Editor: Ben Bond-Lamberty, DOE Pacific Northwest National Laboratory, United States of America

Funding: This work is supported by the National Natural Science Foundation of China (No. 31300378 and No. 31300517). The funders had no role in study design, data collection and analysis, decision to publish, or preparation of the manuscript.

Competing Interests: The authors have declared that no competing interests exist.

* E-mail: hzq@iae.ac.cn

Introduction

While forest studies often focus on tree dynamics, shrubs are also ubiquitous components of the forest community, and their dynamics differ from dominant trees in important ways. Shrubs influence ecosystem properties by maintaining plant diversity, influencing tree seedling regeneration [1,2], and affecting soil nutrient flow [3–5]. However, the demographic traits of understory shrub species are fundamentally different from overstory tall tree species. The dynamics of understory shrubs are mostly driven by vegetative growth and mechanical damage to stems [6]. Also, unlike overstory tall trees, shrubs can spend all their life in the forest understory.

Many shrub species produce multiple stems through vegetative reproduction. The replacement of stems within a shrub has been thought of as an adaptation enabling long persistence of individual shrubs in their environments, as it may allow for increased acquisition of resources in heterogeneous environments [7]. In multi-stemmed shrubs, each stem may experience recruitment, growth, and death semi-independently. Thus, such shrubs usually consist of stems of different ages and sizes [8]. However, interactions among stems in a multi-stemmed plant may be complex. For example, competition among stems within a shrub might restrict height growth relative to single-stemmed trees [9], while more stems can increase overall plant survival compared with individuals of the same species with fewer stems [10].

To understand the dynamics of multi-stemmed understory shrubs, it is important to account for the biotic environment in the understory. Understory shrubs are usually limited by tree competition. Denser neighborhoods may reduce shrub performance due to resource competition. One study that investigated shrub dynamics as a function of the understory neighbor structure suggested that the neighboring understory trees negatively affected the absolute diameter growth and survival of large ramets [11]. Also, large individuals can suppress the performance of smaller individuals [12], as canopy trees usually dominate forest light and root environments [13]. Some researchers have investigated the influence of overstory characteristics on diversity and abundance of shrub species [14,15]. Soil conditions may also affect the performance of understory shrub species. For example, the reproductive modes of an understory shrub was related to soil environment, and the multi-stemmed growth form was preferred in places with abundant litter substrate [16]. Similarly, the dominance and frequency patterns of multi-stemmed plants were driven by soil nutrient status [17,18]. In addition, size strongly affected the absolute diameter growth and survival of large ramets for shrub species [11]. Therefore, soil variables and stem size should be also considered to examine the dynamics of multi-stemmed understory shrubs.

During forest development, floristic composition, diversity, richness, and density of plant communities may change [19]. Hence, the existence or strength of biotic processes, such as neighborhood interactions, can vary at different successional stages, particularly if large changes occur in the vertical structure of plant communities through time [15]. Moreover, soil nutrients are more available in old growth forests owing to less efficiency in nutrient retention than secondary stage in some cases [20]. To our knowledge, no ecological research has been conducted in dynamics of understory shrubs in old growth compared with secondary forests.

The goal of this study is to evaluate the role of stem size, multiple stems, neighbor competition and soil conditions in stem demographic variables of two shrub species in two temperate forests at different successional stages. We investigated the effects of diameter at breast height (dbh), the number of stems in a shrub, overstory and understory neighbors and soil variables. We specifically ask the following questions: (1) How does the number of stems in a shrub affect stem growth and survival of these two shrub species? (2) How do basal area of overstory and understory neighbors and soil variables affect stem growth and survival of these two species? And (3) do the effects of multiple stems, neighbors and soil variables differ between the two study forests? We predict that competition from local neighbors is going to be stronger, and have a more negative impact on growth and survival of these two species in the old growth forest, where there are more larger and older trees [21].

Materials and Methods

Ethics statement

The study forests represent neither privately-owned field, nor endangered or protected species. No specific permits were required for the described field studies.

Study site

We studied two forests in northeastern China, both located in Changbai Mountain Nature Reserve (42°23′N, 128°05′E), which is one of the largest biosphere reserves in China and has been spared from logging and other severe human disturbances. The reserve was established in 1960 and joined the World Biosphere Reserve Network under the Man and the Biosphere Project in 1980 [22,23]. The reserve is about 200,000 ha with elevation ranging from 740 m to 2,691 m [24].

One of our study sites is in broad-leaved Korean pine (*Pinus koraiensis*) mixed temperate forest, which is the dominant vegetation type in northeastern China and well known for high species richness and particular species composition [23,25]. The other is a secondary poplar-birch forest, which is one of the main secondary forest types on Changbai Mountain, resulting from natural regeneration after clear-cutting or fire. The poplar-birch forest is an important stage in the secondary succession of the broad-leaved Korean pine forest [26]. For brevity, we will hereafter refer to these two forests as the 'old growth forest' and the 'secondary forest' respectively. The climate of the study region is characterized by low temperature and high precipitation [24]. Mean annual temperature is 3.3°C (−16.5°C in January and 20.5°C in August). Mean annual precipitation is 672 mm, most of which occurs between June and September (480–500 mm) [24].

Plot censuses

Both these sites have large forest dynamic plots where all individuals with diameter at breast height ≥1 cm have been measured, mapped, tagged, and identified to species. Trees and shrubs are re-measured every five years. In the old forest, the forest dynamics plot is 25 hectare (ha) (500×500 m). It was established in 2004 and recensused in 2009. The elevation in the plot ranges from 791.8 m to 809.5 m. In the 2004 census, there were 59,138 living stems, belonging to 52 species, 34 genera, and 18 families. Mean stand density of living stems was 2,366 stems·ha^{-1} and mean basal area of living trees was 43.75 m^2·ha^{-1}.

In the secondary forest, the forest dynamic plot is 5 ha (250×200 m). It was established in 2005 and recensused in 2009. The elevation ranges from 796.3 m to 800.4 m. In the 2005 census, there were 20,107 living stems, belonging to 50 species, 32 genera, and 18 families. Mean stand density of living stems was 4,021 stems·ha^{-1} and mean basal area of living trees was 28.79 m^2·ha^{-1}. Hence, relative to the old growth forest, in this plot there were much more stems, but these were considerably smaller, suggesting lower canopy height and a denser understory.

Study species

Here we focus on the stem growth and survival dynamics of two shrub species that are most abundant in the understory of both forest types. The first is *Corylus mandshurica* (English common name Manchurian Filbert), an androgynous shrub with edible fruits that are dispersed by gravity. The second species is *Acer barbinerve* (common name Barbed Vein Maple), a dioecious species with wind dispersed seeds. The distributions of these two shrub species in the two forest plot are shown in Figure 1. A large proportion of individuals of these two species have multiple stems and form clumps [27]. Here, we evaluated the dynamics of stems in these clumps.

Statistical analysis

To examine the explanatory variables affecting the dynamics of shrub stems, we used generalized linear mixed models (GLMMs), with gamma errors and log-link function for the growth rate, and with binomial errors and logit-link function for the survival probability. The variables in the models are described below.

Demographic rates. Growth of each shrub stem was calculated as basal area increment (cm^2 yr^{-1}):

$$g = \frac{BA_{t2} - BA_{t1}}{t}$$

where BA_{ti} is the basal area of shrub stems in census i, and t is the inter-census interval. We excluded extremes from the calculation of all shrub stems decreasing in size by >5% yr^{-1} [28], and altered remaining negative growth rates to be zero. To log transform growth, we added 0.01.

Biotic variables. For each shrub, we calculated the following variables: (1) the number of stems in the shrub, (2) basal area of overstory neighbors (dbh≥10 cm) between 5–20 m radii, (3) basal area of understory neighbors (dbh<10 cm) between 5–20 m radii. To eliminate boundary effects, only those shrubs with a distance greater than or equal to 20 m from the forest plot edges were included.

Soil variables. We sampled soils within the two forest plots using a regular grid of points every 30 m. At each grid point, two additional sample points at 2, 5 or 15 m were selected in a random compass direction from the grid point. In total, 967 and 215 soil samples were sampled in the old growth and secondary forest plots respectively. Soil properties (organic matter, total nitrogen, total phosphorus and total potassium) showed significant differences among the two forests (organic matter: t=8.703, P<2.2e-16; total nitrogen: t=−4.551, P=7.331e-06; total phosphorus: t=4.725, P=3.28e-06; total potassium: t=16.254, P<2.2e-16). There were

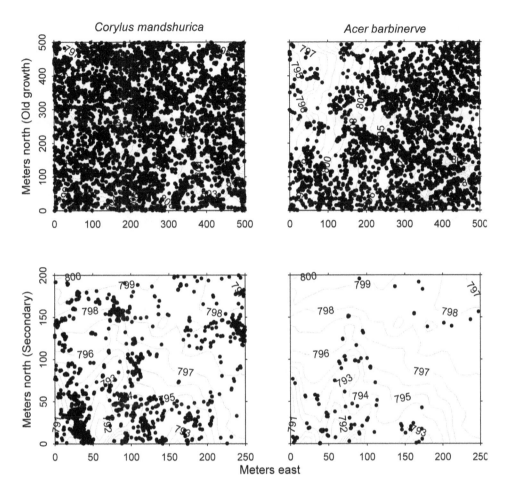

Figure 1. The distribution of *C. mandshurica* **and** *A. barbinerve* **in the old growth and secondary forest plots.** The solid points represent the individuals of the two species. The contour lines refer to the topography of the two forest plots.

higher content of organic matter, total phosphorus and total potassium, but lower content of total nitrogen in the old growth forest (Mean±S.E.: 165.284 ± 2.646 g·kg^{-1} vs. $129.324\pm$ 3.173 g·kg^{-1}; 1.255 ± 0.015 g·kg^{-1} vs. 1.111 ± 0.027 g·kg^{-1}; 16.472 ± 0.086 g·kg^{-1} vs. 13.661 ± 0.150 g·kg^{-1}; $6.370\pm$ 0.087 g·kg^{-1} vs. 7.190 ± 0.158 g·kg^{-1}). Soil organic matter did not show spatial autocorrelation in the secondary forest. We interpolated total nitrogen, total phosphorus and total potassium of the two forest plots to 5×5 m grid by kriging. Then, we converted the calculated values for each grid to z scores [29]. To reduce possibility of overfitting the models [29], we performed a principal components analysis (PCA) on these three soil variables. The first two principle components accounted for 98.8% of the total variance in the three soil variables in the old growth forest and 91.8% in the secondary forest (Table S1 in File S1). In both forests, the first component (PC1) explained variation in total nitrogen and phosphorous, while the second (PC2) represented variation in total potassium.

In all models we included dbh (log-transformed), biotic variables and soil variables (Table 1) as explanatory variables. The values of the explanatory variables were standardized by subtracting the mean value of the variable and dividing by the standard deviation. Because shrub stems from the same individual were not independent replicates of stem number, we included individual as a random effect.

Here we present only results for understory neighbors at 5 m radius and overstory neighbors at 20 m radius because GLMMs under these scales were the most likely models based on AIC and BIC for most cases when testing separately (Table S2 and S3 in File S1). We evaluated spatial autocorrelation in the residuals of the GLMMs (Figure S1 in File S1), and variograms revealed that the models adequately accounted for spatial autocorrelation.

All analyses were carried out in the statistical environment R (version 2.14.2), using the packages 'geoR' [30], 'languageR' [31], 'lme4' [32] and 'sqldf' [33].

Results

Effects of dbh and stem number in a shrub on stem growth and survival

Stem dbh had a strong negative influence on the survival of *C. mandshurica* in the two forests (Figure 2 and 3), such that larger stems had significantly higher mortality in the two forests. For example, stems of *C. mandshurica* of 1 cm dbh had probability of surviving of 0.736 ± 0.002 in the old growth forest and 0.800 ± 0.009 in the secondary forest. This probability declined to 0.696 ± 0.014 and 0.522 ± 0.064 for 3 cm dbh respectively. The effect of stem dbh was small but positive on the growth of *C. mandshurica* in the old growth forest. Stems of 1 cm dbh grew on average 0.200 ± 0.005 cm^2·yr^{-1} in the old growth forest, and this increased to 0.256 ± 0.051 cm^2·yr^{-1} for stems of 3 cm dbh.

Table 1. Parameters included in models.

Variables	Data		
	Range	Mean	Median
Old growth forest			
Corylus mandshurica **(13,498)**			
DBH (cm)	1–10.9	1.580	1.5
Number of stems in a shrub	1–18	3.318	3
Basal area (m²) of overstory neighbors within 20 m	2.217–10.447	5.340	5.284
Basal area (m²) of understory neighbors within 5 m	0–0.044	0.010	0.009
Soil PC1	−2.024–3.279	−0.008	−0.497
Soil PC2	−1.015–1.377	0.140	0.119
Acer barbinerve **(10,216)**			
DBH (cm)	1–13.2	1.931	1.7
Number of stems in a shrub	1–21	5.394	5
Basal area (m²) of overstory neighbors within 20 m	3.066–9.592	5.523	5.470
Basal area (m²) of understory neighbors within 5 m	0–0.042	0.011	0.010
Soil PC1	−2.020–3.299	−0.633	−1.066
Soil PC2	−0.946–1.391	0.034	0.013
Secondary forest			
Corylus mandshurica **(1,269)**			
DBH (cm)	1–4.9	1.551	1.5
Number of stems in a shrub	1–13	2.940	2
Basal area (m²) of overstory neighbors within 20 m	1.705–4.249	3.098	3.180
Basal area (m²) of understory neighbors within 5 m	0.000–0.073	0.033	0.032
Soil PC1	−2.576–3.525	0.579	0.538
Soil PC2	−3.042–2.836	0.133	0.060
Acer barbinerve **(227)**			
DBH (cm)	1–3.7	1.640	1.5
Number of stems in a shrub	1–14	6.366	6
Basal area (m²) of overstory neighbors within 20 m	2.298–4.140	3.247	3.247
Basal area (m²) of understory neighbors within 5 m	0.012–0.062	0.031	0.029
Soil PC1	−1.991–3.340	0.781	1.068
Soil PC2	−1.595–2.430	0.400	0.208

Similarly, larger stems of *A. barbinerve* had faster growth in the two forests (for example, growth increased from 0.204 ± 0.011 and 0.240 ± 0.036 cm²·yr⁻¹ at 1 cm dbh to 0.421 ± 0.035 and 1.099 (only one stem) cm²·yr⁻¹ at 3 cm dbh) (Figure 2 and 3). While in the secondary forest larger stems had a higher survival (increased from 0.821 ± 0.050 to 0.939 ± 0.000, at 1 and 3 cm dbh, respectively) (Figure 3).

The number of stems in an individual positively influenced growth and survival of *C. mandshurica* in the old growth forest (Growth: Effect$\pm$S.E. $= 0.070\pm0.011$, $P = 5.45$e-10; Survival: Effect$\pm$S.E.$= 0.069\pm0.026$, $P= 0.008$), but did not influence in the secondary forest (Figure 2 and 3). For example, average growth of stems of *C. mandshurica* in the old growth forest increased from 0.180 ± 0.003 cm²·yr⁻¹ in shrubs with only one stem, to 0.209 ± 0.003 cm²·yr⁻¹ in shrubs with two or more stems. Similarly, probability of survival of *C. mandshurica* in the old growth forest increased a very small but significant amount, going from 0.711 ± 0.001 in shrubs with only one stem to 0.724 ± 0.001

in shrubs with two or more stems. Moreover, in the old growth forest, there was a significant interaction between stem number and basal area of overstory neighbors in the survival of *C. mandshurica* (Effect$\pm$S.E. $= 0.066\pm0.027$, $P=0.014$; not shown). This interaction suggests that the positive effect of multiple stems on survival was stronger when there were larger trees in the vicinity. The number of stems in a shrub also positively influenced the growth of *A. barbinerve* in the old growth forest (Effect$\pm$S.E. $= 0.051\pm0.014$, $P=0.0003$), but not in the secondary forest, nor did it influence the survival of *A. barbinerve* in either of the two forest plots.

Effects of neighborhood variables and soil variables in old growth and secondary forests

Consistent with our expectations, the effects of neighborhood variables and soil variables in the old growth forest were more apparent. For instance, in the old growth forest, basal area of overstory trees showed significant negative effects on growth and

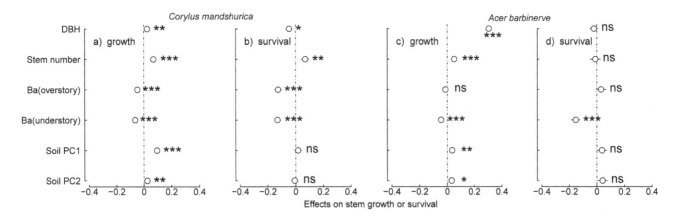

Figure 2. Effects of dbh, stem number in a shrub, neighborhood and soil variables on stem growth or survival of *C. mandshurica* and *A. barbinerve* in the old growth forest plot. *Circles* show the coefficient estimate for each parameter, with 2S.E. indicated by *horizontal lines*. "Ba" represents basal area. Asterisks represent the probability that the estimates are not different from zero: *$P<0.05$, **$P<0.01$, ***$P<0.001$, ns not significant ($P>0.05$).

survival of *C. mandshurica*, such that both decreased when there were more, larger neighbors (Figure 2). This effect of overstory neighbors was not seen for *A. barbinerve* (Figure 2). However, basal area of understory neighbors had significant negative effects for both species (Figure 2). In contrast, in the secondary forest neighborhood variables had little effect on growth or survival of both species, with the exception of a negative effect of basal area of overstory trees for *C. mandshurica* (Figure 3). Growth of the two species in the old growth forest was influenced by soil variables (PC1 and PC2, Figure 2), suggesting that better soil nutrients lead to better growth. In contrast, in the secondary forest only soil PC2 had a positive effect on stem growth of *C. mandshurica* (Figure 3).

Discussion

Our study showed differences in shrub dynamics between the two species and among the two forest sites.

We found that stems of *C. mandshurica* generally had decreased survival as they grew bigger. As older, larger stems probably have better resistance to physical damage and better access to light, decreasing survival can only be explained by stem senescence. We interpret this to mean that established shrubs of this species might shift their investment from surviving long lived stems to producing

new stems. This same behavior was not found for *A. barvinerve*. On the contrary, larger stems of *A. barvinerve* in the secondary forest showed higher, not lower survival. While we cannot explain this with certainty, one possibility is that this population in the secondary forest is much younger. Indeed this species was much less abundant in this plot (45.4 stems·ha^{-1} vs. 408.64 stems·ha^{-1} in the old forest; Table 1), and stems were much smaller than in the old forest (max dbh of 3.7 cm vs. 13.2 cm in the old forest; Table 1). Hence, it is possible that most of the stems we captured were still growing and not senescing. *C. mandshurica* also had lower population density and smaller stems in the secondary forest, but this was not as marked as for *A. barvinerve* (Table 1).

Larger stems had higher basal area growth for *A. barvinerve* in the two forests, and also for *C. mandshurica* in the old growth forest. Similar results were found in recent studies of tree growth [34,35], suggesting that larger stems had more biomass production than smaller ones. This result does not contradict the survival trends. As noted by previous research [12], size affects resource acquisition, as large individuals can obtain a disproportionately larger amount of the resources. Because stems require less diameter growth to produce the same biomass over an increasingly large area, basal area growth may be constant or increased, while diameter growth declines with stem size.

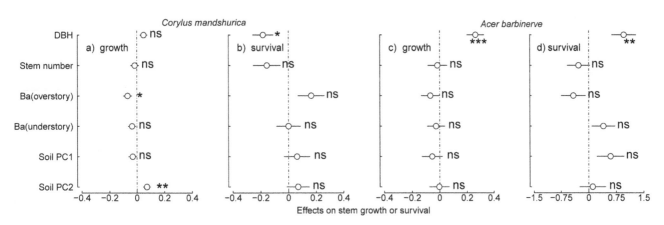

Figure 3. Effects of dbh, stem number in a shrub, neighborhood and soil variables on stem growth or survival of *C. mandshurica* and *A. barbinerve* in the secondary forest plot. *Circles* show coefficient estimate for each parameter, with 2S.E. indicated by *horizontal lines*. See Figure legend 2 for Ba, asterisks and ns.

Multi-stemmed growth form was advantageous in some cases. The presence of multiple stems increased the growth of both species and survival of *C. mandshurica* in the old growth forest. Although the stems within multi-stemmed shrubs have more competition, shrubs with more stems have better resource acquisition, which may benefit stem performance. In multi-stemmed plants, the storage of carbohydrates is generally shared among stems [36], and this may be an effective strategy to buffer plants against stresses and improve survival [37] or growth. However, the advantage of multiple stems was not found in the secondary forest. While this study was unreplicated for stand age, and we can draw no firm conclusions. We hypothesize that there might be stronger competition from local neighbors in the old growth relative to the secondary forest. This conclusion is supported by the interaction between stem number and basal area of overstory neighbors in the survival of *C. mandshurica*. This interaction suggests that the effect of multiple stems is more important in areas of the forest with larger trees, that are probably also darker and more resource-limited. Also, random small-scale gap formation is more common in the old growth forest, and more patchily distributed light levels may be more favorable to multi-stemmed individuals [38].

There were considerable differences between the two species in the effects of overstory neighborhood basal area in the old growth forest. In fact, denser overstory neighborhoods significantly reduced stem growth and survival of *C. mandshurica* in the old growth forest but had little effect on *A. barbinerve*. Also, *A. barbinerve* performed better than *C. mandshurica* in the old growth forest (average of growth rate: 0.280 ± 0.004 cm$^2 \cdot$yr^{-1} vs. 0.198 ± 0.002 cm$^2 \cdot$yr^{-1}; average probability of survival: 0.803 ± 0.001 vs. 0.721 ± 0.001). Since higher basal area of neighbors indicates the presence of larger trees that may produce more shading, these results suggest that *A. barbinerve* may be better adapted to the understory environment in the old growth forest, while *C. mandshurica* is probably more sensitive to suppression with a smaller tolerance for shading. Consistent with previous studies [39–41], *C. mandshurica*, the less shade-tolerant species, also had higher mortality rates in these two forests.

While our data only provides limited information regarding the differences between forests, it was clear that shrub dynamics were different in the two forests. For example, the performance of the two species was clearly influenced by basal area of overstory or understory neighbors in the old growth forest, but not the secondary forest. The old growth forest in our study is characterized by overstory trees with larger diameter and greater basal area relative to the secondary forest (Table 1), which is consistent with previous research [42]. Because of crowding and shading from neighboring large trees, there is less available space in the understory of the old growth than the secondary forest. This may explain why the density of the local neighborhood was more important, and why both our study species had lower percent of surviving stems in the old growth forest relative to the secondary forest. The content of organic matter, total phosphorus and total potassium were significantly higher in the old growth forest, suggesting that soil nutrients are more available in old growth than developing phase. This observation is consistent with previous research in temperate forests [20]. This difference in soil characteristics among forests may explain why both our study species grew more in the old forest and their growth was positively correlated with soil variables in the old growth but not in the secondary forest.

Overall, we found that multi-stemmed growth form was an advantageous trait in the understory for *C. mandshurica* and partly for *A. barbinerve* in the old growth forest. Also, overstory neighbors influenced stem growth and survival for *C. mandshurica* but not for *A. barbinerve* in the old growth forest, which may suggest differences in shade tolerance among these two species. Differences in shade-tolerance among species commonly found in the understory have been reported before [40], but our study improves on previous studies in that we compared their dynamics in two forests with different structure. In our study, the old growth forest is characterized by overstory trees with larger diameter and greater basal area, and also more available soil nutrients relative to the secondary forest. According to the succession theory, the diversity of the secondary forest is not as stable as the old growth forest [43]. There may be more stochastic processes affecting dynamics of the two multi-stemmed understory shrub species in the secondary forest. Hence, we found that the effects of neighborhood and soil variables on growth and survival were not as strong in the secondary forest as in the old forest.

Generally, shrub species have different growth strategies than tall tree species. Many shrub species have more stems through vegetative reproduction, while overstory tree species are able to grow up into the canopy layer and sprout less frequently. Some authors have suggested that multi-stemmed growth form can be beneficial to understory species [10], while others have suggested it can be detrimental to stem dynamics [11]. Here we show that the effects of multiple stems differ for different species. Furthermore, even within the same species the benefits might be facultative, differing among forests and neighborhood compositions.

Supporting Information

File S1 Supporting tables and figure. Table S1. Soil variable loadings for the two PCAs. Table S2. AIC, $\triangle$AIC, BIC and $\triangle$BIC values of stem growth and survival models of the two species for understory neighbors across different scales. The most likely models are shown in bold (red). Table S3. AIC, $\triangle$AIC, BIC and $\triangle$BIC values of stem growth and survival models of the two species for overstory neighbors across different scales. The most likely models are shown in bold (red). Figure S1. Variograms illustrating the spatial autocorrelation in residuals of the GLMMs for stem growth and survival of the two species in the old growth and secondary forest plots.

Acknowledgments

The Smithsonian Tropical Research Institute supported Tania Brenes-Arguedas through most of this collaboration, and provided the facilities for Xuejiao Bai to work on the manuscript with Tania Brenes-Arguedas for one month. We thank Drs. Liza S. Comita and Simon A. Queenborough for helpful suggestions on data analysis. We thank Drs. Ben Bond-Lamberty, Erin Kurten and three anonymous reviewers for critical comments for the manuscript. We also thank Buhang Li and Zhaochen Zhang for their field work and data collection. The publication of this manuscript has been approved by all co-authors.

Author Contributions

Conceived and designed the experiments: XB ZH. Performed the experiments: XB JY XW FL ZY SS DX. Analyzed the data: XB TBA DX. Contributed reagents/materials/analysis tools: ZH JY XW FL ZY DX. Wrote the paper: XB TBA.

References

1. Beckage B, Clark JS, Clinton BD, Haines BL (2000) A long-term study of tree seedling recruitment in southern Appalachian forests: the effects of canopy gaps and shrub understories. Canadian Journal of Forest Research 30: 1617–1631.
2. O'Brien MJ, O'Hara KL, Erbilgin N, Wood DL (2007) Overstory and shrub effects on natural regeneration processes in native Pinus radiata stands. Forest Ecology and Management 240: 178–185.
3. Chen YL, Cui XY, Zhu N, Guan JY (1998) Status and role of the bush layer and main shrub species in nutrient cyclings of Tilia amurensis Pinus koraiensis forest. Journal of Northeast Forestry University 26: 7–13.
4. Guan JY, Chen YL, Zhu N, Zhuo LH, Liu Y (1999) Status and role of the shrub layer and its main shrub species in nutrient cycling of Mongolian oak forest. Bulletin of Botanical Research 19: 100–110.
5. Nilsson M-C, Wardle DA (2005) Understory vegetation as a forest ecosystem driver: evidence from the northern Swedish boreal forest. Frontiers in Ecology and the Environment 3: 421–428.
6. Kanno H, Hara M, Hirabuki Y, Takehara A, Seiwa K (2001) Population dynamics of four understorey shrub species during a 7-yr period in a primary beech forest. Journal of Vegetation Science 12: 391–400.
7. Kroons Hd, Hutchings MJ (1995) Morphological plasticity in clonal plants: the foraging concept reconsidered. Journal of Ecology 83: 143–152.
8. Fujiki D, Kikuzawa K (2006) Stem turnover strategy of multiple-stemmed woody plants. Ecological Research 21: 380–386.
9. Wilson BF (1995) Shrub stems: form and function. In: Garter BL, editor. Plant stems: physiology and functional morphology. New York: Academic Press. pp. 91–102.
10. Tanentzap AJ, Mountford EP, Cooke AS, Coomes DA (2012) The more stems the merrier: advantages of multi-stemmed architecture for the demography of understorey trees in a temperate broadleaf woodland. Journal of Ecology 100: 171–183.
11. Matsushita M, Tomaru N, Hoshino D, Nishimura N, Yamamoto S-I (2010) Factors affecting the production, growth, and survival of sprouting stems in the multi-stemmed understory shrub Lindera triloba. Botany 88: 174–184.
12. Weiner J (1990) Asymmetric competition in plant populations. Trends in Ecology & Evolution 5: 360–364.
13. Wright SJ (2002) Plant diversity in tropical forests: a review of mechanisms of species coexistence. Oecologia 130: 1–14.
14. Gracia M, Montané F, Piqué J, Retana J (2007) Overstory structure and topographic gradients determining diversity and abundance of understory shrub species in temperate forests in central Pyrenees (NE Spain). Forest Ecology and Management 242: 391–397.
15. McKenzie D, Halpern CB, Nelson CR (2000) Overstory influences on herb and shrub communities in mature forests of western Washington, U.S.A. Canadian Journal of Forest Research 30: 1655–1666.
16. Kanno H, Seiwa K (2004) Sexual vs. vegetative reproduction in relation to forest dynamics in the understorey shrub, Hydrangea paniculata (Saxifragaceae). Plant Ecology 170: 43–53.
17. Bellingham PJ, Sparrow AD (2009) Multi-stemmed trees in montane rain forests: their frequency and demography in relation to elevation, soil nutrients and disturbance. Journal of Ecology 97: 472–483.
18. Clarke PJ, Knox KJE, Wills KE, Campbell M (2005) Landscape patterns of woody plant response to crown fire: disturbance and productivity influence sprouting ability. Journal of Ecology 93: 544–555.
19. Finegan B (1984) Forest succession. Nature 312: 109–114.
20. Wirth C (2009) Old-growth forest: function, fate and value: a synthesis. In: Wirth C, Gleixner G, Heimann M, editors. Old-growth forests: Function, fate and value. New York, Berlin, Heidelberg: Springer.
21. Armesto JJ, Carmona M, Celis-Diez JL, Gaxiola A, Gutiérrez AC, et al. (2007) Old-growth temperate rain forests of South America: Conservation, plant-animal interactions, and baseline biogeochemical processes. In: Wirth C, Heimann M, Gleixner G, editors. Old Growth Forests: Function, Fate and Value. Berlin, Germany: Springer Verlag. pp. 367–390.
22. Shao GF, Schall P, Weishampel JF (1994) Dynamic simulations of mixed broadleaved-Pinus koraiensis forests in the Changbaishan Biosphere Reserve of China. Forest Ecology and Management 70: 169–181.
23. Stone R (2006) Ecology - A threatened nature reserve breaks down Asian borders. Science 313: 1379–1380.
24. Yang H, Li D, Wang B, Han J (1985) Distribution patterns of dominant tree species on northern slope of Changbai Mountain. Research of Forest Ecosystem 5: 1–14.
25. Yang X, Xu M (2003) Biodiversity conservation in Changbai Mountain Biosphere Reserve, northeastern China: status, problem, and strategy. Biodiversity and Conservation 12: 883–903.
26. Xu HC (2001) Natural forests of pinus koraiensis in China. Beijing: China Forestry Publishing House.
27. Bai XJ, Li BH, Zhang J, Wang LW, Yuan ZQ, et al. (2010) Species composition, structure and spatial distribution of shrub species in broad-leaved Korean pine (Pinus koraiensis) mixed forest. Chinese Journal of Applied Ecology 21: 1899–1906.
28. Condit R, Ashton PS, Manokaran N, LaFrankie JV, Hubbell SP, et al. (1999) Dynamics of the forest communities at Pasoh and Barro Colorado: comparing two 50-ha plots. Philosophical Transactions of the Royal Society of London Series B: Biological Sciences 354: 1739–1748.
29. Young SS, Carpenter C, Zhi-Jun W (1992) A Study of the Structure and Composition of an Old Growth and Secondary Broad-Leaved Forest in the Ailao Mountains of Yunnan, China. Mountain Research and Development 12: 269–284.
30. Ribeiro PJ, Diggle PJ (2011) geoR: Analysis of geostatistical data. 1.6-34 ed.
31. Baayen RH (2011) languageR: Data sets and functions with "Analyzing Linguistic Data: A practical introduction to statistics". 1.2 ed.
32. Bates D, Maechler M, Bolker B (2010) lme4: Linear mixed-effects models using S4 classes. R Package Version: 0.999375-37 ed.
33. Grothendieck G (2010) sqldf: Perform SQL Selects on R Data Frames. 0.3-5 ed.
34. Nock CA, Baker PJ, Wanek W, Leis A, Grabner M, et al. (2011) Long-term increases in intrinsic water-use efficiency do not lead to increased stem growth in a tropical monsoon forest in western Thailand. Global Change Biology 17: 1049–1063.
35. Anning AK, McCarthy BC (2013) Competition, size and age affect tree growth response to fuel reduction treatments in mixed-oak forests of Ohio. Forest Ecology and Management 307: 74–83.
36. Del Tredici P (2001) Sprouting in temperate trees: A morphological and ecological review. The Botanical Review 67: 121–140.
37. Kobe RK (1997) Carbohydrate allocation to storage as a basis of interspecific variation in sapling survivorship and growth. Oikos 80: 226–233.
38. Koop H (1987) Vegetative reproduction of trees in some European natural forests. Plant Ecology 72: 103–110.
39. Harcombe PA, Marks PL (1983) Five years of tree death in a Fagus-Magnolia forest, southeast Texas (USA). Oecologia 57: 49–54.
40. Yoshida T, Kamitani T (1997) The stand dynamics of a mixed coppice forest of shade-tolerant and intermediate species. Forest Ecology and Management 95: 35–43.
41. Whitney GG (1984) Fifty years of change in the arboreal vegetation of heart's content, an old-growth hemlock-white pine-northern hardwood stand. Ecology 65: 403–408.
42. Ziegler SS (2000) A comparison of structural characteristics between old growth and postfire second-growth hemlock–hardwood forests in Adirondack Park, New York, U. S. A. Global Ecology & Biogeography 9: 373–389.
43. Hao ZQ, Tao DL, Zhao SD (1994) Diversity of higher plants in broad-leaved Korean pine and secondary birch forests on northern slope of Changbai Mountain. Chinese Journal of Applied Ecology 5: 16–23.

2

Carnivore Use of Avocado Orchards across an Agricultural-Wildland Gradient

Theresa M. Nogeire[1]*, **Frank W. Davis**[2], **Jennifer M. Duggan**[1], **Kevin R. Crooks**[3], **Erin E. Boydston**[4]

1 School of Environmental and Forest Sciences, University of Washington, Seattle, Washington United States of America, 2 Bren School of Environmental Science and Management, University of California Santa Barbara, Santa Barbara, California, United States of America, 3 Department of Fish, Wildlife, and Conservation Biology, Colorado State University, Fort Collins, Colorado, United States of America, 4 U.S. Geological Survey, Western Ecological Research Center, Thousand Oaks, California, United States of America

Abstract

Wide-ranging species cannot persist in reserves alone. Consequently, there is growing interest in the conservation value of agricultural lands that separate or buffer natural areas. The value of agricultural lands for wildlife habitat and connectivity varies as a function of the crop type and landscape context, and quantifying these differences will improve our ability to manage these lands more effectively for animals. In southern California, many species are present in avocado orchards, including mammalian carnivores. We examined occupancy of avocado orchards by mammalian carnivores across agricultural-wildland gradients in southern California with motion-activated cameras. More carnivore species were detected with cameras in orchards than in wildland sites, and for bobcats and gray foxes, orchards were associated with higher occupancy rates. Our results demonstrate that agricultural lands have potential to contribute to conservation by providing habitat or facilitating landscape connectivity.

Editor: Mark S. Boyce, University of Alberta, Canada

Funding: Partial funding for this study was provided by The Bren School at UCSB, Orange County Great Park Corporation, and the U.S.G.S. The funders had no role in study design, data collection and analysis, decision to publish, or preparation of the manuscript.

Competing Interests: The authors have declared that no competing interests exist.

* E-mail: tnogeire@gmail.com

Introduction

Land-use change is a leading driver of loss of biological diversity globally [1]. As pressures from habitat loss increase, there is growing interest in agricultural landscapes as potential habitat or movement areas for wildlife. Agricultural landscapes are potentially rich in structure, food, and cover, and many native species forage and reproduce in these landscapes [2]. These lands can support moderate diversity of birds, mammals, arthropods, and plants, depending on the intensity of agriculture [3,4] and on configuration of natural land cover [5].

Mammalian carnivores are frequent targets of conservation efforts [6], and they play a key role in food webs, for example via mesopredator release [7] or trophic downgrading [8]. Because carnivores are typically wide-ranging, it is especially important to consider agricultural landscapes as well as protected areas when forming conservation plans for these species. Wildlife managers and conservation planners currently have little knowledge of carnivore use of agricultural landscapes, but this subject will become increasingly important as agricultural systems continue to expand and protected areas become more isolated. Connectivity between habitat patches is especially critical in human-dominated landscapes [9], but most connectivity models focus on natural vegetation types, not on differences between human-dominated land cover types within such landscapes [10]. When evaluating landscape connectivity for large carnivores, conservation planners have often relied on expert opinion and considered all agriculture as having uniformly low connectivity value (for example [11–13]).

Many members of the order Carnivora are omnivorous and feed, in part, on anthropogenic food sources [14–16]. Scat analysis has identified cultivated fruit in the diets of carnivores, particularly foxes and stone martens [15–17], and Borchert et al. [18] found that at least one orchard type – avocado – was regularly used by carnivores in California. California is a major producer of avocados, with 23,500 hectares of orchards [19] spread across five southern counties. Because avocados grow well on steep slopes, they are planted in a variety of landscape contexts, including hillslopes adjacent to native vegetation as well as valley bottoms adjacent to other types of crops.

We examined the use of avocado orchards by mammalian carnivores across agricultural-wildland gradients in southern California. We assessed whether occupancy of carnivores at motion-activated camera stations was a function of surrounding land cover, and in particular, whether area of orchards influenced carnivore occupancy. If orchards constitute poor quality carnivore habitat relative to natural areas, we would expect to observe carnivores less frequently in orchards than in nearby wildlands.

Methods

Study Area

Coastal southern California is highly urbanized and contains about two thirds of California's 38 million residents; it also has relatively little remaining undeveloped land [20], yet is experiencing rapid population growth [21]. This region has a Mediterranean-type climate, and the dominant natural vegetation types are oak woodland, riparian woodland, sage scrub, and annual grassland. Eleven native members of the order Carnivora occur in this region: American badger (*Taxidea taxus*), American black bear (*Ursus americanus*), bobcat (*Lynx rufus*), coyote (*Canis latrans*), gray fox (*Urocyon cinereoargenteus*), long-tailed weasel (*Mustela frenata*), mountain lion (*Puma concolor*), raccoon (*Procyon lotor*), ringtail (*Bassariscus astutus*), striped skunk (*Mephitis mephitis*), and Western spotted skunk (*Spilogale gracilis*).

Our study area included avocado orchards and wildlands in Santa Barbara and Ventura counties selected for their position in the landscape and for landowner cooperation (Figure 1). Avocado orchards (hereafter "orchards") grew on diverse topographies, from steep mountains to flat floodplains, and were surrounded by natural vegetation including sage scrub (Figure 2), oak woodland, grassland vegetation, other agriculture, or low-density development. Wildland sites with only natural vegetation were located at the University of California's Sedgwick Reserve and Gaviota State Park.

Land-cover Classification

No single existing land cover map met our habitat mapping needs in terms of scale, accuracy, and legend. We therefore created land cover maps from the National Land Cover Database (NLCD) [22], 2005 Southern California Association of Governments (SCAG), California Department of Water Resources, and the California Avocado Commission. Avocado orchards were identified by the California Avocado Association data, along with the SCAG and Department of Water Resources data. We used NLCD land cover to classify natural habitat types, which were not identified in the SCAG layer. Lands in classes which were essentially open space with substantial human activity (e.g., school yards, golf courses, dirt roads, urban parks, low-density development or developed open space) and which had less than 10% impervious surface were classified as 'disturbed'. When land-cover layers from the different sources were inconsistent, we verified classifications with ground visits or visual inspection of air photos (National Agriculture Imagery Program 2005, 1 m natural color).

Camera Stations

We used motion-activated digital cameras (Stealthcam, LLC, Grand Prairie, TX) at 38 sites to detect carnivore species from April 2007 to June 2008, resulting in 1,130 trap nights. Cameras

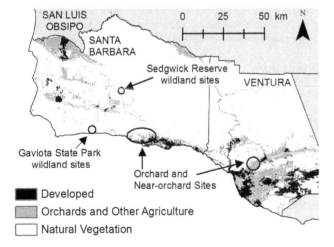

Figure 1. Study area in southern California. Study sites within Santa Barbara and Ventura counties.

Figure 2. Orchard on hillslope. Typical landscape pattern of steep hills with orchards surrounded by wildlands.

were placed in and around 6 orchards (22 sites in orchards and 6 sites in natural vegetation adjacent to orchards) and 2 continuous wildlands (10 sites), with distance to nearest camera between 30–900 meters (mean = 193 meters).

At all sites, cameras were placed along similar-sized dirt roads, near signs of carnivore activity (e.g., scat) or at trail junctions when possible. We placed scent lure (Pred-a-Getter, Murray's Lures and Trapping Supplies, Walker, West Virginia) in front of the camera to encourage animals to approach the camera and to stop long enough to be photographed. For each carnivore species at each camera site, we tallied the number of nights in which the species was detected at least once. We considered each 24-hour trap night to begin at 6:30 am, and cameras were active continuously between 12 and 76 nights (average = 33 nights) at a particular site.

Detections and Occupancy

To determine the difference in carnivore species richness between land cover types, we examined whether the number of native carnivore species differed among cameras situated in orchards, natural vegetation adjacent to orchards, and continuous wildlands using a likelihood-ratio chi-squared test. We also tested for differences between pairs of land-cover categories with a post-hoc Tukey's Honest Significant Differences test.

We assessed the influence of landscape variables on carnivore presence at camera stations using a model selection framework to compare occupancy models. We used program PRESENCE v4.6 (PRESENCE, accessed 8/2/12, http://www.mbr-pwrc.usgs.gov/software/presence.html) to estimate occupancy (ψ) and detection rate (p, the probability of detecting a species if it is present) at each camera site [23] for bobcats, coyotes, and gray foxes (species with sufficient detections to permit analyses). This program uses a likelihood approach and has been used with camera-trap data [24,25], incorporating the effect of both site covariates and sample design. We considered each trap night as a survey. We used Akaike's Information Criterion corrected for small sample size (AICc) to choose the best-performing models [26].

We began the modeling process by selecting the best detection model for each species while holding occupancy rate constant, as in Negrões et al. [25] and Duggan et al. [27]. We expected that season (wet, November – March [28], versus dry) and land cover at the camera site (orchard, natural vegetation adjacent to orchards, or continuous wildland) could affect detection rate (in addition to occupancy) so we included both as covariates in detection models. Detection covariates, as well as predictors of occupancy described below, were standardized by z-score as described in Donovan and Hines [29].

Next, to determine if spatial clustering affected occupancy, and at what scale, we compared models including site (individual orchards at least 1 km from the perimeter of the nearest neighbor), meta-site (2–3 orchards 3–4 km from one another), or county (Ventura versus Santa Barbara) as predictors of occupancy while including any detection variables selected in the previous step. We then included the covariate from the top-ranked model of spatial scale as a predictor in the candidate model set for occupancy of that species.

Finally, for each species we modeled occupancy while including detection covariates from the top-ranked detection model for that species. Potential predictors of occupancy included land cover at the camera site, distance from each camera to the perimeter of continuous wildland (natural areas contiguous with Gaviota State Park, Los Padres National Forest or adjacent wildlands, ranging from 0–3.4 km), and season (wet versus dry). We also evaluated the degree to which area of orchards and other landscape variables in the neighborhood of a camera influenced carnivore occupancy. To do so, we used the land-cover map to quantify the extent (km^2) of orchards and covariates (disturbed, shrub/scrub, grassland/herbaceous, and woodland) within a 1,935 m-diameter circle centered on each camera, approximately the average size of a bobcat home range in this region and intermediate between range sizes of foxes and coyotes [30].

We had 38 sites, and therefore examined only single- and double-factor occupancy models to avoid overparameterization. We included all single- and double-factor models in our candidate model set and then conducted model averaging. We report results for the average model, but also include summaries of the top-ranked model and all models within 2 AICc points of this model, indicating substantial support [26]. To compare the selection support for each predictor variable, we also calculated variable importance weights, which are the sum of the model weights of all models that contain a given variable [26]. Averaged models include only models within 2 AICc points of the best model. Variable importance rates are assessed across all models and therefore each variable has equal representation.

Results

Camera Stations

Cameras were active for a total of 667 trap nights in orchards, 201 in natural vegetation near orchards, and 262 in wildlands. We detected 8 of the 11 native carnivore species in the study region. Seven native species were detected in orchards: coyote (38 detections), striped skunk (28), bobcat (25), gray fox (20), mountain lion (3), black bear (2), and raccoon (2). Eight native species were detected in natural vegetation: coyote (25), bobcat (21), mountain lion (4), gray fox (7), raccoon (2), badger (1), black bear (1), and striped skunk (1). The 3 native carnivore species not detected included ringtail, spotted skunk, and long-tailed weasel.

Detections and Occupancy

On average, the number of native species detected per site differed among land-cover classes ($\chi^2 = 6.69$, df = 2, p = 0.035), but differences between individual classes were not statistically significant (Tukey's HSD, all p>0.12). The number of native species detected was greatest in orchards (mean = 2.1, SE = 0.36), intermediate in sites with natural vegetation adjacent to orchards (mean = 1.8, SE = 0.48), and lowest in wildland sites (mean = 0.8, SE = 0.40).

The top-ranked detection rate models for coyote and gray fox included land cover at the camera station location (Tables 1, 2), with higher detections in avocado orchards (top model: $\beta = 2.62$, SE = 1.09 for coyote and $\beta = 2.21$, SE = 1.10 for fox) or near avocado orchard ($\beta = 3.19$, SE = 1.10 for coyote and $\beta = 1.47$, SE = 1.16 for fox) relative to wildlands. Season was also included in the top-ranked models; for fox, the direction of the effect could not be distinguished from 0 ($\beta = 1.02$; SE = 1.16), while for coyote, detection rate was lower in the dry season than in the wet season ($\beta = -0.78$, SE = 0.29). These detection covariates were included in all subsequent models. For bobcat, the intercept-only model was the top model, so subsequent bobcat models did not include detection covariates. For all three species, the intercept-only occupancy model was the top-ranked model for spatial variation, thus we did not include spatial variables as predictors of occupancy in our final set of candidate models.

Avocado orchards, either at the camera site or in the neighborhood of the camera, were included in at least one competitive occupancy model for all three carnivore species (Tables 1, 2, 3). The area of avocado orchard in the neighborhood of a camera was the most important predictor of bobcat occupancy (model average: $\beta = 0.56$, SE = 0.32; Table 4) and was included in all top four models for bobcat occupancy (Table 3). The area of avocado orchards in the neighborhood of a camera was the third most important predictor for gray fox occupancy ($\beta = 0.17$, SE = 0.14). For coyote, the area of orchard in the neighborhood had a weak negative effect ($\beta = -0.25$, SE = 0.20), but both avocado orchard and 'near orchard' at the camera site had a positive effect (avocado orchard: $\beta = 13.41$, SE = 6.73; near orchard: $\beta = 13.23$, SE = 6.67). Land cover at the camera site was not included in any competitive bobcat or gray fox occupancy models. Distance to continuous wildland was the most important variable for predicting gray fox occupancy (model average: $\beta = -1.16$, SE = 0.93) and third most important variable for bobcats ($\beta = -0.16$, SE = 0.14; Table 4), with occupancy increasing closer to or within wildland habitat. Distance to continuous wildland was not, however, included in any competitive coyote models. The area of disturbed land in the neighborhood of the camera was included in competitive models for both coyote ($\beta = 4.81$, SE = 2.20) and gray fox ($\beta = 0.31$, SE = 0.39) occupancy (Tables 1, 2), but large standard error values for fox occupancy suggested a weak influence. Disturbed land was not included in models for bobcat occupancy (Table 3). Woodland, shrub, and grassland/herbaceous vegetation in the neighborhood of a camera had a positive effect on occupancy in all models for all species, except that woodland had a negative effect on gray fox occupancy.

Discussion

Carnivores were detected with surprising frequency in avocado orchards. We detected most carnivore species native to coastal southern California in avocado orchards, and these orchards were used frequently by bobcats, coyotes and gray foxes. Further, we detected more carnivore species in orchards than in wildland sites. Although orchards are often adjacent to wildlands, the presence of carnivores in orchards does not appear to be simply an artifact of landscape context. If this were the case, we would expect to find more carnivores in wildlands than in orchards, which we did not. We would also expect to find that distance to continuous wildland was a more consistently important predictor in our models; although it was the strongest predictor of occupancy for gray fox, it was present in only one competitive model for bobcat occupancy and no competitive models for coyote.

Table 1. Top-ranked models of site occupancy (ψ) and detection rate (p) for gray fox.

Model I.D.	K	−2*log-likelihood	ΔAICc	Relative likelihood	ω	$\hat{\psi}$ (SE)	$\hat{p}$ (SE)
Detection (p) ~							
Land cover + season	5	230	0.00	1.00	0.45	0.47 (0.13)	0.018 (0.025)
Land cover	4	234	0.55	0.76	0.34	0.41 (0.11)	0.017 (0.033)
Season	3	238	2.89	0.24	0.11	0.45 (0.13)	0.015 (0.029)
Intercept only	2	241	2.98	0.23	0.10	0.38 (0.10)	0.059 (0.012)
Occupancy (ψ) ~							
Distwild	6	227	0	1.00	0.14	0.47 (0.14)	0.009 (0.010)
Distwild + woodland	7	224	0.40	0.82	0.11	0.61 (0.09)	0.066 (0.019)
Intercept only	5	230	0.67	0.72	0.10	0.47 (0.13)	0.067 (0.020)
Distwild + Avocado orchard	7	225	0.98	0.61	0.08	0.39 (0.17)	0.068 (0.021)
Shrub	6	228	1.33	0.51	0.07	0.56 (0.12)	0.066 (0.019)
Distwild + disturbed	7	225	1.51	0.47	0.06	0.55 (0.14)	0.065 (0.018)
Averaged model					*0.50*	*0.51 (0.13)*	*0.049 (0.016)*

Footnote: All models with ΔAICc <2.0, plus the intercept-only models, are reported. K is the number of parameters, ΔAICc is the difference between the AICc of the model and the lowest-AICc model, ω is the AICc model weight (summed for the averaged model), ψ is the predicted occupancy at a site and p is the probability of detecting the species at a given site. Covariate abbreviations: distwild is distance to continuous wildland, land cover is land cover (avocado orchard, near orchard, or wildland) at the camera site, and woodland, avocado orchard, shrub and disturbed refer to the area of that land cover in the neighborhood of the camera site.

The food subsidy value of avocados may explain why omnivores such as bears, coyotes, and raccoons were present in orchards. Indeed, remote cameras have recorded these species eating avocados in southern California (M. Borchert, U.S. Forest Service, personal communication), but why obligate carnivores like mountain lions and bobcats would be present in orchards is less clear. Orchards may provide good cover for carnivores; many of these species are habitat generalists, and orchards often replace oak woodlands with structurally similar vegetation. Irrigation lines in orchards act as a rare source of perennial water in arid landscapes. In our study, we did not find an effect of wet versus dry season on occupancy, as might be expected if carnivores were attracted by water sources. However, irrigation lines, combined with abundant avocados, might simulate year-round wet-season conditions for small mammals, perhaps leading to bottom-up effects in these agricultural systems. Future research could assess whether orchards are providing more food and water for small mammals than native vegetation, and whether a relative increase in prey might help explain the use of these lands by carnivores [31]. Finally, further study could evaluate whether the presence of infrequently-used dirt roads in orchards might appeal to animals moving across densely vegetated landscapes.

Table 2. Top-ranked models of site occupancy (ψ) and detection rate (p) for coyote.

Model I.D.	K	−2*log-likelihood	ΔAICc	Relative likelihood	ω	$\hat{\psi}$ (SE)	$\hat{p}$ (SE)
Detection (p) ~							
Land cover+season	5	440	0.00	1.00	0.86	0.68 (0.10)	0.079 (0.019)
Land cover	4	447	4.12	0.13	0.11	0.71 (0.10)	0.069 (0.014)
Season	3	452	6.93	0.03	0.03	0.55 (0.095)	0.10 (0.019)
Intercept only	2	486	38.70	0.00	0.00	0.44 (0.10)	0.056 (0.0068)
Occupancy (ψ) ~							
Disturbed	6	433	0.00	1.00	0.18	0.76 (0.08)	0.056 (0.012)
Grass/herbaceous	6	434	0.68	0.71	0.13	0.75 (0.10)	0.055 (0.012)
Avocado orchard + disturbed	7	432	1.24	0.54	0.10	0.77 (0.10)	0.055 (0.012)
Land Cover + grass/herbaceous	8	429	1.42	0.49	0.09	0.52 (0.10)	0.070 (0.027)
Intercept only	5	440	3.74	0.15	0.03	0.68 (0.10)	0.055 (0.013)
Averaged model					*0.53*	*0.72 (0.09)*	*0.06 (0.015)*

Footnote: All models with ΔAICc <2.0, plus the intercept-only models, are reported. K is the number of parameters, ΔAICc is the difference between the AICc of the model and the lowest-AICc model, ω is the AICc model weight (summed for the averaged model), ψ is the predicted occupancy at a site and p is the probability of detecting the species at a given site. Covariate abbreviations: distwild is distance to continuous wildland, land cover is land cover (avocado orchard, near orchard, or wildland) at the camera site, and grass/herbaceous, avocado orchard and disturbed refer to the area of that land cover in the neighborhood of the camera site.

Table 3. Top-ranked models of site occupancy (ψ) and detection rate (p) for bobcat.

Model I.D.	K	-2*log-likelihood	ΔAICc	Relative likelihood	ω	$\hat{\psi}$ (SE)	$\hat{p}$ (SE)
Detection (p) ~							
Intercept only	2	368	0	1.00	0.48	0.44 (0.10)	0.080 (0.012)
Season	3	367	1.26	0.53	0.26	0.52 (0.14)	0.066 (0.019)
Land cover	4	365	2.10	0.35	0.17	0.47 (0.11)	0.071 (0.019)
Land cover+season	5	364	3.27	0.20	0.09	0.55 (0.14)	0.060 (0.022)
Occupancy (ψ) ~							
Avocado orchard + grass/herbaceous	4	363	0	1.00	0.13	0.44 (0.15)	0.081 (0.012)
Avocado orchard + woodland	4	363	0.17	0.92	0.12	0.26 (0.17)	0.081 (0.012)
Avocado orchard	3	365	0.20	0.90	0.12	0.42 (0.13)	0.081 (0.012)
Avocado orchard + distwild	4	363	0.29	0.87	0.11	0.42 (0.15)	0.081 (0.012)
Intercept only	2	368	0.34	0.84	0.11	0.44 (0.10)	0.080 (0.012)
Woodland	3	366	1.11	0.57	0.07	0.28 (0.16)	0.080 (0.012)
Woodland + grass/herbaceous	4	365	1.88	0.39	0.05	0.35 (0.15)	0.079 (0.012)
Averaged model					0.71	0.38 (0.14)	0.080 (0.012)

Footnote: All models with ΔAICc <2.0, plus the intercept-only models, are reported. K is the number of parameters, ΔAICc is the difference between the AICc of the model and the lowest-AICc model, ω is the AICc model weight (summed for the averaged model), ψ is the predicted occupancy at a site and p is the probability of detecting the species at a given site. Covariate abbreviations: distwild is distance to continuous wildland, land cover is land cover (avocado orchard, near orchard, or wildland) at the camera site, and woodland, grass/herbaceous, shrub, avocado orchard and disturbed refer to the area of that land cover in the neighborhood of the camera site.

There is growing interest in managing for movement of wild animals through agricultural areas [10,32]. Knowing the value of different land-cover types for habitat or movement can inform conservation decisions regarding which lands should be purchased or put under easements, or which areas are most suitable for the placement of highway crossings [12]. Avocado orchards appear to serve as both foraging and movement habitat for most carnivore species in California, and conservation easements or other incentives to keep land in orchards could offer a cost-effective conservation strategy. Such alternative conservation strategies are particularly important when considering agriculture (including avocados) in Mediterranean-type ecosystems, which are highly threatened [33].

Table 4. Variable importance weights (ω) for predictors of occupancy for bobcats, coyotes, and gray foxes.

Bobcat		Coyote		Gray fox	
Covariate	**∑ ω**	**Covariate**	**∑ ω**	**Covariate**	**∑ ω**
Avocado orchard	0.58	Disturbed	0.50	Distance to wildland	0.51
Woodland	0.29	Grass/herbaceous	0.47	Shrub	0.21
Grass/herbaceous	0.28	Avocado orchard	0.15	Avocado orchard	0.18
Distance to wildland	0.13	Shrub	0.12	Woodland	0.18
Season	0.09	Land cover at site	0.10	Disturbed	0.13
Shrub	0.07	Woodland	0.10	Grass/herbaceous	0.11
Disturbed	0.06	Season	0.08	Season	0.09
Land cover at site	0.03	Distance to wildland	0.06	Land cover at site	0.05

Acknowledgments

M. Borchert, B. Kendall, P. Jantz, S. Riley, J. Orrock, and S. Rothstein contributed ideas during early phases of the project. E. Fleishman, B. McRae, J. Yee, J. Hilty and M. Ricca provided helpful comments on the manuscript. The use of trade, product, or firm names in this publication is for descriptive purposes only and does not imply endorsement by the U.S. Government.

Author Contributions

Conceived and designed the experiments: FD TN. Performed the experiments: TN. Analyzed the data: TN JD FD KC EB. Contributed reagents/materials/analysis tools: FD TN. Wrote the paper: TN FD EB JD KC.

References

1. Sala OE, Chapin FS, Armesto JJ, Berlow E, Bloomfield J, et al. (2000) Global Biodiversity Scenarios for the Year 2100. Science 287: 1770–1774.
2. Brosi BJ, Daily GC, Davis FW (2006) Agriculture and Urban Landscapes. In: Scott JM, Goble DG, Davis FW, editors. The Endangered Species Act at 30: Volume 2. Washington, D.C.: Island Press. 256–274.
3. Benton TG, Vickery JA, Wilson JD (2003) Farmland biodiversity: is habitat heterogeneity the key? Trends in Ecology & Evolution 18: 182–188.
4. Flynn DFB, Gogol-Prokurat M, Nogeire T, Molinari N, Richers BT, et al. (2009) Loss of functional diversity under land use intensification across multiple taxa. Ecology Letters 12: 22–33.
5. Daily GC, Ceballos G, Pacheco J, Suzán G, Sánchez-Azofeifa A (2003) Countryside Biogeography of Neotropical Mammals: Conservation Opportunities in Agricultural Landscapes of Costa Rica. Conservation Biology 17: 1814–1826.
6. Ray JC (2005) Large carnivores and the conservation of biodiversity. Washington, D.C.: Island Press.
7. Crooks KR, Soulé ME (1999) Mesopredator release and avifaunal extinctions in a fragmented system. Nature 400: 563–566.
8. Estes J, Terborgh J, Brashares J (2011) Trophic downgrading of planet earth. Science 333: 301–306.
9. Crooks KR, Sanjayan MA (2006) Connectivity conservation. Cambridge Univ Pr.
10. Cosentino BJ, Schooley RL, Phillips CA (2011) Connectivity of agroecosystems: dispersal costs can vary among crops. Landscape ecology 26: 1–9.

11. Beier P, Penrod K, Luke C, Spencer W, Cabañero C (2006) South Coast Missing Linkages: restoring connectivity to wildlands in the largest metropolitan area in the United States. In: Crooks KR, Sanjayan MA, editors. Connectivity Conservation. New York: Cambridge Univ Pr. 555–586.

12. Beier P, Garding E, Majka D (2008) Arizona Missing Linkages: Linkage Designs for 16 Landscapes. Phoenix, AZ: Arizona Game and Fish Department.

13. Singleton P, Gaines W, Lehmkuhl J (2002) Landscape permeability for large carnivores in Washington: a geographic information system weighted-distance and least-cost corridor assessment. Portland, Oregon: US Department of Agriculture Forest Service.

14. Fedriani J, Fuller T, Sauvajot R (2001) Does availability of anthropogenic food enhance densities of omnivorous mammals? An example with coyotes in southern California. Ecography 24: 325–331.

15. Padial J, Avila E, Sanchez J (2002) Feeding habits and overlap among red fox (*Vulpes vulpes*) and stone marten (*Martes foina*) in two Mediterranean mountain habitats. Mammalian Biology 67: 137–146.

16. Shapira I, Sultan H, Shanas U (2007) Agricultural farming alters predator–prey interactions in nearby natural habitats. Animal Conservation 11: 1–8.

17. López-Bao J V, González-Varo JP (2011) Frugivory and spatial patterns of seed deposition by carnivorous mammals in anthropogenic landscapes: a multi-scale approach. PloS one 6: e14569.

18. Borchert M, Davis F, Kreitler J (2008) Carnivore use of an avocado orchard in southern California. California Fish and Game 94: 61–74.

19. California Avocado Commission (2010). Available: www.avocado.org.

20. Landis JD, Reilly M (2003) How We Will Grow: Baseline Projections of the Growth of California's Urban Footprint through the Year 2100.

21. Conservation International (2010) Biodiversity Hotspots. Available: www.biodiversityhotspots.org.

22. Homer C, Huang C, Yang L (2004) Development of a 2001 national landcover database for the United States. Photogrammetric Engineering and Remote Sensing 70: 829–840.

23. MacKenzie DI, Nichols JD, Lachman GB, Droege S, Andrew Royle J, et al. (2002) Estimating site occupancy rates when detection probabilities are less than one. Ecology 83: 2248–2255.

24. Linkie M, Dinata Y, Nugroho A, Haidir IA (2007) Estimating occupancy of a data deficient mammalian species living in tropical rainforests: Sun bears in the Kerinci Seblat region, Sumatra. Biological Conservation 137: 20–27.

25. Negrões N, Sarmento P, Cruz J, Eira C, Revilla E, et al. (2010) Use of camera-trapping to estimate puma density and influencing factors in central Brazil. Journal of Wildlife Management 74: 1195–1203.

26. Burnham KP, Anderson DR (2002) Model selection and multimodel inference: a practical information-theoretic approach. New York: Springer Verlag.

27. Duggan J, Schooley R, Heske E (2011) Modeling occupancy dynamics of a rare species, Franklin's ground squirrel, with limited data: are simple connectivity metrics adequate? Landscape ecology 26: 1477–1490.

28. Keeley JE (2000) Chaparral. In: Barbour M, Billings W, editors. North American terrestrial vegetation. 203–254.

29. Donovan TM, Hines J (2007) Exercises in Occupancy Estimation and Modeling. Available: http://www.uvm.edu/envnr/vtcfwru/spreadsheets/occupancy/occupancy.htm.

30. Crooks K (2002) Relative sensitivities of mammalian carnivores to habitat fragmentation. Conservation Biology 16: 488–502.

31. Thibault K, Ernest S, White E, Brown J, Goheen J (2010) Long-term insights into the influence of precipitation on community dynamics in desert rodents. Journal of Mammalogy 91: 787–797.

32. Muntifering JR, Dickman AJ, Perlow LM, Hruska T, Ryan PG, et al. (2006) Managing the matrix for large carnivores: a novel approach and perspective from cheetah (*Acinonyx jubatus*) habitat suitability modelling. Animal Conservation 9: 103–112.

33. Cox RL, Underwood EC (2011) The importance of conserving biodiversity outside of protected areas in mediterranean ecosystems. PloS ONE 6: e14508.

3

Genetic Population Structure of Cacao Plantings within a Young Production Area in Nicaragua

Bodo Trognitz[1]*, **Xavier Scheldeman**[2], **Karin Hansel-Hohl**[1], **Aldo Kuant**[3], **Hans Grebe**[3], **Michael Hermann**[4]

1 Austrian Institute of Technology, Seibersdorf, Austria, **2** Bioversity International, Cali, Colombia, **3** Pro Mundo Humano, Managua, Nicaragua, **4** Crops for the Future, Serdang, Selangor, Malaysia

Abstract

Significant cocoa production in the municipality of Waslala, Nicaragua, began in 1961. Since the 1980s, its economic importance to rural smallholders increased, and the region now contributes more than 50% of national cocoa bean production. This research aimed to assist local farmers to develop production of high-value cocoa based on optimal use of cacao biodiversity. Using microsatellite markers, the allelic composition and genetic structure of cacao was assessed from 44 representative plantings and two unmanaged trees. The population at Waslala consists of only three putative founder genotype spectra (lineages). Two (B and R) were introduced during the past 50 years and occur in >95% of all trees sampled, indicating high rates of outcrossing. Based on intermediate allelic diversity, there was large farm-to-farm multilocus genotypic variation. GIS analysis revealed unequal distribution of the genotype spectra, with R being frequent within a 2 km corridor along roads, and B at more remote sites with lower precipitation. The third lineage, Y, was detected in the two forest trees. For explaining the spatial stratification of the genotype spectra, both human intervention and a combination of management and selection driven by environmental conditions, appear responsible. Genotypes of individual trees were highly diverse across plantings, thus enabling selection for farm-specific qualities. On-farm populations can currently be most clearly recognized by the degree of the contribution of the three genotype spectra. Of two possible strategies for future development of cacao in Waslala, i.e. introducing more unrelated germplasm, or working with existing on-site diversity, the latter seems most appropriate. Superior genotypes could be selected by their specific composite genotype spectra as soon as associations with desired quality traits are established, and clonally multiplied. The two Y trees from the forest share a single multilocus genotype, possibly representing the Mayan, 'ancient Criollo' cacao.

Editor: Thomas Mailund, Aarhus University, Denmark

Funding: Funded by the Austrian Development Agency, ADA (NICACAO Project), (http://www.entwicklung.at/en/) and the Austrian Institute of Technology, AIT (1.U4.00077) (http://www.ait.ac.at). The funders had no role in study design, data collection and analysis, decision to publish, or preparation of the manuscript. The publication content is solely the responsibility of the authors and does not necessarily represent the official views of ADA nor AIT.

Competing Interests: The authors have declared that no competing interests exist.

* E-mail: bodo.trognitz@ait.ac.at

Introduction

In Central America, the cacao tree (*Theobroma cacao* L.), a plant of the humid neotropics, was already being cultivated by the Olmecs and early Mayas, 3000 years ago. Recent investigations on the origin of the ancient Central American cacao, traditionally referred to by its morphogeographic name 'Criollo', suggest that it may have been introduced from an area now in Venezuela, adjacent to the center of highest diversity of *Theobroma cacao* L. in upper Amazonia [1]. However, Criollo cacao represents only a small part of the allelic bandwidth of cultivated and natural cacao populations occurring in Amazonian forests where the species originated. Today's descendants of the Mayan ancient Criollo cacao can therefore be considered as the products of multigenerational selection by Amerindian farmers [2,3]. Hybrids of Criollo and some Forastero accessions, known as Trinitario or modern Criollo [2,4], and as 'Trinidad Selected Hybrids' (TSH), are renowned for their distinct aroma making them a preferred raw material for fine cocoa chocolate [5]. Therefore, remaining sources of ancient Criollo that can still be found in Central America, including Nicaragua, contain potentially valuable germplasm for future breeding of high quality cacao.

Types of cacao are distinguished by several partly overlapping naming schemes. There is the traditional recognition of morphogeographical groups or cultivars (Criollo from Central America, Forastero from Amazonian South America, Amelonado, a Forastero with distinct fruit shape, Trinitario from Trinidad and Tobago, and Refractario from Ecuador originally selected for its resistance to witches' broom disease, *Crinipellis perniciosa* (see also http://sta.uwi.edu/cru/icgt/types.asp). Traditional traders' 'varieties' are recognized by the trade quality (e.g., Trinitario, Criollo, Amelonado, Catongo, Nacional) [4], and cocoa and chocolate are frequently graded and marketed under the name of the country (or region) of production, e.g. Amazonia, Belize, Ecuador, Ivory Coast, or Venezuela. Although modern plantations are often composed of clones (grafted trees or rooted cuttings), propagation by seed has been the simple traditional method for the multiplication of cacao trees. Cacao possesses poorly characterized sexual self-incompatibility, but many trees under cultivation are sufficiently self-fertile [6] to allow for secure yields, and to give rise to inbreeding. The use of clonally propagated, bred and selected cultivars, as are widely used with many horticultural fruit crops in temperate zones, is only just beginning.

The gourmet chocolate sector makes up 4% of the total world chocolate market (S. Vervliet, Puratos/Belcolade, 2007, pers.

comm.) but is growing quickly. 'Fine-flavor cocoa' fetches a considerable price premium, up to four times of the price of standard commodity cocoa. The manufacture of gourmet chocolate depends to a large extent on intrinsic cocoa qualities which are determined by genotype, and on-farm processing including the selection of pods, fermentation, and drying of beans. This offers good opportunities for quality differentiation and value addition that would benefit the growers (S. Petchers, CATIE, 2004, pers. comm.). However, cacao is predominantly produced by smallholder farmers whose level of training and organization in the production chain is often insufficient to maximize the benefits from the production of high quality cocoa.

In Nicaragua, one of the largest cocoa production areas is found in the municipalities of Waslala and Rancho Grande, towns in the central northern part of the country. In pre-Colombian times, this area was under cultural and linguistic influences from the Mayas, and from the Aztecs further north [7]. Waslala is equidistant between the Pacific and Atlantic oceans, at an elevation of 200–740 m in a south-east facing depression adjacent to the Peñas Blancas massif. The average annual temperature ranges between 21.3 and 24.9°C, and mean annual rainfall is between 2170 and 2660 mm (Worldclim database, www.worldclim.org [8]). The beginnings of commercial cocoa cultivation date back to 1961 (E. Rios, first president of the cocoa producers cooperative Cacaonica, 2007, pers. comm.). During and after the civil war in the 1980s, refugees and migrants from all over Nicaragua arrived, and cocoa production has greatly expanded since 1991 with the establishment of the non-governmental organization Pro Mundo Humano, and the foundation of Cacaonica. Cocoa has since become a popular cash crop. The area planted with cacao is now some 1700 ha, having increased during the past five years due to the attractive prices. Typically a household cultivates 0.7–1 ha of cacao, containing 300–600 trees. Plot sizes rarely exceed 2 ha because cacao cultivation is labor intensive, in particular pruning, manual removal of diseased fruits, and continual harvesting and processing. Plantings are distributed on steep slopes that are not useful for cattle pasture. Individual farms rarely yield more than 0.5 t/ha of dried cocoa beans per year, but together, the municipality's total annual crop contributes considerably to the national cocoa production of 2650 t (in 2009). Farmers can obtain higher prices for high quality cocoa grades, especially if organically certified. There is also potential for adding value from quality differentiation based on characteristics imparted through locality-dependent (environmental), management, and genetic factors. Several commercial cocoa and chocolate companies source their raw material in Waslala, including Ritter (Germany), Cocoa S.A. (Costa Rica), Daarnhouver (the Netherlands), and Zotter (Austria).

It is believed that only a limited number of introductions contributed to the present-day germplasm in Waslala cocoa plantings, although few records are available. The Tropical Agricultural Research and Education Center (CATIE, Turrialba, Costa Rica) distributed seed (beans) in the 1980s to Central American countries including Nicaragua (W. Phillips, CATIE, 2007, pers. comm.), some of which arrived at Waslala. In addition, several farmers interviewed during this research claimed to have occasionally brought in seed and scions from other regions, and others reported finding rare pre-existing cacao trees when they arrived at their new farmland in the 1970s and 1980s. Cacao has been predominantly introduced to Waslala as seedlings, and to a lesser extent through grafted clones. The farmers themselves propagate cacao mostly through the use of seedlings.

This paper explores the genetic composition and structure of cacao populations, as a prerequisite for varietal certification and denomination of Waslala cocoa. It also assesses optimal means to improve cocoa yield and quality in this area for the benefit of the farmers and cocoa producers. For the cacao research community, it is of interest to understand how cacao populations are shaped by germplasm introductions and management. We have representatively sampled the municipality and surveyed the genotype of trees by a number of well-defined simple sequence repeat (SSR) markers. It addressed questions related to the extent of allelic diversity and the possibility of discerning the genetic structure of population, with the objective of identifying specific genetic backgrounds which can be related to geographic areas, farmers' degree of access to germplasm, and specific environmental conditions.

Results

Descriptive statistics and genetic diversity

The 15 microsatellite primer pairs detected 116 individual alleles (with 7.73 alleles per locus on average) across all samples collected in Waslala municipality. There were no null-alleles apparent. When only one allele was detected, the individual was considered homozygous at this locus. Two trees had three individual alleles at only two SSR loci for unknown reasons. For the analyses of population genetics, the rarer alleles, relative to the entire data set, were considered in these exceptional cases. Six groups of trees sharing an identical multilocus genotype were found, and two of these genotypes were frequent (Table 1, group E with 10 members, and group D with 7 members).

Considering individual farms as separate, independent entities with individual compositions of genotypes, the average number of effective alleles present within all trees sampled at a single farm was 3.38 (range 1.0–5.4). Private alleles [9] occurred within only ten trees from seven farms (Table 2), including 8 of the 15 SSR loci investigated. The degree of expected heterozygosity (He) averaged over all 45 sites and 15 SSR loci was 0.476 (range 0–0.688). This is equivalent to an average of almost 50% (47.6%) of all loci being heterozygous. The rate of fixed loci was less than 30%, indicating a moderate degree of inbreeding at the current state.

Estimation of the genetic diversity among farms

The degree of genetic diversity was calculated as the percentage of significant differences in all pairwise comparisons among farms, for every SSR locus in separate. Of a total of 14,864 comparisons by the G test using Shannon's mutual information index ($^{S}H_{UA}$) as implemented in GenAlEx [10], 39.4% were significant. This can be interpreted as showing considerable among-farm differences in frequency and composition of alleles at the 15 loci under study. The existence of large among-farm differences is further supported by the large differences in the frequencies of individual alleles by farm (e.g. Figure 1).

Tracing the population structure across all cacao plantings in Waslala

Several simulations were performed in the program *Structure* [11] on all individuals and markers with and without consideration of the individual farms, as a factor contributing to the distribution of 'farm subpopulations' (LOCPRIORS option on or off, respectively). Simulations for up to K = 20 clusters were made. Each cluster was considered to represent one distinct group of ancestral genetic backgrounds that are referred to in this paper as a 'genotype spectrum' or lineage (known as 'formenkreis' in German). In contrast, a genotype as represented by a single individual can be made up entirely of just one genotype spectrum, or from parts of several such genotype spectra.

Table 1. Trees with matching multilocus genotypes across 15 SSR markers.

Sample	Farm/Location	No. Matches	Label
W042	FBBSB	2	A
W041	FBBSB		A
W356	FJM	3	B
W161	F178		B
W366	F178		B
W327	F165	2	C
W207	F003		C
W305	F084	7	D
W153	F018		D
W102	F022		D
W105	F022		D
W106	F022		D
W108	F022		D
W141	F027		D
W049	FBBlandon	10	E
W132	F006		E
W309	F083		E
W201	F174		E
W290	F195		E
W299	F225		E
W330	F227		E
W043	FBBlandon		E
W044	FBBlandon		E
W046	FBBlandon		E
W359	F166	2	F
W325	F166		F

(For codes see Table S1).

Table 2. Private alleles at the farm level for 15 SSR loci across all 317 cacao trees sampled in Waslala, Nicaragua.

Sample (Tree)	Farm	Locus
W187	F005	mTcCIR24–193
W158	F018	mTcCIR33–350
W104	F022	mTcCIR26–281
W265	F029	mTcCIR7–147
W264	F029	mTcCIR11–307
W264	F029	mTcCIR18–346
W264	F029	mTcCIR33–273
W264	F029	mTcCIR37–144
W221	F143	mTcCIR22–291
W224	F143	mTcCIR22–291
W328	F165	mTcCIR37–186
W11	FERPozolera	mTcCIR18–333
W11	FERPozolera	mTcCIR18–343

The most probable number of populations (genotype spectrum clusters) was 3, as determined by a graphical method [12] as well as by the method applying Bayes' rule [13]. The partitioning of individuals across the three clusters was stable both with and without taking into consideration the location (farm). These three groups of genotype spectra, were denominated Blue, Red, and Yellow (B, R, and Y), for further investigation. Individuals within any of the three genotype groups contained different degrees of admixture from one or both of the other lineages (Figure 2). The Y group consisted of only three trees, namely the two FBBSB orphan trees from the forest (W041 and W042), and tree W357 from farm F204 (for identities, see supplementary Table S1). Tree W357 included a 14% admixture with components from the B lineage, and 23% from the R lineage. Another 12 trees, labelled as the BY admixture group, consisted of 30–50% Y, 30–50% B, and up to 20% R shared genotype spectra. There was also a BR admixture group of inferred genotype spectra (27–66% B, 33–65% R, 0–34% Y) consisting of 81 trees. A further 107 trees possessed a majority of B lineage components (39–99% B, 0–33% R, 0–32% Y), and 114 samples were mainly R (0–33% B, 42–99% R, 0–32% Y). Subsets of samples corresponding to the B or R clusters defined in this way were subjected to clustering simulations in *Structure*, but all attempts to detect sub-clusters within the B or the R genotype spectrum failed, and no further

separation by the genotype spectrum was applicable within this set of data.

The average genetic distance between the genotype spectra B, R, and Y was estimated by Nei's Genetic Distance and Genetic Identity and Wright's Fst as implemented in Genalex. Groups BR and BY with large admixtures were excluded. The results are presented in Figure 3. The closest related groups were B and R, with a Genetic Distance of 0.303, corresponding to a Genetic Identity of 73.8% and an Fst of 0.121. The Y group was most distant (Genetic Distance; 1.743 to group B and 1.141 to R), although this result must be taken with caution due to the small sample size of Y. The indices of relatedness were also calculated on reduced sets of samples restricting the portion of admixture genotypes. Allowing a minimum of 75, 85, or 95% presence of the B, Y, or R genotype spectrum (by removing samples with more than 25, 15, or 5% admixture, correspondingly), Genetic Distance increased and Genetic Identity shrunk as expected (Figure 3). This indicates that the clustering in the *Structure* program was successful in the detection of distinct genotype spectra.

The allelic diversity is largest in the R genotype spectrum group, followed by B, whereas the three samples representing the Y group possess very few different alleles per marker locus (Table 3). In fact, the two pure Y trees, W041 and W042, have perfectly matching alleles. Lineage R is also separated from B by having a larger number of private alleles. With increasing purity, i.e., virtually selecting for higher percentages of the prospected founder genotype spectra, the allelic diversity and expected heterozygosity decline, and the numbers of private alleles increase (Table 3).

Analysis of molecular variance (AMOVA as implemented in GenAlEx) on the B and R genotype clusters (assuming they represent founder genotype spectra) revealed 65% variation within and 35% among these clusters, and a Φ_{PT} value of 0.354 (P<0.001). A relatively small among-cluster variation was expected due to the fact that both lineages share the same alleles and possess large Genetic Identity values.

Distribution of the three prospected founder genotype spectra at farm level

As a measure of relatedness between different farms by genotype spectra composition, the average genetic distance over

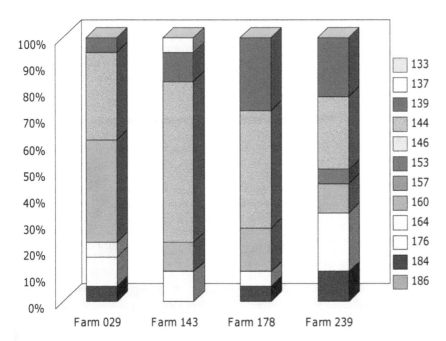

Figure 1. Frequency of SSR alleles (marker mTcCIR37) at four farms with samples from nine trees. A farm is represented by a single column showing cumulative frequencies of individual alleles. Alleles are indicated by their size in base pairs. Farms are labelled by their PCC numbers (also see Acknowledgement).

all 15 marker loci quantified by Shannon's index was applied. The results are summarized via principal coordinates analysis in Figure 4. Except FBBSB, the two orphan trees W041 and W042 near the forest, most farms were not well separated from each other by this method. This reflects the genetic composition of farms; with every farm having trees possessing genotypes of various states of admixture, considering the lineages as detected by the *Structure* program. That is illustrated by the pie diagrams on the map of Waslala municipality (Figure 5), each pie plot representing the proportion of the three genotype lineages contributing to an individual farm. The majority of farms are represented by tree genotypes made up of two (B and R) or three (B, R, and Y) lineages. Only a few farms consist of nearly exclusively B genotype spectrum partitions, and only the closely spaced south-eastern plantings F083 and F084, both owned by

the same single farmer, contain nearly pure B lineage trees (Figure 5).

Association among geographic characteristics of the sample sites, genotype spectra and geographic features

Pairs of the 5 continuous variables; distance to the road, altitude (m above sea level), elevation relative to nearest road (calculated as the difference between the altitude of the sample tree and that of the nearest point of the road), mean annual precipitation, and mean annual temperature, were subjected to correlation and regression analyses in a descriptive approach (Table 4). There was a small but highly significant correlation between the trees' distance to the road, and the elevation relative to the road, altitude, and temperature. A stronger correlation (R = 0.721) existed for distance to the road and mean annual precipitation.

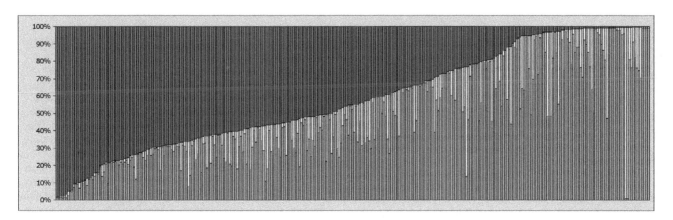

Figure 2. Bayesian clustering of cacao trees in Waslala. Best fit was achieved with three clusters representing three hypothetical founder genotype spectra, Blue, Red, and Yellow, with varying degrees of admixture within single individuals. A single column represents one of 317 individuals, with its proportions of the genetic lineages B, Y, and R. The two trees from forest remnants, W041 and W042, representing the ancient Criollo genotype spectrum, are indicated by the two entirely yellow columns.

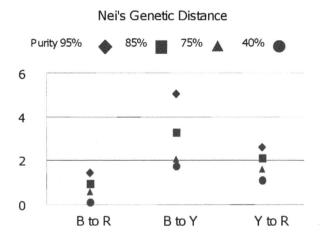

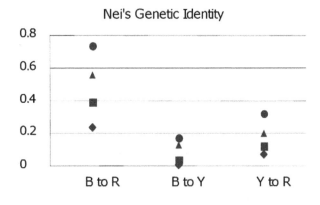

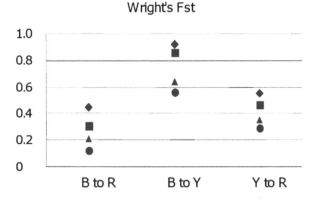

Figure 3. Nei's Genetic Distance and Genetic Identity, and Wright's Fst. Genetic distances between B (Blue), R (Red), and Y (Yellow) genotype spectra comprising groups of cacao trees whose admixed genotypes have certain minimum degrees of purity of the corresponding genotype spectrum (complementary to the maximum degrees of admixture with other genotype spectra, as shown in Figure 2). The increase or reduction of the parameter values throughout different degrees of purity in alignment support the clustering results shown in Figure 2.

The elevation relative to the nearest road and absolute altitude were highly positively correlated (R = 0.799), meaning that trees at higher locations frequently grow on steep hills high above the neighboring roads. Consequently, the negative correlation of elevation relative to the road, and temperature, reflects the expected negative correlation between altitude and mean annual temperature (R = 0.946). This data (Table 4) also suggests that in

this location, although mean annual precipitation tends to increase with increasing elevation as expected, some areas at low elevation receive much precipitation which may produce a cooling effect.

The discrete genotype spectra were used as a factor to compare geographical and climatic characteristics that they may be preferentially associated with (Table 5), in an exploratory approach. To avoid sampling bias due to grossly differing sample sizes, the under-represented groups Y (3 individuals) and BY (12 individuals) were excluded from these analyses. There were well-supported associations of individual genotype groups with the geographic distance to the nearest road, and mean annual precipitation (Table 5). The B genotype spectrum occurred more frequently at locations far from main roads (average 4.5 km) and the R and BR groups were frequently found nearer to roads (average 2.0–2.5 km). The group B was found in areas receiving the highest mean annual precipitation (2452 mm), whereas R and BR were not distinguishable in areas of 2409 mm mean annual rainfall. The elevation of R genotype spectrum trees above the nearest road was marginally but significantly above average. It is worth noting that replication, i.e., the individual trees at their given locations, also made a significant contribution to the total variance.

In summary, genotype spectrum B occurred more frequently further from the road than the R genotype spectrum. Genotype spectrum B is more frequent at lower elevations with higher mean annual rainfall, whereas R occurs preferentially at higher elevations with lower mean annual rainfall, but R is more frequent than B higher above the closest road. This could be interpreted in the way that the R lineage is found preferentially in the mountainous part of Waslala municipality, where it is planted on slopes that steeply descend from the roads. The map (Figure 5) supports this notion. This also means that in the higher elevations (the mountainous south-west), the farms are located higher above the roads than in the lowlands. These higher altitudes with slightly increased mean annual rainfall experience lower temperatures, as is suggested by the strong, negative correlation (Table 4).

The two orphan trees, FBBSB, representing the pure Y genotype lineage, are located at an average altitude of 373 m at a relatively dry area (mean annual precipitation; 2333 mm; within the lower one sixth of the range for all trees sampled), where it is relatively warm (mean annual temperature 23.8°C; compared to the maximum temperature for all sampled farms being 24.9°C). Similarly, the 12 trees representing the BY group all grew in low, relatively dry and warm places (average for this group; 266 m elevation, 2396 mm mean annual rainfall, 24.5°C mean annual temperature).

Examination of the spatial distribution revealed that several single SSR alleles occur most frequently or exclusively in locations close to the main road (Figure 6). A total of 17 alleles are unique to a buffer zone of 2 km either side of the roads. As an example, allele mTcCIR292 occurs 18 times exclusively in these farms. In contrast, only four alleles were found uniquely in the area 2–15 km away from the nearest road. The number of effective alleles is also higher close to the road (3.2 within the 2 km corridor, relative to 2.36 further away; with the degree of expected heterozygosity, He, being 0.652 vs. 0.548, respectively). These increased levels of allelic diversity nearer to the roads suggest possibly more intense introduction of genetic materials along access roads. There is, however, a possible bias in sampling frequency (206 trees near, and 104 far from, the roads) that could interfere with part of these differences.

Genetic lineage and fruit type

Assignment to one of three morphological fruit types, Acriollado, Común, and Híbrido, was achieved for 250 trees.

Table 3. Allelic frequencies and parameters of clustered SSR multilocus genotypes among cacao trees in Waslala, excluding trees with extremely admixed (<39% purity) genotypes.

Minimum purity[1]	No. different alleles (Na)			
(15 loci)	n	B[2]	Y	R
39%	224	5.067	1.600	7.133
75%	116	4.400	1.600	6.267
85%	70	2.867	1.000	5.333
95%	32	1.533	1.000	3.600
	No. alleles with min. 5% frequency (Na>=0.05)			
39%	224	2.733	1.600	4.867
75%	116	1.867	1.600	4.800
85%	70	1.467	1.000	3.933
95%	32	1.200	1.000	3.600
	No. effective alleles (Ne)			
39%	224	1.593	1.246	4.060
75%	116	1.379	1.246	3.778
85%	70	1.239	1.000	3.297
95%	32	1.159	1.000	2.273
	No. private alleles			
39%	224	0.267	0.000	2.267
75%	116	0.333	0.000	2.533
85%	70	0.400	0.067	3.333
95%	32	0.667	0.533	2.667
	Unbiased expected heterozygosity (UHe)			
39%	224	0.344	0.196	0.723
75%	116	0.246	0.196	0.711
85%	70	0.154	0.000	0.686
95%	32	0.094	0.000	0.574
	Shannon's Information Index (I)			
39%	224	0.693	0.268	1.523
75%	116	0.505	0.268	1.441
85%	70	0.297	0.000	1.326
95%	32	0.153	0.000	0.955

[1]Increased purity (complementary to reduced admixture) simulates increased strength of selection for pure genotype spectra.
[2]Genotype cluster, B; Blue, Y; Yellow, R; Red inferred founder genotype spectra.

Although all types were presented in all different locations, their ratios were not equal across the genetic lineages. The Común type was confined to the B lineage, except for a single individual in the R group (Table 6). For the two large groups B and R, whose members possess at least a two thirds share of the Blue and Red genotype spectra, respectively, the ratios of the frequent Acriollado and Híbrido trees were checked with the Chi-square test for goodness-of-fit. The B group had 34 trees assigned to the Acriollado type, and the R group had 8 of these. In total, in the Acriollado and Híbrido types, 85 B and 80 R individuals were recorded, therefore, 42.5 (85/2) B and 40 (80/2) R trees were expected to be encountered with the assumption of unbiased distribution of genetic lineages across the two fruit types (Table 6). Testing the observed frequencies of 34:8, B:R individuals to fit the expected ratio of 42.5:40, revealed an unequal distribution or departure from homogeneity (Chi square = 14.28, P<0.001; ***).

This allows the conclusion that the Acriollado morphotype is highly significantly underrepresented in the R lineage, and overrepresented in the B lineage. This was the most pronounced biased distribution found; with 81% of the Acriollado type within all B and R samples being present among the B lineage trees. Likewise, the Híbrido fruit type, albeit outnumbering the other varieties, was cumulated at the 5% level of marginal significance to the R genotype spectrum. Testing the observed frequency of 34:51 Acriollado:Híbrido individuals within the B genotype spectrum, and 8:72 within R, to fit the expected ratio of 21:61.5, revealed a similar bias at P<0.01 (**) in both comparisons.

Discussion

The genetic structure of smallholder cacao plantings in Waslala was investigated. This is an economically significant Nicaraguan

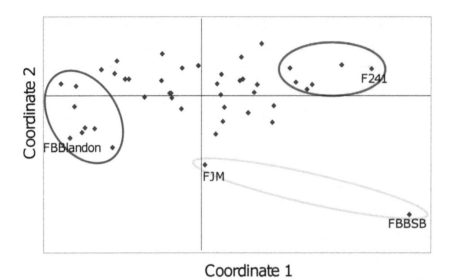

Figure 4. Principal Coordinates Analysis (PCA) on mean Shannon (sHua) values for pairwise farm comparisons. Plot of the first two main PCA axes. Comparisons included 15 SSR loci and 45 sites in Waslala, Nicaragua, represented by 317 cacao trees (first axis 39.9% and second axis 21.8% of total information). Circles indicate sites and farms with large portions (>75%) of the Blue, Red, and Yellow genotype spectra. Sites with largest shares of the genotype spectra are indicated by their code (compare with Figures 2 and 4, and Table S1).

area of production, where this crop has been grown since 1961. The majority of these cacao plantings appear to possess a large diversity of tree genotypes that seems to originate from a limited number of genotype spectra. Notwithstanding, the differences in allele and genotype composition at the farm level are important.

Markers used and allelic diversity

The 15 microsatellite loci sampled in this study are dispersed across nine of the ten linkage groups (chromosomes) of *Theobroma cacao*. These loci were selected as robust, informative markers for cacao and have been characterized in detail [14]. The 15 markers have been widely used to assess the genetic diversity and redundancy among new cacao collections and within clonal collections held at genebanks [1,3,5,15]. Therefore, these markers were considered appropriate to assess the cacao genepool present at the municipalities of Waslala/Rancho Grande, Nicaragua. The markers are anonymous and unlikely to target specific expressed genes, therefore they can be considered as neutral, i.e. not under selection and thus are unbiased markers for this investigation of population structure.

To assess the allelic diversity in Waslala, the total number of alleles, and private alleles, can be used. The 116 individual alleles found within the samples are almost exactly one-half of the number of 231 alleles observed for the same loci among 548 accessions with distinct genotypes that were sampled by Zhang et al. [15] at the live cacao genebank in CATIE, Costa Rica. This means that the allelic richness in Waslala of 7.73 alleles per microsatellite locus, is approximately 50% of the richness within the CATIE collections that have 15.4 alleles per locus. The collection of the USDA-ARS Tropical Agricultural Research Station at Mayaguez, Puerto Rico, holds at the same SSR loci in total 132 alleles with 8.8 alleles per locus [16], being comparable to Waslala, although actual differences in the individual alleles are likely to exist. The level of allelic richness in Waslala is also comparable to that of a collection of semi-natural cacao from the upper Amazon, held at Universidad Nacional Agraria de la Selva, Tingo Maria, Peru [17], with allelic richness levels comparable to that of the USDA-ARS Mayaguez collection

[18]. A subgroup of Ecuadorian cacao collections recognized as being the genetically narrow 'Refractario', had in total 63 alleles and 4.2 alleles per locus [19]. Again, the identities of the alleles may be different although the same microsatellite loci were investigated.

Cacao population structure across plantings in Waslala

Of the 13 private alleles detected by the rarefaction method (Table 2), 8 are dispersed among only 7 trees from three farms. This supports the notion of the wide dispersal of a comparatively small set of common alleles across Waslala, although there is much diversity at the genotype level (a specific combination of alleles at all loci). Evidence for this arises from the occurrence of only a few highly similar SSR genotypes. There are only 7 groups of trees with matching genotypes (Table 1), pointing towards sufficient genetic recombination, probably achieved through planned crosses. The small number of matching genotypes also indicates that during the sampling, trees of clonal origin were successfully omitted. The main method of tree propagation in Waslala is by seed, although in recent years, grafting of scions onto established rootstocks of trees that are cut due to low productivity, has become an alternative method.

The experimental station and germplasm distribution unit in Nicaragua, El Recreo, receives cacao germplasm from CATIE, and apparently, seed from crosses at El Recreo were distributed to Nicaragua's production zones including Waslala. During 1991–96, considerable dispersal of seed from controlled Trinitario×Forastero crosses and from clonal propagation of superior Trinitario genebank accessions was recorded in Waslala (S. Thienhaus, FADCANIC, Centro Agroforestal Sostenible, Wawashang, Nicaragua, 2010, pers. comm.), and the Cacaonica cooperative was involved in the distribution of this germplasm to farms sampled in this study. Nonetheless, the data on alleles and genotypes shows that the material used may have been selected from certain parts of the genotype spectra available in cacao.

The considerable differences were observed in the frequency and on-farm composition of genotypes across farms, as witnessed

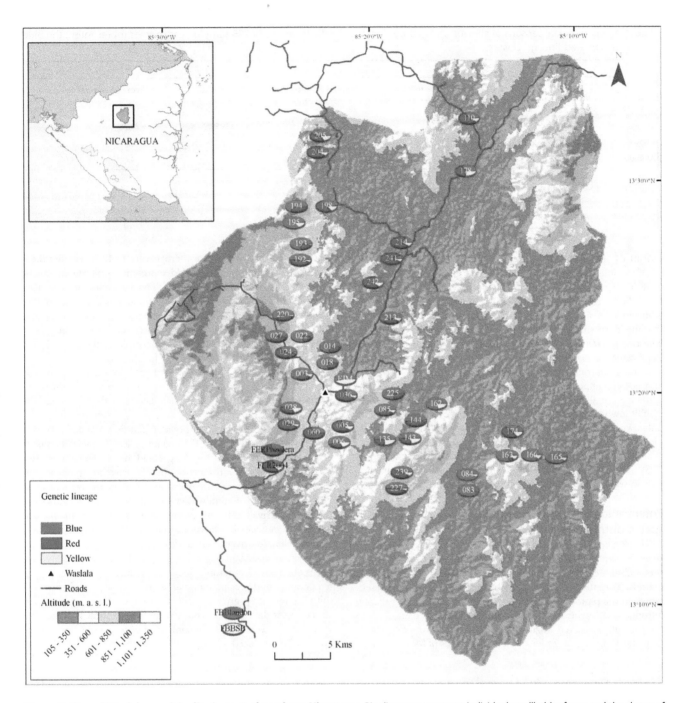

Figure 5. Map of Waslala municipality in central northern Nicaragua. Pie diagrams represent individual smallholder farms and the shares of putative founder genotype spectra, B, R, and Y, totalled over all cacao trees sampled.

by the spatial distribution of genotypes with widely differing degrees of lineage admixture (Figure 5). This may reflect seed trade activities of the past. Nonetheless, neither differentiation-based diversity (principal coordinates analysis on mean Shannon values, Figure 4), nor probabilistic inference of population structure [20] revealed any indication of more than three distinct genotype spectra within all samples from Waslala. Likely causes for this include the preference by farmers for only a few sources of genetic material for unknown reasons, newly introduced trees of <20 years of age may not yet be among the high-yielding trees and were thus not sampled, or the parents used for the crosses were closely related. The inference of population structure applied

here can only give information on the number of genotype spectra that are discernible in the existing data set. However, it cannot assess the absolute magnitude of diversity any of these single genotype spectra consists of. Likewise, at this stage it is problematic to trace any individual donors of the B and R genotype spectra due to the large number of choices that are available at the genebanks (e.g. SSR fingerprints of clonal accessions offered by the International Cocoa Germplasm Database; www.icgd.reading.ac. uk/index.php). This can be achieved by integrating the current data on the Nicaraguan populations with information on particular parental material that may have contributed to this genepool.

Table 4. Pairwise comparisons of climatic and geographic data for the locations of 295 sampled cacao trees representing the Blue, Red, and Blue-Red lineage clusters.

	Elevation relative to road	Annual precipitation	Altitude	Average annual temperature
Distance to road	−28.4***	72.1***	−25.6***	20.2***
Elevation relative to road		−13.5*	79.9***	−78.7***
Annual precipitation			12.1*	−15.7**
Altitude				−94.6***

Coefficients of correlation (in percent) and levels of significance (as determined by F tests in regression analyses) are shown.
*; P<0.05,
**; P<0.01,
***; P<0.001.

Origin of the Y genotype

Of the inferred three founder genotype spectra, two, B and R, were frequent and widespread, whereas only the two cacao trees from the forest, W041 and W042 represented the pure, non-admixed Y lineage. Several instances point toward the assumption that the Y trees may indeed represent the ancient Criollo lineage. The two forest trees were growing in a wild state, and appeared significantly older than all the managed plantation trees. Farmers do not harvest fruit from such forest trees because of their low yield and small, unpigmented seed. Criollo is known to possess extremely small allelic diversity, small unpigmented seed, and exhibit low yields. The majority of Criollo trees were killed by an unknown incident in 1727 [21], and only a few plants apparently escaped by chance, with rare trees to be found at sheltered sites near ancient settlement places in this Central American region [2,3]. However, confirmation of the two orphan trees being Criollo will require additional comparative studies.

Potential identity of the B and R genotypes and their spatial distribution patterns

The B and R lineages are present predominantly in admixed states (Figure 2), and residues of the Y lineage were detected by the probabilistic clustering method within a minority of the BR hybrids. Y-admixture could mean hybridization with Y representatives in the past, but it could also mean that intercrosses among introduced BY hybrids could have split the putative Y lineage into the presently observed fragmentary levels. Such parental hybrids could be Trinitario accessions which are hybrids of Mesoamerican (Criollo) and Amazonian (Forastero) cacaos [3,4,5]. Whether the B or the R lineages, or both, could represent Trinitario cannot be discerned with the data available. The fact that no clear R-Y hybrids were found among the 317 samples could be the result of insufficient time for this hybridization to take place. It also suggests that R may represent most recent introductions that have been intentionally hybridized with B, for example in the crossing and propagation programs conducted in the early 1990s at the Cacaonica cooperative and by other organizations. The hypothesis that the R lineage was only recently introduced is shown by its preferential distribution near to main access roads and around the town of Waslala itself (Figures 5 and 6).

The B lineage is more widespread in plantings situated relatively further from main roads. This could reflect farmers' habits of distributing their plants or, they could be a remnant of two successive periods of introduction, the B lineage being older. It is, however, somewhat remarkable that the R genotypes have not found a wider distribution within the purported 15–30 years since their likely arrival, as the maximum distances from the main roads within the municipality rarely exceed 15 km (Figure 5).

Microclimate-driven spatial distributions of individual genotypes within wild plant (including grass and tree) populations have been observed. In nature, subtle differences of shading [22], temperature and precipitation variation [23], or precipitation and soil alkalinity [24], are sufficient to strongly influence population structure. At the relatively young plantings found in Waslala, that vary from 13 years old to maximum of 49 years old, single trees are quickly replaced when they are unproductive, affected by

Table 5. Summary results of general linear models for analysis of variance of climatic and geographic factors for three abundant, inferred cacao genotype spectra in Waslala municipality.

Factor	Genotype spectrum P (F test)	Replication P (F test)	Multiple means comparison	Corresponding mean values		
Distance to road	***	**	B BR R	4492	2534	1905 m
Elevation relative to road	*	-	R BR B	96	51	34 m
Annual precipitation	**	*	B BR R	2452	2411	2407 mm
Altitude	-	-	B BR R	428	438	458 m
Average temperature	-	-	B BR R	23.5	23.4	23.4°C

Levels of significance of dependent variables Genotype spectrum and Replication (representing individual trees within a genotype, used for calculation of the error term). Multiple means comparisons were made with the Waller-Duncan function in SAS-GLM; items connected by an underscore are not significantly different. Number of samples included by genotype cluster; 107 B, 111 R, 77 BR.
*; P<0.05,
**; P<0.01,
***; P<0.001; –; (not significant).

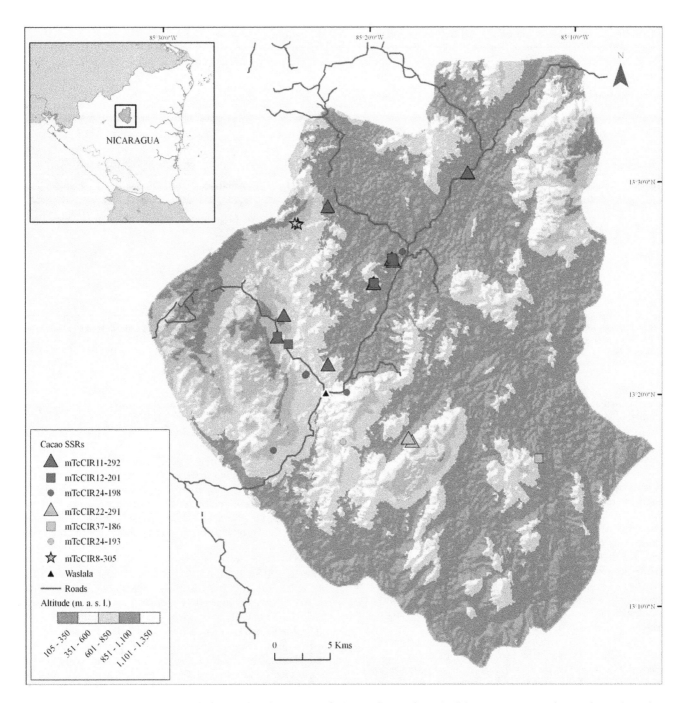

Figure 6. Map of trees possessing alleles confined to zones relative to the roads. Pink; alleles unique to areas close to the road (0–2 km; shared among 44 trees). Amber; alleles occurring far from the road (2–15 km; 7 trees).

diseases, or when more promising planting material become available. Under these circumstances, and because the majority of locally available material belong to only two basic genotype spectra, it cannot be excluded at present that microclimatic variations, in particular precipitation, may be a factor that partially determines the spatial distribution of these genotypes, alongside management practices. Again, clarity can only be obtained through additional experiments.

Remarkably, the Y lineage putatively representing the ancient Criollo type has a narrow distribution in an area that experiences relatively low annual precipitation and relatively higher mean annual temperatures. This may point to the preferred environ-

mental conditions that facilitate the survival of this lineage under unmanaged conditions, and may help to elucidate the nature of the unknown incident that wiped out the Criollo crop in 1727 [21]. However, these findings must be treated with caution due to the small number of Y individuals.

Distribution of fruit types

Despite the great variability of morphological characteristics, the distribution of types identified by fruit shape, seed color and size (Acriollado and Común varieties) and in addition, to a limited extent the technology of production (for the Híbrido type), was unequal across the three inferred genotypes B, Y, and R. The B

Table 6. Frequencies of morphological fruit types relative to inferred genotype spectrum group.

All trees identified

Fruit type\genotype	B[1]	BY	BR	R	Total
Acriollado	34	2	22	8	66
Común	9	0	0	1	10
Híbrido	51	8	43	72	174
Total	94	10	65	81	250

B and R groups only[2]	B	R	Total	observed	expected	Chi squ	P
Acriollado	34	8	42	34:8	42.5:40	14.28	***
Híbrido	51	72	123	51:72	42.5:40	5.27	*
Total	85	80					
observed	34:51	8:72					
expected	21:61.5	21:61.5					
Chi square	9.47	10.06					
P	**	**					

[1]For codes of genotype spectrum (or founder genetic lineage) clusters, compare legend of Table 3. BY; cluster containing individuals with admixed Blue-Yellow, BR; Blue-Red, genotypes at equal proportions.
[2]Bottom section; fitness-to-homogeneity tests of frequencies across genetic lineage and fruit types showing observed and expected numbers of individuals, Chi square value, and corresponding probability level, P (*; P<0.05, **; P<0.01, ***; P<0.001).

genotype contained nine of the ten Común-type trees distinguished by their melon shaped fruits. Among the two main genetic lineages, B and R, B represented most of the Acriollado type trees (Table 6). The Híbrido type is the only vernacular 'variety' that is applied to trees based on a mix of categories; fruit morphology and recorded technique of their production by controlled crosses. Accordingly, trees recognized as Híbrido occurred in all inferred genetic lineages at high frequencies, although R, the lineage with the largest distribution along main roads and more influenced by new introductions, had slightly more Híbrido individuals than B, at the marginal significance level of P<0.05. Therefore, fruit and seed morphology are, at least in part, genetically determined, and can be selected for by visual examination. Exploring the features that lead to the identification of vernacular varieties as is demonstrated with Híbrido trees, is recommended. However, as the designation to this type is based on a mixture of natural and technical criteria, its usefulness is limited.

In conclusion, the multilocus genotypes as detected by the 15 microsatellite markers can be used directly to denominate and recognize individual cacao trees and farms. This opens a means to select and breed for further enhancement of the crop and diversification of cocoa quality, both within the entire area and at the farm level. Of the two scenarios for future breeding, enhancement using the existing germplasm, or hybridization with superior imported material, the latter could likely disturb the already established and valued site-specific cocoa quality based on existing alleles and genotypes. Multi-year measurements of the culinary quality of cocoa and chocolate from the sampled trees are under way, and if these experiments reveal distinct features of the lineages, this could open up opportunities for breeding and selecting genotypes conferring elite quality.

Materials and Methods

Forty four cacao plantings in smallholder farms were selected to represent 14 climatic zones within the municipality of Waslala, Nicaragua. Two naturally occurring orphan cacao trees remaining from recently cleared forest were also included. This group is referred to as derived from "farm FBBSB".

A total of 315 trees identified were selected as consistently high yielding by their owners, and two low-yielding FBBSB trees, on average 7 trees per location (range 2–20). Eight locations were represented by less than 5 trees. High yield was defined as the stable production of many fruits year-round. This 'high yield' of individual trees as observed by the farmer may depend on the degree of stylar self-compatibility, distance from neighboring cacao and shade trees, and degree of fertilization, rather than on the genotype, and the principle of random sampling was therefore adhered to. Care was taken to sample non-grafted seedlings. New, fully expanded adult leaves were dried on silicagel in sealed plastic bags and shipped and stored at room temperature until use. Total genomic DNA was extracted from dry leaf tissue with the Dneasy Plant Mini Kit (Qiagen) according to the manufacturer's protocol.

Three types, mainly defined by morphological characteristics of the fruit (pod) and seed (beans) were identified. Acriollado has white beans, and pulp color and fruit shape with some resemblance to the original Criollo type. Común was used to describe trees producing fruits of one Forastero morphotype, namely Amelonado, possessing spherical pods similar in shape to honey melons (*Cucumis melo*). Finally, Híbrido was used to describe plants producing pods of intermediate shape and characteristics, as they occur frequently after hybridizing crosses of Forastero and Criollo. These pods often are elongated with pronounced acuminate tips and reduced seed size. The Híbrido classification was also applied to trees reported to be obtained from seed programs by the Nicaraguan genebank, El Recreo, or by the Honduran Foundation of Agricultural Research (FHIA), that are creating varietal hybrids through controlled crosses.

Primers for 15 simple sequence repeat (SSR or microsatellite) markers [25] specified in Table 7 were purchased from Sigma. For each marker, one of the primers was labelled with a fluorescent dye (FAM or HEX), and the PCR amplicon was separated on ABI Prism 3100 and ABI Prism 3130xl capillary sequencers to visualize the microsatellite alleles. The data generated in the Sequencher 4

Table 7. Cacao microsatellite (simple sequence repeat; SSR) primers [27] used to fingerprint trees from plantings in Waslala, Nicaragua, 2007–2009.

SSR code[1]	EMBL No	5'-Primer	3'-Primer	Chr	Size (bp)	AT °C
mTcCIR1	Y16883	GCAGGGCAGGCTCAGTGAAGCA	TGGGCAACCAGAAAACGAT	8	128–146	59
mTcCIR6	Y16980	TTCCCTCTAAACTACCCTAAAT	TAAAGCAAAGCAATCTAACATA	6	225–247	48
mTcCIR7	Y16981	ATGCGAATGACAACTGGT	GCTTTCAGTCCTTTGCTT	7	147–162	53
mTcCIR8	Y16982	CTAGTTTCCCATTTACCA	TCCTCAGCATTTTCTTTC	9	286–305	50
mTcCIR11	Y16985	TTTGGTGATTATTAGCAG	GATTCGATTTGATGTGAG	2	287–337	48
mTcCIR12	Y16986	TCTGACCCCAAACCTGTA	ATTCCAGTTAAAGCACAT	4	186–220	55
mTcCIR15	Y16988	CAGCCGCCTCTTGTTAG	TATTTGGGATTCTTGATG	1	231–257	50
mTcCIR18	Y16991	GATAGCTAAGGGGATTGAGGA	GGTAATTCAATCATTTGAGGATA	4	330–354	53
mTcCIR22	Y16995	ATTCTCGCAAAAACTTAG	GATGGAAGGAGTGTAAATAG	1	272–291	48
mTcCIR24	Y16996	TTTGGGGTGATTTCTTCTGA	TCTGTCTCGTCTTTTGGTGA	9	185–202	51
mTcCIR26	Y16998	GCATTCATCAATACATTC	GCACTCAAAGTTCATACTAC	8	281–306	46
mTcCIR33	AJ271826	TGGGTTGAAGATTTGGT	CAACAATGAAAATAGGCA	4	271–350	53
mTcCIR37	AJ271942	CTGGGTGCTGATAGATAA	AATACCCTCCACACAAAT	10	133–186	50
mTcCIR40	AJ271943	AATCCGACAGTCTTTAATC	CCTAGGCCAGAGAATTGA	3	258–294	51
mTcCIR60	AJ271958	CGCTACTAACAAACATCAAA	AGAGCAACCATCACTAATCA	2	186–211	53

[1]The code of the SSR and corresponding EMBL accession number, PCR primers, number of the cacao chromosome (Chr), fragment size (Size), and PCR annealing temperature (AT) are indicated.

software (Gene Codes Corp., Ann Arbor, USA) was analyzed with the aid of Genotyper, Peak Scanner 1 (ABI), or Genemapper programs. The individual alleles were labelled by the size in bases of their largest repeat. The PCR was replicated to up to five times to eliminate uncertainties. Together with newly shipped samples, previously analyzed control samples were included to provide the correct assignment of allele sizes. For each sampled tree, DNA was isolated once or twice. Trees were sampled during three years, from 2007 to 2009. For several trees, a second leaf was sampled in a different year.

Basic parameters on the samples' genetic composition and allele frequencies were calculated using the GenAlEx application [10] in Microsoft Excel. Principal coordinates analysis (PCA) and analysis of molecular variance were also performed in GenAlEx. For PCA, the mean Shannon mutual information indices (sHua) for pairwise farm comparisons were calculated as the fraction of Total Information index across each pair of populations, which were comprised of the weighted Allele Information indices of both populations in the pair, for each locus (compare www.anu.edu.au/BoZo/GenAlEx/new_version.php, GenAlEx Tut1, p. 35). The genotypes were further analyzed with Bayesian statistical methods in the program *Structure* [11] to attempt to trace the number and genetic composition of founder populations or kinships in Waslala cacao plantings. Settings for the simulations in *Structure* were 100,000 permutations during the burnin phase and 50,000 to 100,000 during simulations under a model allowing for genotype admixture.

Spatial climate data were extracted from Worldclim (www.worldclim.org). This database provides detailed information on climate characteristics at 1 km×1 km-resolutions, and its estimated tolerance of annual precipitation values is 10–25 mm for this part of Central America [8].

Geographic information system (GIS) analyses and maps were made with the DIVA-GIS software (www.diva-gis.org). Administrative and access information was based on maps by MARENA, the Nicaraguan Ministry of Environment and Natural Resources [26].

Planned potential associations among geographic and climate variables and inferred genotypes were tested by correlation and regression analyses in Excel or by general linear models in SAS (SAS Institute Inc., Cary, USA), of the type $Y = \beta X + e$, where X is the discrete genotype, e the error represented by the replication dependent variance, and Y the individual factor of influence, where appropriate. The individual trees within one genotype group were considered as replications for this genotype.

Supporting Information

Table S1 List of the 317 cacao trees. Owner; farmer's name. Comarca; rural district. Lineage; inferred genotype spectrum. Climate zone; defined by average temperature and precipitation. CIR1–CIR60; SSR fingerprint. The two alleles at each SSR locus are listed in two columns within one row.

Acknowledgments

Dapeng Zhang and Liz Johnson, USDA-ARS, gave advice on the choice and application of appropriate SSR markers. We thank Eduardo Somarriba, Carlos Astorga, Wilberth Phillips, and Luis Orozco at CATIE for their support with this research. We based our selection of farms on a survey of cocoa producing households in Waslala conducted in 2007 by the CATIE-led PCC (Proyecto Cacao de Calidad) project. Susanne Thienhaus, FADCANIC RAAS, coordinator of the Innovation Program at the Centro Agroforestal Sostenible, Wawashang, Nicaragua, shared information on the distribution of cacao germplasm. Nora Castañeda, Bioversity/CIAT, made valuable contributions to the final layout of Figures 5 and 6.

This study has been conducted in the framework of a research and development project entitled 'Sustainable futures for indigenous smallholders in Nicaragua: harnessing the high-value potential of native cacao diversity'. The project aims at assisting farmers in Waslala to make a gradual transition from their current production of only average quality cocoa to the production of differentiated high value cocoa based on the optimized use of cacao biodiversity, choice of locality and appropriate postharvest procedures.

Author Contributions

Conceived and designed the experiments: BT XS MH. Performed the experiments: KH-H BT AK HG. Analyzed the data: BT XS. Contributed reagents/materials/analysis tools: BT XS AK. Wrote the paper: BT XS.

References

1. Motamayor JC, Lachenaud P, Wallace da Silva e Mota J, Loor R, Kuhn DN, et al. (2008) Geographic and genetic population differentiation of the Amazonian chocolate tree (*Theobroma cacao* L). PloS ONE 3(10): e3311. Available: http://www.plosone.org/article/info%3Adoi%2F10.1371%2Fjournal.pone.0003311.

2. Motamayor JC, Risterucci AM, Lopez PA, Ortiz CF, Moreno A, et al. (2002) Cacao domestication I: the origin of the cacao cultivated by the Mayas. Heredity 89: 380–386.

3. Motilal LA, Zhang D, Umaharan P, Mischke S, Mooleedhar V, et al. (2009) The relic Criollo cacao in Belize – genetic diversity and relationship with Trinitario and other cacao clones held at the International Cocoa Genebank, Trinidad. Plant Genetic Resources: Characterization and Utilization. doi: 10.1017/S1479262109990232.

4. Motamayor JC, Risterucci AM, Heath M, Lanaud C (2003) Cacao domestication II: progenitor germplasm of the Trinitario cacao cultivar. Heredity 91: 322–330.

5. Johnson ES, Bekele FL, Brown SJ, Song Q, Zhang D, et al. (2009) Population structure and genetic diversity of the Trinitario cacao (*Theobroma cacao* L.) from Trinidad and Tobago. Crop Sci 49: 564–572.

6. Lanaud C, Sounigo O, Amefia YK, Paulin D, Lachenaud P, et al. (1987) Nouvelles données sur la fonctionement du systéme d'incompatibilité du cacaoyer et ses consequences pour la selection. Café Cacao Thé 31(4): 267–277.

7. Merrill T (1994) Pre-Colonial Period. In: Merrill T, ed. Nicaragua: a country study (3rd edition). Washington DC: Library of Congress, Available: http://lcweb2.loc.gov/frd/cs/nitoc.html#ni0013.

8. Hijmans RJ, Cameron SE, Parra JL, Jones PG, Jarvis A (2005) Very high resolution interpolated climate surfaces for global land areas. Intl J Climatol 25: 1965–1978.

9. Kalinowski ST (2004) Counting alleles with rarefaction: Private alleles and hierarchical sampling designs. Conservation Genet 5: 539–543.

10. Peakall R, Smouse PE (2006) GenAlEx 6: Genetic analysis in Excel. Population genetic software for teaching and research. Molecular Ecology Notes 6: 288–295.

11. Pritchard JK, Stephens M, Donnelly P (2000) Inference of population structure using multilocus genotype data. Genetics 155: 945–959.

12. Evanno G, Regnaut S, Goudet J (2005) Detecting the number of clusters of individuals using the software STRUCTURE: a simulation study. Mol Ecol 14: 2611–2620.

13. Pritchard JK, Wen X, Falush D (2010) Documentation for structure software: Version 2.3. Chicago, USA: Univ Chicago. pp 15–17. Available: http://pritch.bsd.uchicago.edu/structure_software/release_versions/v2.3.3/structure_doc.pdf.

14. Motilal LA, Zhang D, Umaharan P, Mischke S, Boccara M, et al. (2008) Increasing accuracy and throughput in large-scale microsatellite fingerprinting of cacao field germplasm collections. Tropical Plant Biol online 2(1). Available: http://www.springerlink.com/content/120913/?k=Motilal.

15. Zhang D, Mischke S, Johnson ES, Phillips-Mora W, Meinhardt L (2009) Molecular characterization of an international cacao collection using microsatellite markers. Tree Genet & Genomes 5: 1–10.

16. Irish BM, Goenaga R, Zhang D, Schnell R, Brown JS, et al. (2010) Microsatellite fingerprinting of the USDA-ARS Tropical Agriculture Research Station cacao (*Theobroma cacao* L.) germplasm collection. Crop Sci 50: 656–667.

17. Zhang D, Arevalo-Gardini E, Mischke S, Zúñiga-Cernades L, Barreto-Chavez A, et al. (2006) Genetic diversity and structure of managed and semi-natural populations of cocoa (*Theobroma cacao*) in the Huallaga and Ucayali valleys of Peru. Ann Bot, doi: 10.1093/aob/mcl146. Available: http://www.aob.oxfordjournals.org.

18. Zhang D, Mischke S, Goenaga R, Hemeida AA, Saunders JA (2006) Accuracy and reliability of high-throughput microsatellite genotyping for cacao clone identification. Crop Sci 46: 2084–2092.

19. Zhang D, Boccara M, Motilal L, Butler DR, Umaharan P, et al. (2008) Microsatellite variation and population structure in the "Refractario" cacao of Ecuador. Conserv Genet 9: 327–337.

20. Falush D, Stephens M, Pritchard JK (2003) Inference of population structure using multilocus genotype data: linked loci and correlated allele frequencies. Genetics 164: 1567–1587.

21. Wood GAR, Lass RA (1985) Cocoa. 4th edn. London: Longman. 620 p.

22. Li Y-C, Röder MS, Fahima T, Kirzhner VM, Beiles A, et al. (2002) Climatic effects on microsatellite diversity in wild emmer wheat (*Triticum dicoccoides*) at the Yehudiyya microsite, Israel. Heredity 89: 127–132.

23. Hübner S, Höffken M, Oren E, Haseneyer G, Stein N, et al. (2009) Strong correlation of wild barley (*Hordeum spontaneum*) population structure with temperature and precipitation variation. Mol Ecol 18: 1523–1536.

24. Muir G, Fleming CC, Schlötterer Ch (2000) Species status of hybridizing oaks. Nature 405: 1016.

25. Saunders JA, Mischke S, Leamy EA, Hemeida AA (2004) Selection of international molecular standards for DNA fingerprinting of *Theobroma cacao*. Theor Appl Genet 110: 41–47.

26. MARENA (2005) Atlas de la Reserva de la Biosfera Bosawas. Managua, Nicaragua: Ministerio del Ambiente y los Recursos Naturales (MARENA). 65 p.

27. Risterucci AM, Grivet L, N'Goran JAK, Pieretti I, Flament MH, et al. (2000) A high-density linkage map of *Theobroma cacao* L. Theor Appl Genet 101: 948–955.

4

Regeneration of *Betula albosinensis* in Strip Clearcut and Uncut Forests of the Qinling Mountains in China

Yaoxin Guo[1], Gang Li[2], Youning Hu[3], Di Kang[3], Dexiang Wang[3], Gaihe Yang[1]*

1 College of Agronomy, Northwest A&F University, Yangling, Shaanxi, China, 2 College of Life Science, Northwest A&F University, Yangling, Shaanxi, China, 3 College of Forestry, Northwest A&F University, Yangling, Shaanxi, China

Abstract

To contribute to a better understanding of the regeneration strategy of *Betula albosinensis* forests and the likely reasons behind either the successful recovery or failure after strip clearcutting, we compared the population structures and spatial patterns of *B. albosinensis* in eight *B. albosinensis* stands in Qinling Mountains, China. Four cut and four uncut stands were selected, and each sampled using a single large plot (0.25 ha). Results indicated that, on the one hand, *B. albosinensis* recruitment was scarce (average of 48 stems ha^{-1}) in the uncut stands, relative to the mature population (average of 259 stems ha^{-1}), suggesting a failure of recruitment. On the other hand, the subsequent regeneration approximately 50 years after the strip clearcutting showed that the density of the target species seedlings and saplings has increased significantly, and the current average density of seedlings and saplings was 156 stems ha^{-1}. The clumped spatial pattern of *B. albosinensis* suggested that their regeneration was highly dependent on canopy disturbance. However, recruitment remained poor in the uncut stands because most gaps were small in scale. The successful regeneration of sunlight-loving *B. albosinensis* after strip clearcutting was attributed to the exposed land and availability of more sunlight. Bamboo density did not influence *B. albosinensis* recruitment in the uncut stands. However, stand regeneration was impeded after strip clearcutting; thus, removing bamboo is essential in improving the competitive status of *B. albosinensis* at the later stage of forest regeneration after clearcutting. The moderate severity of disturbance resulting from strip clearcutting reversed the degeneration trend of primary *B. albosinensis* stands. This outcome can help strike a balance between forest conservation and the demand for wood products by releasing space and exposing the forested land for recruitment. Life history traits and spatiotemporal disturbance magnitude are important factors to consider in implementing effective *B. albosinensis* regeneration strategies.

Editor: Ben Bond-Lamberty, DOE Pacific Northwest National Laboratory, United States of America

Funding: This research was funded by the Key Program of the National Basic Sci-Tech Special Fund of China (2007FY110800). The funders had no role in study design, data collection and analysis, decision to publish, or preparation of the manuscript.

Competing Interests: The authors have declared that no competing interests exist.

* E-mail: ygh@nwsuaf.edu.cn

Introduction

Betula albosinensis, a deciduous hardwood, is a tree species endemic to China, distributed in the mid-high mountains of warm temperate regions. As one of the most important species in the Qinling Mountains, *B. albosinensis* thrives over a wide elevation range of 1950 m to 2750 m [1,2]. The community (i.e., species composition, spatial structure, and gap characteristics) [3–5] and seed germination characteristics [6,7] of *B. albosinensis* forests have been studied. No direct study, however, has reported on the regeneration of *B. albosinensis* trees in forests where they dominate after natural disturbance or artificial management treatments. This type of ecological knowledge is essential in implementing conservation strategies and ensuring the sustainable utilization of forests [8,9].

Tree regeneration is influenced by many factors, such as the life history attributes and disturbance of species and the competitive interactions among them. Disturbances play an important role in the regeneration dynamics of many mature hardwood forests [8–12]. For pioneer tree species, natural disturbance is regarded as an important measure of population persistence [13]. In China, the natural regeneration of *B. albosinensis* forests in the Qinling

Mountains is poor and may soon be replaced by other stable species [5,14–15]. In response to the increasing demand for forest products through regeneration, strip clearcutting was conducted in several *B. albosinensis* forests in the Qinling Mountains as a sustainable alternative for forest regeneration [7,16–17]. Nevertheless, further studies from ecological and silvicultural perspectives on the regeneration dynamics of *B. albosinensis* after strip clearcutting are required.

Understory bamboos in temperate and tropical subalpine forests are particularly effective in reducing seedling recruitment and tree regeneration when they reach a high degree of dominance [18–21]. In the Qinling Mountains, *Fargesia qinlingensis* is a common understory bamboo in *B. albosinensis* forests and dominates the understory in most sites. Therefore, understanding the role of *F. qinlingensis* in *B. albosinensis* forests may be critical in determining the regeneration dynamics of the latter. In this study, we analyzed the population structures and spatial patterns of *B. albosinensis* populations in strip clearcut and uncut *B. albosinensis* forests under different bamboo covers. The objectives included the following: (1) identify the regeneration patterns of *B. Albosinensis*, (2) to examine whether *B. albosinensis* regeneration after strip clearcutting was adequate to grow a new forest, and (3) to determine the influence

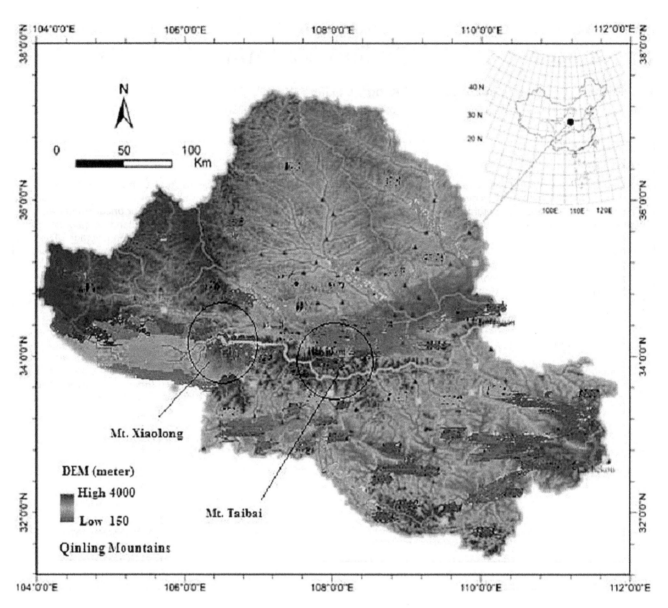

Figure 1. Location of the study areas in the Qinling Mountains of China.

of understory bamboo on the regeneration and community structure of *B. albosinensis*.

Methods

Study Area

This study was conducted at the Mt. Taibai National Nature Reserve (33°49 to 34°10′N, 107°19′ to 107°58′E, Shaanxi Province) and Mt. Xiaolong National Nature Reserve (33°35′ to 34°06′N, 106°13′ to 106°34′E, Gansu Province), located in the middle and western areas of the Qinling Mountains in China, respectively (Fig. 1). The Qinling Mountains run east–west and act as an important watershed divider between two great rivers of China, the Yangtze River and the Yellow River, which constitute a transitional zone between northern subtropical zone and warm-temperate zone. Mt. Taibai is the highest mountain in the Qinling Mountains, which spans an altitudinal gradient of 530 to 3767 m. Mean annual rainfall is 750 to 1100 mm, primarily falling in June through August, which are also the warmest months with mean

monthly temperature of 13.9 and 12.1°C, and December and January are the coldest months with monthly temperature of −5.7 and −4.4°C [15,22]. Elevation in the Mt. Xiaolong ranges from 704–3300 m. Annual precipitation ranges from 460–850 mm, most of which falls between July and September. Annual mean temperature ranges from 7 to 13°C [2]. The Nature Reserves were established for multiple-uses, including research, animal protection and forest production. Research activities were conducted under the scientific use permits issued respectively by Forestry Department of Shaanxi Province and Forestry Department of Gansu Province. Our field study did not involve any endangered or protected species in the Nature Reserves.

The numbers in parentheses are the estimated seedlings and saplings established in canopy gaps.

B. albosinensis forest is an important type of forest vegetation of Qinling Mountains, distributed from 1950 to 2750 m in Mt. Taibai and from 2000 to 2600 m in Mt. Xiaolong. *B. ablosinensis* forests in Mt. Xiaolong were strip clearcut in 1960 s and 1970 s to promote regeneration [2]. In Mt. Taibai, however, human

Table 1. Characteristics of stands and *B. ablosinensis* population in strip clearcut and uncut *B. ablosinensis* forest of Qinling Mountains.

Stands	I	II	III	IV	V	VI	VII	VIII
	Uncut	Uncut	Uncut	Uncut	Strip clearcut	Strip clearcut	Strip clearcut	Strip clearcut
Altitude (m)	2368	2418	2463	2536	2387	2397	2468	2486
Slop (°)	30	28	17	15	32	25	28	26
Aspect	NW	N	NW	N	N	NE	N	W
Bamboo cover (%)	45	15	8	5	5	10	20	50
Density (DBH ≥ 5.0 cm,ha^{-1})	244	249	267	276	529	693	590	465
Basal area (m^2 ha^{-1})	18.1	17.6	22.8	19.3	24.8	24.7	27.6	23.8
Seedlings (<1.5 m tall, ha^{-1})	16 (11)	8 (6)	16(12)	8 (7)	72	88	58	46
Saplings(<5.0 cm DBH and ≥ 1.5 m tall, ha^{-1})	36 (31)	32 (26)	48(39)	28 (24)	106	86	92	76
Observed no. of species	13	16	20	18	15	18	16	11
Shannon diversity (including *F. qinlingensis*)	0.79	1.45	1.63	1.76	1.58	1.68	1.38	0.49

activities are rare due to the relatively difficult accessibility, and some of the only remaining intact forests occur in the region. Thus, the study chooses the stands in Mt. Taibai and Mt. Xiaolong to examine regeneration of *B. ablosinensis* comparatively.

Field Sampling

After reconnaissance, we selected four cut stands (V–VIII) in the strip clearcut area where the primary trees and undergrowth were felled from the base in the 1960 s, except some large individuals as mother trees. Four stands (I–IV) without cut disturbance were also sampled. Stands were selected if they met the following criteria: (i) *B. albosinensis* dominated the stands and represented the typical forest structure at each site; (ii) the trees in the stands were 100 years, thus classifying them as mature stands; and (iii) sampled stands included observable variations in bamboo cover.

In the current study, each stand was sampled with a large plot (50 m×50 m). Within each plot, all trees with diameter at breast height (or DBH, i.e., 1.3 m above ground level) longer than 5.0 cm were measured. Trees in the stands with DBH less than 5.0 cm and other woody plants with height taller than 1 m were included in five subplots measuring 5 m×5 m, with the trees distributed in the middle and at the four corners of each large plot. Bamboo cover in the stands was estimated by tallying the bamboo coverage in the subplots upon the culms.

Data Analyses

Woody plant diversity was computed for each stand using Shannon's formula [23]. Only woody plants that reached 1.0 m in height were considered in the diversity estimation. Species richness in a plot was computed as the number of woody species with stems taller than 1.0 m. Size-structure diagram for *B. albosinensis* was prepared to depict the frequency of different-sized individuals and to interpret the trends in population dynamics.

The spatial pattern of the *B. albosinensis* population was identified using Morisita's index [24]:

$$I_\delta = q \sum n_i(n_i - 1)/N(N - 1)$$

Where q is the number of quadrats, n_i is the number of individuals in the *i*th quadrat, and N is the total number of individuals in all quadrats. I_δ equals 1.0 when the population is randomly distributed, $I_\delta > 1$ if a population is clumped, and $I_\delta < 1$ if a population is regularly distributed. The intensity of pattern was interpreted from the magnitude of the index value. The greater the index value the greater the intensity of clumping. The scale of pattern (m^2) was identified by computing Morisita'index values for quadrats of varying size. Block sizes were computed using quadrats of 5×5, 5×10, 10×10, 10×15, 15×15, 15×20, 20×20, 20×25, 25×25, and 25×30 units. All of the statistical analyses were conducted using SPSS 11.5 software. Figures were plotted by Origin7.5 software.

Results

Stand Characteristics

The basal areas and densities of *B. albosinensis* in different stands varied (Table 1). Basal areas ranged from 18.1 m^2 ha^{-1} to 22.8 m^2 ha^{-1} and from 23.8 m^2 ha^{-1} to 27.6 m^2 ha^{-1} in the uncut and cut stands, respectively. Densities ranged from 244 stems ha^{-1} to 276 stems ha^{-1} in the uncut stands, whereas *B. albosinensis* were more abundant (range 465 stems ha^{-1} to 693 stems ha^{-1}) and had an approximate twofold increase in the cut stands. The bamboo cover ranged from 5% to 50% among the stands. In addition to *F. qinlingensis*, other woody species were also found in *B. albosinensis* forests (Table 2). The other woody plants in the uncut stands were more abundant than those in the cut stands because of strip clearcutting.

Understory Vegetation and Stand Structure

Except the evergreen bamboo (*F. qinlingensis*), deciduous species characterize understory vegetation in terms of number of species (Table 2). Differences in *B. albosinensis* seedling and sapling densities that were observed in the uncut stands seemed were unrelated to bamboo cover (Table 1). In the cut stands, however, *B. albosinensis* density was lower when bamboo coverage in the stands reached 50%. Moreover, seedling and sapling densities were negatively correlated (r = −0.98, p<0.01, Spearman rank correlation coefficient) with bamboo coverage (Table 1). Reduction in other woody plant abundance was also evident when bamboo coverage increased to 50%.

Table 2. Number of woody plants (>1.0 m tall) ha^{-1} in the *B.ablosinensis* stands in Qinling Mountains, China.

Speices	I	II	III	IV	V	VI	VII	VIII
	Uncut	Uncut	Uncut	Uncut	Strip clearcut	Strip clearcut	Strip clearcut	Strip clearcut
Pinus armandii	48	84	132	104	32	16	28	56
Betula utilis		16	48	96	24	56	112	48
Tilia latevirens	48	80	46	14	48	32	8	4
Acer robustum	56	64	36	48	6	16	32	
Abies fargesii			32	82	8	14	84	96
Ribes glaciale	68	105	76	38	16	44	80	
Populus davidiana		48	38	16	40	16	32	16
Pterocarya stenoptera						56	84	16
Rose omiensis	96	102	136	12	56	82		30
Lonicera fangutica		64	102	44	84	46	34	
Sorbus discolor	64	72	84	40				
Meliosma cuneifolia		16	32		16	32	64	
Rhododendron purdomli		32	64	96				
Salix matsudana					32	32	64	
Salix caprea	32	48	28	24				
Cerasus tomentosa					48	16	32	24
Populus purdomii	24	16	32	8	8	16		
Quercus liaotungensis	28	32	8					
Cornus controversa	24		16				44	16
Litsea tsinlingensis					8	16	32	
Sorbus koehneana	8	32		6	16		32	
Acer coesium ssp.giraldii			8	40				
Corylus mandshurica	35	24			8	4		
Acer miaotaiense					6		6	34
Populus purdomii		8	32	4				
Acer shenlauensis				36	12	6	4	
Ilex yunnanensis		14	58					
Pertya sinensis			6	28				
Total	531	857	1014	736	468	500	772	340
Fargesia qinlingensis	1040	790	560	320	470	580	820	1250

Blanks represent no individuals found in the plot.

Bamboo density reduced woody plant diversity (Table 1). There was also a negative correlation ($r = -0.76$, $p = 0.032$, Spearman rank correlation coefficient) between species richness and bamboo cover. When *F. qinlingensis* densities were included in the Shannon computation, a negative correlation ($r = -0.87$, $p = 0.005$, Spearman rank correlation coefficient) was also found between species diversity and bamboo cover (Table 1).

Size Structure

The population of *B. albosinensis* showed a bell-shaped diameter distribution pattern in the uncut stands (Figure 2). *B. albosinensis* trees mainly thrived in the middle and larger diameter classes, whereas young *B. albosinensis* with DBH less than 15 cm were scarce and accounted for just 11.9% of the total individuals, suggesting poor recruitment over the past decades. *B. albosinensis* seedlings and saplings were sparser, and only several were found in each uncut stand. The diameter class distribution for *B. albosinensis* in strip clearcut stands had a reverse J shape (Figure 2), with young

stems (<20 cm) accounting for 52.8%. Recent recruitment was also abundant, suggesting successful regeneration over the past 50 years after strip clearcutting.

Spatial Patterns

Spatial distributions of small (DBH<10 cm) and larger (DBH >10 cm) *B. albosinensis* differed between the cut and the uncut stands (Figure 3). In the uncut stands, small *B. albosinensis* were clumped at almost all scales (25 m^2 to 750 m^2), and the clumping magnitude fluctuated severely among different scales, suggesting a heterogeneously natural disturbance. In comparison, small and larger *B. albosinensis* clumps of highest intensity both occurred at small scales (25 m^2 to 150 m^2), suggesting that recruitment in single-tree gaps was common. In the cut stands, the high clumping of small *B. albosinensis* occurred at small to intermediate scales, whereas taller *B. albosinensis* trees were clumped only at intermediate scales (200 m^2 to 400 m^2). Such results suggested that

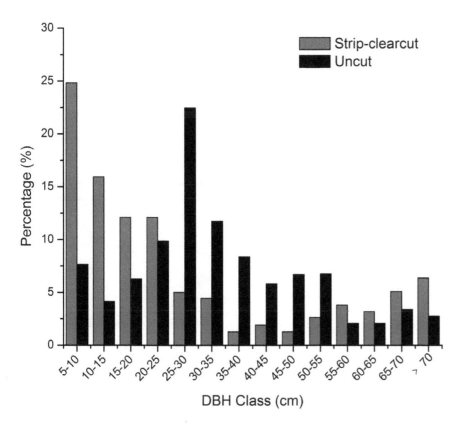

Figure 2. Size class distribution of the *B. ablosinensis* trees in strip clearcut and uncut *B. ablosinensis* forests of Qinling Mountains, China.

B. albosinensis recruitment in the past period was initiated by large disturbance.

Discussion

Seedling and sapling pools of *B. albosinensis* in the uncut stands were scarce, suggesting a degeneration trend. Poor natural regeneration of *B. albosinensis* forests elsewhere in China has also been reported [25,26]. The absence of *B. albosinensis* recruitment generally indicates an unfavorable environment for regeneration. Thick forest litter is a main factor influencing regeneration because of restrictive seed germination in deciduous forests. Based on laboratory simulation experiment results, the germination rate of *B. albosinensis* seeds declined when covered with mulch, especially broad-leaf mulch [6,7]. Similar effects of forest litter on germination have also been observed in *Betula alleghaniensis* in Canada [27] and *Betula maximowicziuna* in Japan [28]. General field observations demonstrate that *B. albosinensis* are prone to germination in places with less litter fall (e.g., under open-canopy or near roads). Even after a successful germination, *B. albosinensis* seedlings beneath closed canopies easily die as a result of shade sensitivity, although fast-growing seedlings can compensate for less shade-tolerance.

Disturbances drive the regeneration dynamics of most closed-canopy forests by creating opportunities that facilitate the establishment of new individuals through canopy opening [8,11,29]. In the present work, poor *B. albosinensis* regeneration was observed in the uncut stands, and these mainly occurred in the canopy gaps produced by disturbances. This finding suggests that *B. albosinensis* are dependent on gaps for regeneration. The intensely clumped distributions of small (<10 cm DBH) and larger (>10 cm DBH) *B. albosinensis* at small scales (25 m^2 to 150 m^2) are consistent with the canopy gap sizes (20 m^2 to 100 m^2) surveyed in the forests of the Qinling Mountains [4]. Larger *B. albosinensis* were found clumped with lower intensity, indicating a characteristic of thinned population [11]. *B. albosinensis* possess life history traits associated with the gap characteristics, and are less shade-tolerant. In addition, *B. albosinensis* produce more frequent seed crops, have smaller seeds that disperse farther, and have seedlings that grow faster. These life history traits promote the rapid colonization and early dominance of *B. albosinensis* in the gaps. Therefore, gap disturbance seems critical for the maintenance of *B. albosinensis* populations.

However, *B. albosinensis* recruitment in canopy gaps in this report was very limited in number, which may be related to the disturbance scale. *Betula* in other subalpine forests [30–32] exhibit great dependency on relatively large gaps for regeneration. *B. albosinensis* also appear to require large gaps for persistence. Natural disturbances that have occurred in *B. albosinensis* forests, including standing death and snapping of canopy trees caused by heavy snow, diseases, and climber twining, are frequently small-scale and rarely large-scale [4,14]. As a result of these disturbances, light and ground layers are only slightly changed over a small area. Most gaps are filled by either vegetation growth of the surrounding adults or replacement of shade-tolerant species that are already present as suppressed individuals. These frequent small-scale disturbances may not fulfill the demand of *B. albosinensis* regeneration for an environment with much light and exposed soil. Therefore, without large-scale disturbances to clear space and expose the covered land, it may be difficult for *B. albosinensis* stands to persist for generations to come.

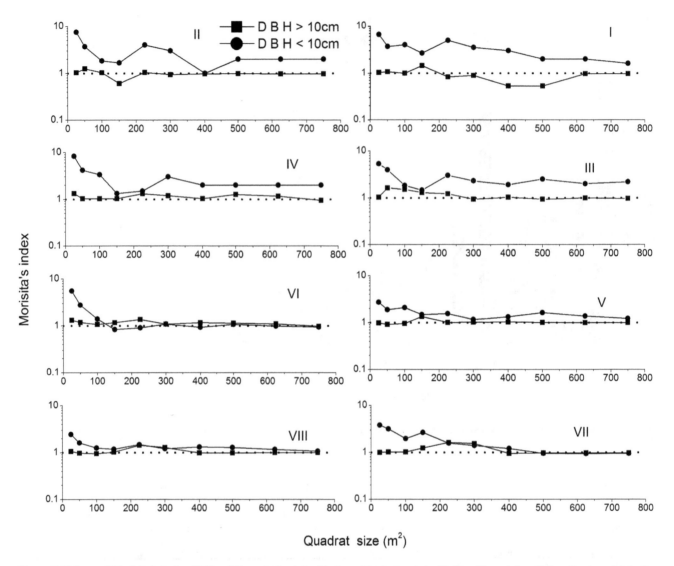

Figure 3. Values of Morisita's index (I_δ) for different-sized _B. ablosinensis_ in 8 stands in Qinling Mountains, China. Random distribution ($I_{\delta} = 1.0$) is shown by the dot line.

Bamboo, as a common understory plant in subalpine forests, restricts tree regeneration and species diversity [20–21,29,33]. Where bamboo fully occupies a forest understory, the frequency and number of tree seedlings and shrubs are lower because of intense competition with bamboo for space and resources. In our study, a significant negative correlation was found between understory woody plant diversity (richness and Shannon's formula) and _F. qinlingensis_ cover, suggesting that dense bamboo reduced plant diversity. This finding is consistent with published reports on other subalpine forests where bamboos dominate the forest understory [18,34,35–37]. However, no significant correlation was found between _B. albosinensis_ recruitments and _F. qinlingensis_ cover despite the wide range (5% to 45%) in the uncut stands. Previous studies in mixed hardwood-conifer forests have suggested that understory bamboos impede _Betula_ regeneration, although it did so with less intensity than conifer [36,37].

It is likely that there are other restricting factors (e.g., forest litter, closed canopy, and dense shrubs) that weaken the impeding effect of bamboo alone on _B. albosinensis_ regeneration. In addition, the ability of _B. albosinensis_ to disperse into gaps as well as go through a fast-growing juvenile stage may more or less help them

break the shade of bamboo layers. Therefore, bamboo density has no significant effect on the distribution and establishment of _B. albosinensis_ seedlings and saplings. However, understory bamboo may contribute greatly to the persistence of _B. albosinensis_ population by restricting coexisting conifer species [32,37]. Furthermore, bamboo flowering usually creates a favorable environment for _B. albosinensis_ regeneration as a large forest disturbance.

Contrary to failed natural regeneration, a large number of recruitments were found in stands with subsequent regeneration approximately 50 years after strip clearcutting. _B. albosinensis_ dominated the regenerating stands as a pioneer species. After strip clearcutting, the sudden exposure of previously forested lands and more sunlight reaching the forest floor seemed responsible for the successful regeneration of sunlight-loving _B. albosinensis_. The establishment of _B. albosinensis_ in large numbers after strip clearcutting suggests that disturbance magnitude is important for understanding the regeneration strategy of _B. albosinensis_. However, we found that high bamboo coverage in the cut stands reduced _B. albosinensis_ seedling and sapling recruitment. With vegetation restoration after the cut, _B. albosinensis_ regeneration became prone to bamboo restriction when they attained a high degree of dominance. As such, other artificial silvicultural methods, such as bamboo removal, may be necessary in improving the competitive status of *B. albosinensis* at the later stage of forest regeneration after clearcutting.

The above results confirm that strip clearcutting can prevent primary *B. albosinensis* stands from degenerating. The moderate severity of disturbance caused by strip clearcutting may be the best approach to achieve the dual objectives of forest conservation and timber production. Palynological evidence suggests that *Betula* forests have existed as zonal forests in the geological period and in modern times in the Qinling Mountains [38]. Naturally, *B. albosinensis* population may be capable of maintaining their stability as a whole despite the particularly poor regeneration. Thus, the aggregation of different-spatiotemporal cohorts driven by disturbances may be the pattern and strategy of natural *B. albosinensis*

population stability. However, persistent and periodic artificial disturbance by strip clearcutting is necessary from a forest production perspective.

Acknowledgments

Great thanks to the Mt. Taibai National Nature Reserve and the Mt. Xiaolong National Nature Reserve for their helps in field work and logistical support. We are also grateful to Junjie Yang and Lingtong Kong for the assistance in field work.

Author Contributions

Provided direction and suggestions for the manuscript: GL DW GY. Conceived and designed the experiments: YG GL. Performed the experiments: YG DK YH. Analyzed the data: YG. Wrote the paper: YG.

References

1. Zhang YQ (1989) Shaanxi Forest. Xi'an: Xi'an Science and Technology Press. (In Chinese.).
2. Li ZY, Zhang YC, Li J (2002) The forest of Xiaolong Mountain. Tianshui: Xiaolongshan Forest Experiment Bureau. (In Chinese.).
3. Fu ZJ, Guo JL (1994) Preliminary studies of *Betula albosinensis* forest in Taibai Mountain. Chinese Journal of Plant Ecology 18(3): 261–270. (In Chinese with English abstract.).
4. Su JW, Yue M, Wang YJ (2006) Gap characteristics of *Betula albosinensis* forest in Taibai Mountains. Chinese Journal of Applied and Environmental Biology 12(2): 195–199. (In Chinese with English abstract.).
5. Guo YX, Kang B, Li G, Wang DX, Yang GH, et al. (2011) Species composition and point pattern analysis of standing trees in secondary *Betula albosinensis* forest in Xiaolongshan of west Qinling Mountains. Chinese Journal of Applied Ecology 22(10): 2574–2580. (In Chinese with English abstract.).
6. Ren JY, Lin Y, Yue M (2008) Seed germination characteristics of *Betula albosinensis* at Mountain Taibai, China. Chinese Journal of Plant Ecology 32(4): 883–890. (In Chinese with English abstract.).
7. Wu Y, Liu Q, He H, Liu B, Yin H (2004) Effects of light and temperature on seed germination of *Picea asperata* and *Betula albosinensis*. Chinese Journal of Applied Ecology 15(12): 2229–2232. (In Chinese with English abstract.).
8. Cullen LE, Stewart GH, Duncan RP, Palmer JG (2001) Disturbance and climate warming influences on New Zealand Nothofagus tree-line population dynamics. Journal of Ecology 89: 1061–1071.
9. Antos JA, Parish R (2002) Structure and dynamics of a nearly steady-state subalpine forest in south-central British Columbia, Canada. Oecologia 130: 126–135.
10. Yahner RH (2000) Eastern Deciduous Forest: Ecology and Wildlife Conservation. Minneapolis, MN: University of Minnesota Press.
11. Veblen TT, Donoso C, Schlegel FM, Escobar B (1981) Forest dynamics in south-central Chile. Journal of Biogeography 8: 211–247.
12. Pebyelas J, Ogaya R, Boada M, Jump AS (2007) Migration, invasion and decline: changes in recruitment and forest structure in a warming-linked shift of European beech forest in Catalonia (NE Spain). Ecography 30, 829–837.
13. Takahashi A, Koyama H, Takahashi N (2008) Habitat expansion of *Robinia pseudoacacia L.* and role of seed banks in the Akagawa River basin. Journal of the Japanese Forest Society 90: 1–5.
14. Lin Y, Ren JY, Yue M (2008) Population structure and spatial analysis of Betula albosinensis at Taibai Mountain, northwestern China. Chinese Journal of Plant Ecology 32(6): 1335–1345. (In Chinese with English abstract.).
15. Li JJ, Zhu ZC, Min ZL (1989) Comprehensive Suvey of the Taibai Mountain Preserve. Xi'an: Shaanxi Normal University Press. (In Chinese.).
16. Basnet K (1992) Effect of topography on the pattern of trees in Tabonuco (*Dacryodes excelsa*) dominated rain forest of Puerto Rico. Biographica 24: 31–42.
17. Allison D, Art W, Cunningham E, Teed R (2003) Forty-two years of succession following strip clearcut in a northern hardwoods forest in northwestern Massachusetts. Forest ecology and management 182: 285–301.
18. Taylor AH, Shiwei J, Lianjun Z, Chunping L, Changjin M, et al. (2006) Regeneration patterns and tree species coexistence in old-growth Abies-Picea forests in southwestern China. Forest Ecology and Management 223: 303–317.
19. Gonzalez ME, Veblen TT, Donoso C, Valeria L (2002) Tree generation response in a lowland Nothofagus-dominated forest after bamboo die-back in south-central Chile. Plant Ecology 161: 59–73.
20. Narukawa Y, Yamamoto S (2002) Effects of coniferous and broad-leaved trees in a Japanese temperate mixed forest. Journal of Vegetation Science 2,413–418.
21. Dang HS, Zhang YJ, Zhang KR, Jiang MX, Zhang QF (2010) Age structure and regeneration of subalpine fir (*Abies fargesii*) forests across an altitudinal range in the Qinling Mountains, China. Forest Ecology and Management 259: 547–554.
22. Ren Y, Liu MS, Tian LH, Tian XH, Li ZJ (2006) Biodiversity, conversation and management of Taibaishan Nature Reserve. Beijing: China forestry Publishing House. (In Chinese.).
23. Shannon CE, Weaver W (1949) The Mathematical Theory of Communication. Urbana, IL: University of Illinois Press.
24. Morisita M (1959) Measuring of the dispersion of individuals and analysis of the distributional patterns. Memoirs of the Faculty of Science Kyushu University, Series E 2: 215–235.
25. Liu SJ, Su ZX (2004) T-square study on spatial pattern and regeneration of Betula albosinensis population on west slope of Jiuding Mountain. Chinese Journal Applied Ecology 15(1): 1–4. (In Chinese with English abstract.).
26. Miao N, Liu SR, Shi ZM, Yu H, Liu XL (2009) Spatial pattern of dominant tree species in subpine Betula-Abies forest in west Sichuan of China. Chinese Journal of Applied Ecology 20(6), 1263–1270. (In Chinese with English abstract.).
27. Peterson CJ, Facelli JM (1992) Contrasting germination and seedling growth of *Betula allenghaniensis Britton* and *Rhus typhina L.* subjected to various amounts and type of plant litter. American Journal of Botany 79: 1209–1216.
28. Osumi K, Sakurai S (1997) Seedling emergence of *Betula maximowicziana* following human disturbance and the role of buried viable seeds. Forest Ecology and Management 93: 235–243.
29. Taylor AH, Qin Z, Liu J (1996) Structure and dynamics of subalpine forests in the Wang Lang Natural Reserve, Sichuan, China. Vegetation 124: 25–38.
30. Koyama T (1984) Regeneration and coexistence of two *Abies* species dominating subalpine forests in central Japan. Oecologia 62: 156–161.
31. Nakamura T (1985) Forest succession in the subalpine region of Mt, Fuji, Japan. Vegetation 64: 15–27.
32. Taylor AH, Qin Z (1988) Tree replacement patterns in subalpine *Abies-Betula* forests, Wolong Natural Reserve, China. Vegetation 78: 141–149.
33. Takahashi K, Kohyama T (1999) Size-structure dynamics of two conifers in relation to understorey dwarf bamboo: a simulation study. Journal of Vegetation Science 10: 833–842.
34. Veblen TT, Veblen AT, Schlegel FM (1979) Understorey patterns in mixed evergreen-deciduous Nothofagus forests in Chile. Journal of Ecology 67: 809–823.
35. Peet RK (1981) Forest vegetation of Colorado Front Range: composition and dynamics. Vegetation 45: 3–75.
36. Taylor AH, Qin Z (1988) Regeneration patterns in old-growth *Abies-Betula* forests in the Wolong National Reserve, Sichuan, China. Journal of Ecology 76: 1204–1218.
37. Wang W, Franklin SB, Ren Y, Ouellette JR (2006) Growth of bamboo Fargesia qinlingensis and regeneration of trees in a mixed hardwood-conifer forest in the Qinling Mountains, China. Forest Ecology and Management 234: 107–115.
38. Zhu ZC (1991) Stability of the Betula forest in the Taibai mountain of Qinling mountain range. Journal of Wuhan Botanical Research 9: 169–175. (In Chinese with English abstract.).

5

Factors Affecting Spatial Variation of Annual Apparent Q_{10} of Soil Respiration in Two Warm Temperate Forests

Junwei Luan[1], Shirong Liu[1]*, Jingxin Wang[2], Xueling Zhu[3]

1 The Research Institute of Forest Ecology, Environment and Protection, Chinese Academy of Forestry, Key Laboratory of Forest Ecology and Environment, China's State Forestry Administration, Beijing, PR China, **2** West Virginia University, Division of Forestry and Natural Resources, Morgantown, West Virginia, United States of America, **3** Baotianman Natural Reserve Administration, Tuandong, Chengguan Town, Neixiang County, Henan Province, PR China

Abstract

A range of factors has been identified that affect the temperature sensitivity (Q_{10} values) of the soil-to-atmosphere CO_2 flux. However, the factors influencing the spatial distribution of Q_{10} values within warm temperate forests are poorly understood. In this study, we examined the spatial variation of Q_{10} values and its controlling factors in both a naturally regenerated oak forest (OF) and a pine plantation (PP). Q_{10} values were determined based on monthly soil respiration (R_S) measurements at 35 subplots for each stand from Oct. 2008 to Oct. 2009. Large spatial variation of Q_{10} values was found in both OF and PP, with their respective ranges from 1.7 to 5.12 and from 2.3 to 6.21. In PP, fine root biomass (FR) (R = 0.50, P = 0.002), non-capillary porosity (NCP) (R = 0.37, P = 0.03), and the coefficients of variation of soil temperature at 5 cm depth (CV of T_5) (R = −0.43, P = 0.01) well explained the spatial variance of Q_{10}. In OF, carbon pool lability reflected by light fractionation method (L_{LFOC}) well explained the spatial variance of Q_{10} (R = −0.35, P = 0.04). Regardless of forest type, L_{LFOC} and FR correlation with the Q_{10} values were significant and marginally significant, respectively; suggesting a positive relationship between substrate availability and apparent Q_{10} values. Parameters related to gas diffusion, such as average soil water content (SWC) and NCP, negatively or positively explained the spatial variance of Q_{10} values. Additionally, we observed significantly higher apparent Q_{10} values in PP compared to OF, which might be partly attributed to the difference in soil moisture condition and diffusion ability, rather than different substrate availabilities between forests. Our results suggested that both soil chemical and physical characters contributed to the observed large Q_{10} value variation.

Editor: Ben Bond-Lamberty, DOE Pacific Northwest National Laboratory, United States of America

Funding: This study was jointly funded by the Ministry of Finance (numbers 200804001 and 201104006), China's National Natural Science Foundation (30590383; 31200370), the Ministry of Science and Technology (2011CB403205, 2008DFA32070, 2006BAD03A04), and CFERN & GENE Award Funds on Ecological Paper. The funders had no role in study design, data collection and analysis, decision to publish, or preparation of the manuscript.

Competing Interests: The authors have declared that no competing interests exist.

* E-mail: liusr@caf.ac.cn (SL)

Introduction

Soils are the largest carbon pool in the terrestrial ecosystem, estimated to contain almost three times as much carbon as the atmosphere between the depths of 0–300 cm of soil [1,2]. This value is much higher if northern permafrost regions are also considered [3]. Annual CO_2 efflux from soil respiration (R_S), the second largest terrestrial carbon flux, is ten times higher than CO_2 efflux from fossil burning [4,5]. R_S is also probably the least well constrained component of the terrestrial carbon cycle [6]. Thus, the response of R_S to climate change, which usually is called apparent temperature sensitivity of R_S (Q_{10} value) and estimated based on empirical functions, is of importance in predicting possible feedbacks between the global carbon cycle and the climate system [7]. Recently, the efficiency and accuracy of R_S estimation based on apparent Q_{10} values and the method used to estimate Q_{10} values [7,8], has been widely debated [9]. Nevertheless, empirical response functions are still a valid method to derive annual estimates of R_S based on specific field measurements (e.g. Savage et al. [10]), particularly when it is not limited by water content and the simulation is made through interpolation rather than extrapolation [11].

The Q_{10} of R_S has been a focus of R_S research and is widely reported in the literature. Soil moisture condition has been suggested to be a factor that affects Q_{10} [12–14]. However, a positive [14] or a topographic position dependent [13] relationship between soil moisture and Q_{10} has been reported. Davidson and Janssens [15] pointed out that soil moisture could exert a secondary effect on apparent Q_{10} due to its interaction with substrate availability [16]. The seasonal change in autotrophic respiration, which is driven by the strong seasonality in tree below ground C allocation, could also influence the variability in apparent Q_{10} values [17,18]. Thus annual and seasonal variations of Q_{10} values have been widely reported [14,19]. Furthermore, the relationship between soil organic matter (SOM) quality and temperature sensitivity of organic matter decomposition has been extensively studied recently [7,8]. Whether SOM of different quality has similar [20–22] or different temperature sensitivities has also been debated [23–25].

The variability of temperature sensitivity among ecosystems has been reported, accounting for substrate quality [23], climate factors [26], or different range of temperature used to estimate Q_{10} values [27]. Mahecha et al. [28] found a global convergence in the temperature sensitivity of respiration at the ecosystem level, but high spatial variation of temperature sensitivity exists within plots

[14,29]. Spatial variation of R_S has been discussed, e.g., in boreal forest [30]; tropical rainforest [31]; as well as savanna ecosystem [32]. However, direct field evidence of factors affecting the spatial variation of apparent Q_{10} values within plots has not been fully investigated, and it is still ambiguous whether variation is attributed to the spatial distribution of SOM quality or soil microclimate.

In this study, both a natural regenerated oak forest (OF) and a nearby artificially regenerated pine plantation (PP) were chosen in warm temperate China, to determine characteristics of spatial variability of apparent Q_{10} values within plot at locations in a 10 m×10 m grid based on R_S field measurements. Our specific objectives were to 1) identify the spatial variation of Q_{10} values in both OF and PP; and 2) determine factors correlated with spatial variability of Q_{10} values within each plot.

Materials and Methods

Study Sites and Experimental Design

The study sites were located at the Forest Ecological Research Station in the Baotianman Natural Reserve (111°47′–112°04′E, 33°20′–33°36′N), Henan Province, PR. China. Baotianman Natural Reserve Administration (Neixiang County, Henan Province) issued the permission for our experimental sites. The average elevation is 1400 m, with an annual mean precipitation and air temperature of 900 mm and 15.1°C, respectively. Precipitation occurs mainly in summer, accounting for 55–62% of the annual total [33]. Upland soils are dominated by mountain yellow brown soils (Chinese classification). The OF stand was dominated by *Quercus aliena* var. *acuteserrata*, while the nearby PP stand was dominated by *Pinus armandii* Franch (for detailed information of these two stands see Luan et al. [34]). No intensive management was conducted in the PP since its establishment. One 40 m×60 m study plot was delineated in each stand with an average slope of <8°. Within each plot, a 10 m×10 m square grid was then placed and 35 subplots (1 m×1 m) were positioned at each intersection of the grid. PVC collars (19.6 cm inside diameter) were installed at each subplot in September 2008 and were kept on the site throughout the study period.

Soil Respiration, Microclimate Measurements, and Q_{10} Calculation

Soil respiration measurements were conducted for a total of 12 (OF, measurement on 19 May, 2009 was canceled due to rain event) and 13 (PP) measurement campaigns using a Li-8100 soil CO_2 flux system (LI-COR Inc., Lincoln, NE, USA), from October 2008 to October 2009 avoiding snow cover period (9 and 17 Oct., 1 and 11 Nov. of 2008; 19 Mar., 7 and 17 Apr., 19 May., 2 and 23 Jun., 2 Aug., 19 Sept., and 19 Oct. of 2009). Sampling was performed between 9:00 and 15:00 (GMT +8:00). Soil temperature at 5 cm (T_5) was measured adjacent to each respiration collar with a portable temperature probe provided with the Li-8100. Soil volumetric water content (SWC) at 0–5 cm was measured with a portable time domain reflectometer MPKit-B soil moisture gauge (NTZT Inc., Nantong, China) at three points close to each chamber. We avoided early morning and post-rain measurements to reduce the possible effect of rapid transition on the soil respiration rate during the observations.

An exponential equation (Eqn (1)) was used to describe the temporal relationship between R_S and T_5 for each subplot (n = 12 for OF; or 13 for PP):

$$R_S = \alpha e^{\beta T_5} \qquad (1)$$

where R_S is soil respiration; T_5 is the soil temperature at 5 cm depth; and α and β are fitted parameters. The temperature sensitivity parameter, Q_{10} of each subplot was calculated as:

$$Q_{10} = e^{10\beta} \qquad (2)$$

Our analysis showed that one measurement fewer for OF compared to PP do not have significant impact on Q_{10} estimation (data were not shown).

The number of samples required to estimate the Q_{10} of R_S of each stand at the 10% or 20% of its actual value at the 95% probability level was obtained using Eq. 3 described by Hammond and McCullagh [35]:

$$n = \left(\frac{t_\alpha CV}{D} \right)^2$$

where t_α is Student's t with degrees of freedom ($\alpha = 0.05$), CV is the sample coefficient of variation derived from data obtained for this study, and D is allowable error of field sampling process.

Soil Properties, Root Biomass, and Carbon Pool Lability

Five soil samples were collected from the top 5 cm depth of the mineral soil next to each chamber using 100 ml (50.46 mm diameter, 50 mm height) sampling cylinders in August, 2009. Three soil samples were combined and used for mass-based measurements of soil organic carbon (SOC), total nitrogen (TN), and light fraction organic carbon (LFOC). The remaining two cylinder samples were used for analyses of bulk density (BD), total soil porosity (TP), capillary porosity and non-capillary porosity (NCP) on the basis of soil water-retention capacity [36]. Light fraction soil organic matter at a depth of 0–10 cm was obtained by the density fractionation method proposed by Six et al. [37], but with a modification using $CaCl_2$ solution (density of 1.5 g ml^{-1}; Garten et al. [38]). Bulk-soil and light-fraction organic carbon contents were determined by the wet oxidation method with 133 mM $K_2Cr_2O_7$ at 170–180°C [39]. In August 2009, roots were extracted from 0–30 cm fresh soil samples by two cores (10 cm diameter) located close to the collars. The samples were washed; coarse (>5 mm), medium (2–5 mm), and fine (<2 mm) roots were manually separated and then their dry biomass (70°C, 24 hours) was measured. We found that stand structure parameters (total basal area, maximum DBH for trees within 4 m (radius) of the measurement points) well explained the spatial distribution of fine root biomass [34], which indicated that the spatial pattern of fine root biomass is comparably stable, because stand structure is relatively stable for an ecosystem in a given time. The leaf area index (LAI) was measured above each subplot using hemispherical photographs with WinSCANOPY (Regent Instruments Inc., Quebec, Canada) in August 2009.

The term 'lability' of SOC was defined as the ratio of the oxidized to non-oxidized SOC [40]. We applied this definition to the density fractionation method, and calculated subplot carbon pool lability (L_{LFOC}) as described by Luan et al. [41]:

$$L_{LFOC} = \frac{LFOC}{SOC - LFOC} \qquad (3)$$

LFOC is the light fraction organic carbon and SOC is the soil organic carbon.

Statistical Analysis

Descriptive statistics (mean, range, standard deviation (SD) and coefficient of variation (CV)) were used to show the characteristics of the spatial variability of R_S, Q_{10}, and soil parameters. Variogram computations were also performed to determine the strength and scale of the spatial variability of Q_{10} and soil parameters. The spatial variability was quantified by the semivariance (γ (h)). The semivariance of any parameter z is computed as:

$$\gamma(h) = \frac{1}{2n(h)} \sum_{x=1}^{n} (z_x - z_{x+h})^2 \qquad (4)$$

where n (h) is the number of lag pairs at distance intervals of h and z_x and z_{x+h} are the values of the variable z at x and $x+h$, respectively. Plotting γ(h) against h gives the semivariogram, which will exhibit either purely random behavior or systematic behavior described by a theoretical model (linear, spherical, gaussian or power law distribution). The nugget, sill, range and structural variance (Q) parameters were obtained from the model with the best fit to the semivariance data. Geostatistical analyses were performed with GS+ (Geostatistics for the Environmental Sciences, v.5.1.1, Gamma Design Software, Plainwell, MI).

Pearson correlations were performed to assess factors (soil moisture, seasonal CV of T_5 and SWC, LFOC, L_{LFOC}, FR, NCP) controlling spatial variation of Q_{10} values among subplots for each forest (n = 35) or pooled data of two forests (n = 70). Geostatistical analyses showed that Q_{10} values and soil parameters were spatially independent (Fig. 1). This allowed us to treat our measurement locations as independent samples for inferential statistics. Therefore, general linear models (GLM) were employed to examine the effect of forest type on Q_{10} values, where L_{LFOC}, FR, SWC (averaged over 12 or 13 measurement campaigns), and NCP were included in the model as co-variables, respectively. Statistical analyses were conducted using SPSS version 13.0 (SPSS Inc., Chicago, USA).

Results

Microclimate and Soil Parameters Variance within Plots

All the subplots experienced similar seasonal fluctuations of T_5 and SWC (Fig. 2). High spatial variation of SWC was found in all measurement campaigns (Fig. 2), with the CV of SWC ranging from 10.7% to 27.2% for PP and from 10.7% to 26% for OF (Fig. 2). Soil carbon and nitrogen contents at 5 cm depths, the C/N ratio, soil bulk density, light fraction organic carbon, fine root biomass and soil carbon pool lability (L_{LFOC}) for the OF and PP showed high spatial variation in the stand (Table 1). The semivariograms of L_{LFOC}, FR, and NCP showed no change in semivariance with distance, indicating that they had no spatial autocorrelation in this scale (Fig. 1 a, b, d, f, g, i). Although averaged SWC had moderate spatial dependency, the ranges and sills observed were not precisely determined because the ranges were larger than the effective range of 43.27 m, which is equal to 60% of the maximum lag in the 10-m grids (Fig. 1 c, h).

Spatial Variation of Q_{10} Values

Exponential equation well described the relationship between R_S and T_5 for each subplot, and all the correlations were significant at the P<0.05 (R^2>0.34) level. The Q_{10} values varied considerably among subplots, ranging from 1.7 to 5.12 and 2.3 to 6.21 for the OF and the PP, respectively (Table 1). Among the Q_{10} values, 37.1% and 48.6% of them were between 4 and 5 for the

OF and the PP, respectively. Spatial distribution of Q_{10} values for both forests are shown in Figure 3. According to our power calculation, the number of measurements required to estimate the Q_{10} of R_S per stand within 10% or 20% of its actual value at the 0.05 probability level are 26 and 6 for OF, respectively, and 15 and 4 for PP. Geostatistical analyses showed that Q_{10} values had no spatial autocorrelation (Fig. 1e, 2j). The absence of autocorrelations among Q_{10} values and soil parameters allowed us to treat our measurement locations as independent samples for inferential statistics.

Controls on Q_{10} Variation

In PP, both FR and NCP were positively correlated with the Q_{10} values, while CV of T_5 was negatively correlated with the Q_{10} values (Table 2). In OF, we found a significantly positive correlation between L_{LFOC} and the Q_{10} values ($P=0.038$; Table 2). Regardless of forest type, L_{LFOC} and NCP were positively correlated, while SWC was negatively correlated with Q_{10} values (Table 2). No significant correlations between seasonal CV of SWC and Q_{10} were found for either forest or pooled data of two forests (Table 2). Significantly different Q_{10} values between forests was found ($F=4.517$, $P=0.037$; Table 3). However, significant difference in Q_{10} values between OF and PP disappeared when SWC or NCP was included as a co-variable in the GLM (Table 3).

Discussion

Spatial Variation of Q_{10} Values within Plots

Although the average Q_{10} values (3.80 and 4.25 for the OF and the PP) in this study was within the range of Q_{10} values reported in other temperate forests [42,43], there was a large variation in Q_{10} values between subplots, such as 1.7–5.12 for the OF and 2.3–6.21 for the PP (see Table 1). Spatial variability in Q_{10} was also reported in a managed Ponderosa pine *(Pinus ponderosa)* forest (1.2–2.5; Xu and Qi [14]) and in a Japanese cedar *(Cryptomeria japonica)* plantation (1.3–3.2; Ohashi and Gyokusen [29]). This large variation of Q_{10} values among subplots suggests a potential risk of bias estimation of the soil respiration at a plot scale, which has not been adequately addressed. Similar estimates for soil respiration sampling have also been made in other studies. It was recommended to measure at least eight locations to stay within 20% of its actual value at the 95% confidence level in a mature beech forest [44]. Saiz et al. [45] also suggested that the sampling strategy of 30 sampling points per stand was adequate to obtain an average rate of soil respiration within 20% of its actual value at the 95% confidence level in four Sitka spruce stands.

Controlling Factors on Q_{10} Variance

High spatial variance in soil moisture was found in both stands for most sampling dates (Figure 2), which could be attributed to the microtopography, the high spatial variability of soil organic matter content [34] and of root distribution (e.g. we found a significant negative correlation between SWC and fine root biomass $R^2 = 0.16$, $P = 0.021$, n = 35). Such a short scale soil moisture spatial variation have also been reported in other forests [29,46,47]. We even found a slight spatial autocorrelation for soil moisture (Fig. 1 d, i). It was reported that the high spatial variance of soil moisture exerted significant negative impact on soil respiration rate [34]. However, spatially, no significant impacts of soil moisture on Q_{10} values were found for PP and OF (Table 2).

In our study, all the subplots experienced similar seasonal fluctuations of soil temperature and moisture even though their magnitudes were different (Fig. 2). So we expect that there could be no obvious influence of different microclimate fluctuation on

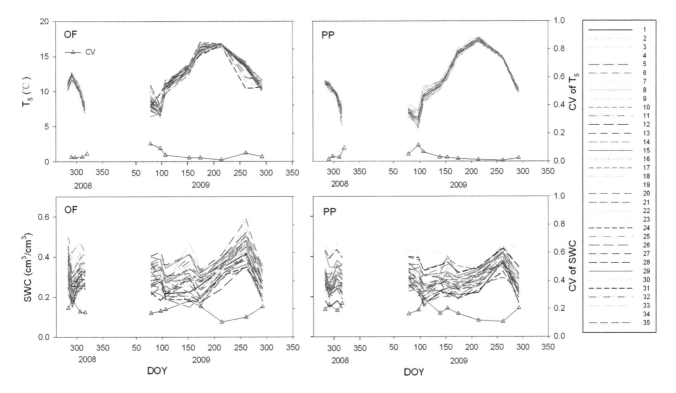

Figure 1. Semivariograms of L_{LFOC} (a, f), FR (b, g), SWC (c, h), NCP (d, i), and Q_{10} (e, j) in 10-m grid squares of OF (left panel) and PP (right panel), respectively. Model for SWC are exponential models. The SWC were averaged over the 12 (OF) or 13 (PP) measurement campaigns.

Q_{10} calculation at a given plot level in this study. However, the above mentioned influence was still found in PP where seasonal CV of T_5 correlated significantly with Q_{10} values (Table 2). Nevertheless, microclimate fluctuation difference can not fully

explain the spatial variability of Q_{10} values since no similar significant correlations were found in OF or when we pooled data together for all measurements regardless of forest types (Table 2). Therefore, we posit that the spatial variation of Q_{10} values among

Table 1. Statistical analysis of soil parameters, fine root biomass, soil respiration rate, Q_{10} values, and carbon pool lability (L_{LFOC}) for the oak forest and pine plantation.[a]

Parameters	Oak forest				Pine plantation			
	mean	S.D.	Range	CV	mean	S.D.	Range	CV
R_S ($\mu mol\, m^{-2} s^{-1}$)	2.12	0.58	1.16–4.17	0.27	2.01	0.44	1.07–3.16	0.22
Q_{10}	3.80	0.95	1.7–5.12	0.25	4.25	0.81	2.30–6.21	0.19
SOC (g/kg soil)	78.90	18.49	47.50–117.58	0.23	77.94	24.63	45.88–153.89	0.32
TN (g/kg soil)	6.03	1.38	3.65–9.26	0.23	5.17	1.28	3.27–8.82	0.25
C:N (g/g)	13.08	0.61	11.76–15.45	0.05	14.92	1.30	12.69–18.02	0.09
BD (g/cm³)	0.71	0.138	0.42–0.96	0.19	0.69	0.121	0.49–1.00	0.17
LAI (m²/m²)	3.50	0.60	2.60–4.90	0.17	2.96	0.30	2.41–3.68	0.10
Averaged SWC (cm³ cm⁻³)	0.31	0.0495	0.233–0.437	0.16	0.28	0.045	0.215–0.421	0.16
Seasonal CV of T_5	0.27	0.02	0.22–0.30	0.08	0.32	0.02	0.28–0.38	0.07
Seasonal CV of SWC	0.21	0.04	0.14–0.30	0.20	0.17	0.05	0.07–0.30	0.29
LFOC (g/kg soil)	30.55	12.22	16.85–64.17	0.40	28.57	20.53	7.53–101.17	0.72
L_{LFOC} (g/g)	0.69	0.43	0.31–2.58	0.62	0.64	0.49	0.13–2.12	0.77
FR (g/m²)	223.40	76.80	31.04–330.94	0.34	164.45	61.07	69.45–298.32	0.37
NCP (m³/m³)	0.084	0.031	0.015–0.14	0.365	0.097	0.032	0.045–0.18	0.325

[a]S.D.: standard deviation; CV: coefficient of variance; R_S: soil respiration; SWC: soil water content; TOC: total organic carbon; TN: total nitrogen; LFOC: light fraction organic carbon; FR: fine root biomass; BD: bulk density; LAI: leaf area index; NCP: non-capillary porosity. $n = 35$. The soil respiration rates R_S and SWC in this table were averaged over the 12 (OF) or 13 (PP) measurement campaigns.

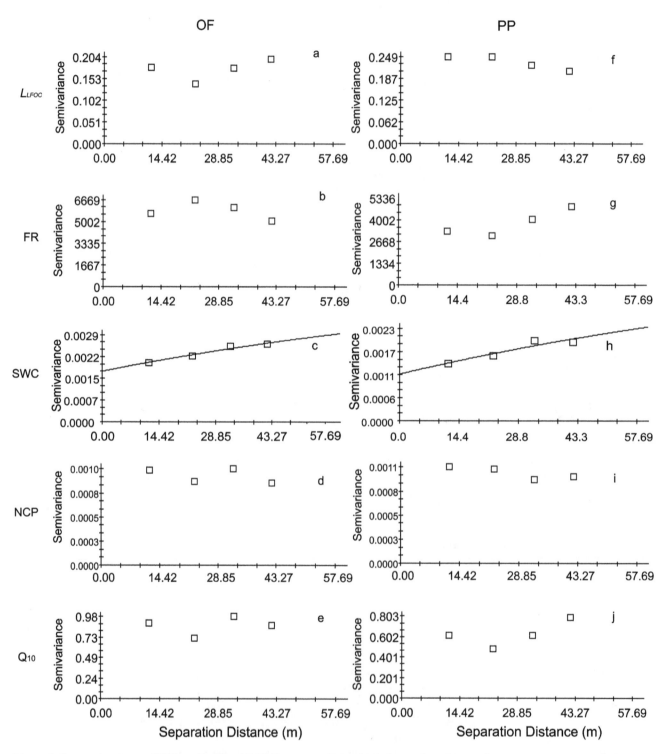

Figure 2. Seasonal pattern of T_5 (up panel) and SWC (lower panel) for OF (left panel) and PP (right panel) for each subplot, as well as the seasonal pattern of the CV (up triangle) of T_5 and SWC among subplots.

subplots should be associated with other inherent characteristics of each subplot, i.e, spatial differences in substrate availability as suggested by [15]. Gershenson et al. [16] also found a positive relationship between substrate availability and temperature sensitivity.

In our study, fine root biomass well explained the Q_{10} variance in PP, and was marginally significantly correlated with Q_{10} when we pooled data of all forest types (Table 2). Since fine roots are

associated with the fast turnover carbon pool [48–50], the positive linear correlation between Q_{10} and FR implied the positive relationship between Q_{10} and lability of the substrate. It was also reported that Q_{10} values may be related to seasonal change in autotrophic respiration [17]. The correlations between fine root biomass and Q_{10} may also imply there exists a connection between Q_{10} and autotrophic respiration, i.e., the higher autotrophic respiration was coincided with the higher fine root biomass in the

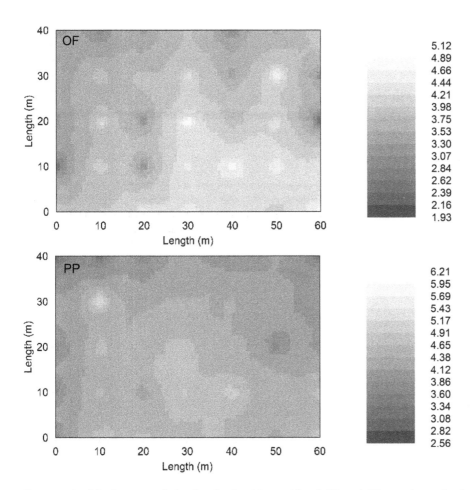

Figure 3. Isarithmic maps of the Q_{10} in the 10-m grids of OF and PP are shown in the top and bottom panels respectively, interpolations were done by the inverse distance weighting method. White areas indicate high values and dark areas indicate low values.

subplots. This inference was supported by our previous study as we found a similar positive correlation between FR and R_S [34].

Light fraction organic carbon (LFOC), which has been widely recognized as a labile carbon indicator [51,52], is comprised largely of incompletely decomposed organic residues with turnover times of years to decades [53], thus the concentration of LFOC can indicate substrate supply quantity to some extent [34,54,55].

There was no correlation found between Q_{10} and labile organic carbon concentration (LFOC) as reflected by light fractionation (Table 2). Nevertheless, significant correlations between carbon pool lability (L_{LFOC}) and Q_{10} were found in OF as well as when we pooled data together from all forest types (Table 2). This demonstrated that the carbon pool lability as reflected by light fractionation, which can partly stand for SOM quality [41], may

Table 2. Pearson correlation coefficients between Q_{10} and variables in spatially.

Independent Variables	Pine plantation		Oak forest		Pooled data of two forests	
	R	Sig. (2-tailed)	R	Sig. (2-tailed)	R	Sig. (2-tailed)
LFOC	0.178	0.306	0.161	0.355	0.142	0.241
L_{LFOC}	0.290	0.091	**0.351**	**0.038**	**0.293**	**0.014**
FR	**0.497**	**0.002**	0.240	0.165	0.207	0.086
SWC	−0.213	0.219	−0.246	0.155	**−0.290**	**0.015**
NCP	**0.369**	**0.029**	0.282	0.101	**0.355**	**0.003**
CV of T_5	**−0.426**	**0.011**	−0.245	0.157	−0.010	0.932
CV of SWC	−0.053	0.762	−0.112	0.521	−0.169	0.161

Abbreviations see Table 1. n = 35 for each forest, n = 70 for pooled data of two forest types. The SWC in this table were averaged over the 12 (OF) or 13 (PP) measurement campaigns.

Table 3. General Linear Models for examine forest type effect on Q_{10} values, where F test was conducted. L_{LFOC}, FR, SWC (averaged over 12 or 13 measurement campaigns), and NCP were taken as co-variables of the GLM respectively to examine which factor could exert influence on Q_{10} value difference between forest.

Variable type	Variables	F values	Sig.
Co variable	None	–	–
Fixed variable	Forest type	4.517	**0.037**
Co variable	L_{LFOC}	7.539	**0.008**
Fixed variable	Forest type	5.689	**0.020**
Co variable	FR	8.965	**0.004**
Fixed variable	Forest type	10.548	**0.002**
Co variable	SWC	3.8	**0.055**
Fixed variable	Forest type	2.14	0.148
Co variable	NCP	7.7	**0.007**
Fixed variable	Forest type	2.62	0.11

Abbreviations see Table 1. None: No co-variable.
For all tests, df = 1 for fixed variable and co variables, and df = 67 for error.

exert more impact on Q_{10} values compared to the concentration of LFOC. This indicates a connection between the spatial distribution of SOM quality and the apparent Q_{10} as we speculated.

Multi-pool soil C models have been employed to simulate changes in soil C stocks as a single, homogeneous soil C pool [56–58], but the same Q_{10} value for different carbon fractions have still been applied. With increasing the understanding of temperature sensitivity of different soil organic carbon fractions [7–9]. Our findings on the connection between Q_{10} values and C availability among subplots suggest that different Q_{10} values corresponding to carbon fractions with different turn over times should be incorporated into soil carbon models.

Q_{10} Values between Stands

In our study, Q_{10} values were significantly higher in the PP than that in the OF (Table 3), which is consistent with Wang et al.'s [59] findings in Korean pine plantation vs. *Mongolian* oak forest. Although we found significant correlations between L_{LFOC} and FR with Q_{10} values, GLM showed that both L_{LFOC} and FR can not explain why the higher Q_{10} occurred in the PP rather than in the OF (Table 3). No significant difference in Q_{10} values was found between PP and OF when averaged SWC was included as co-variables in GLM, but GLM showed a marginally significant correlation between averaged SWC and Q_{10}. This implied that different soil moisture conditions accounted for different apparent Q_{10} values in the studied forests. Higher water content could impede O_2 diffusion, thereby reducing decomposition rates and microbial production of CO_2. In this case, the temperature response of CO_2 efflux would be lower (i.e. a lower Q_{10} value) in wetter subplots than in dryer subplots, implying that the temperature response of CO_2 efflux would be lesser in wet years than in dry years as Davidson et al [43] reported.

Furthermore, we speculate that effects of soil moisture conditions on Q_{10} may be partly attributed to different soil physical characteristics, such as the soil non-capillary porosity, which is an important factor in relation to soil gas diffusion. This was confirmed by our analysis, which showed that there was no significant difference in Q_{10} values between PP and OF when NCP was included as a co-variable, while there was a significant positive correlation between the spatial distribution of NCP and Q_{10} values (Table 3). This also indicated that the difference in NCP between two forests resulted in the difference in Q_{10} values. Similarly, a weak spatial correlation between hardness (related to soil porosity) of the A layer and Q_{10} variation was reported by Ohashi et al. [29]. Conant et al. [9] recently also suggested that the physico-chemical protection from decomposition of organic matter (OM) will affect temperature response of SOM. A negative correlation between averaged SWC and NCP ($R = -0.306$, $P = 0.01$) in this study regardless of forest type also suggested that there was an interaction between soil moisture and porosity. Soil porosity could exert intense impacts on temperature sensitivity of R_S in combination with soil moisture condition. Therefore, lower Q_{10} values in the OF compared to that in the PP may have been partly caused by the higher soil moisture or lower NCP.

In contrast, Xu and Qi [14] reported a positive correlation between Q_{10} values and soil moisture, with SWC values range from 10% to 24%. In our study, however, SWC values were 0.23–0.389 m/m^3 for the PP and 0.241–0.451 m/m^3 for the OF, respectively, which was higher than that reported by Xu and Qi [14]. This implies that there is a complex relationship between Q_{10} and soil moisture, which may result in contrasting effects. A marginal critical soil moisture condition may exist which determines a positive or negative relationship between Q_{10} and soil moisture.

Conclusions

High spatial variances in apparent Q_{10} values were found for both forests. Parameters related to substrate availability and gas diffusion both exerted significant impact on the spatial variation of Q_{10} values within each stand. Higher Q_{10} values in the PP compared to the OF were also found, which could be attributed to the difference in soil moisture conditions or NCP, rather than substrate availability. Our results suggested that the R_S estimation at stand level could be improved through considering the spatial variation of Q_{10} values and its influencing factors.

Acknowledgments

We gratefully acknowledge the support of Yin Wu, Ye Tian, Xinfang Yang, Xiaojing Liu, and Jiguo Liu of the Baotianman National Nature Reserve for their assistance in field monitoring and sampling. We also thank Mr. Damon Hartley of West Virginia University for his valuable comments on the earlier versions of this manuscript. Special thanks also go to Dr. Ben Bond-Lamberty for his constructive comments and suggestions used to revise the manuscript.

Author Contributions

Conceived and designed the experiments: JL SL JW XZ. Performed the experiments: JL XZ. Analyzed the data: JL SL JW XZ. Contributed reagents/materials/analysis tools: JL SL XZ. Wrote the paper: JL SL JW XZ.

References

1. Jobbágy EG, Jackson RB (2000) The Vertical Distribution of Soil Organic Carbon and its Relation to Climate and Vegetation. Ecological Applications 10: 423–436.

2. Schimel DS (1995) Terrestrial ecosystems and the carbon cycle. Global Change Biology 1: 77–91.

3. Tarnocai C, Canadell JG, Schuur EAG, Kuhry P, Mazhitova G, et al. (2009) Soil organic carbon pools in the northern circumpolar permafrost region. Global Biogeochemical Cycles 23: GB2023, doi:2010.1029/2008GB003327.

4. Raich JW, Potter CS (1995) Global Patterns of Carbon Dioxide Emissions from Soils. Global Biogeochem Cycles 9: 23–36.

5. Hashimoto S (2012) A New Estimation of Global Soil Greenhouse Gas Fluxes Using a Simple Data-Oriented Model. PLoS ONE 7: e41962.

6. Bond-Lamberty B, Thomson AM (2010) A global database of soil respiration data. Biogeosciences 7: 1915–1926.

7. Davidson EA, Janssens IA, Luo Y (2006) On the variability of respiration in terrestrial ecosystems: moving beyond Q_{10}. Global Change Biology 12: 154–164.

8. Kirschbaum MUF (2006) The temperature dependence of organic-matter decomposition–still a topic of debate. Soil Biology & Biochemistry 38: 2510–2518.

9. Conant RT, Ryan MG, Ågren GI, Birge HE, Davidson EA, et al. (2011) Temperature and soil organic matter decomposition rates – synthesis of current knowledge and a way forward. Global Change Biology 17: 3392–3404.

10. Savage K, Davidson EA, Richardson AD (2008) A conceptual and practical approach to data quality and analysis procedures for high-frequency soil respiration measurements. Functional Ecology 22: 1000–1007.

11. Tang J, Bolstad PV, Martin JG (2009) Soil carbon fluxes and stocks in a Great Lakes forest chronosequence. Global Change Biology 15: 145–155.

12. Jassal RS, Black TA, Novak MD, Gaumont-Guay D, Nesic Z (2008) Effect of soil water stress on soil respiration and its temperature sensitivity in an 18-year-old temperate Douglas-fir stand. Global Change Biology 14: 1305–1318.

13. Craine JM, Gelderman TM (2010) Soil moisture controls on temperature sensitivity of soil organic carbon decomposition for a mesic grassland. Soil Biology and Biochemistry 43: 455–457.

14. Xu M, Qi Y (2001) Spatial and Seasonal Variations of Q_{10} Determined by Soil Respiration Measurements at a Sierra Nevadan Forest. Global Biogeochem Cycles 15: 687–696.

15. Davidson EA, Janssens IA (2006) Temperature sensitivity of soil carbon decomposition and feedbacks to climate change. Nature 440: 165–173.

16. Gershenson A, Bader NE, Cheng W (2009) Effects of substrate availability on the temperature sensitivity of soil organic matter decomposition. Global Change Biology 15: 176–183.

17. Högberg P (2010) Is tree root respiration more sensitive than heterotrophic respiration to changes in soil temperature? New Phytologist 188: 9–10.

18. Högberg MN, Briones MJI, Keel SG, Metcalfe DB, Campbell C, et al. (2010) Quantification of effects of season and nitrogen supply on tree below-ground carbon transfer to ectomycorrhizal fungi and other soil organisms in a boreal pine forest. New Phytologist 187: 485–493.

19. Chen B, Liu S, Ge J, Chu J (2010) Annual and seasonal variations of Q_{10} soil respiration in the sub-alpine forests of the Eastern Qinghai-Tibet Plateau, China. Soil Biology & Biochemistry 42: 1735–1742.

20. Fang C, Smith P, Moncrieff JB, Smith JU (2005) Similar response of labile and resistant soil organic matter pools to changes in temperature. Nature 433: 57–59.

21. Reichstein M, Kätterer T, Andrén O, Ciais P, Schulze ED, et al. (2005) Does the temperature sensitivity of decomposition vary with soil organic matter quality? Biogeosciences Discuss 2: 737–747.

22. Reichstein M, Subke J-A, Angeli AC, Tenhunen JD (2005) Does the temperature sensitivity of decomposition of soil organic matter depend upon water content, soil horizon, or incubation time? Global Change Biology 11: 1754–1767.

23. Fierer N, Craine JM, McLauchlan K, Schimel JP (2005) Litter Quality and the temperature Sensitivity of Decomposition. Ecology 86: 320–326.

24. Conant RT, Drijber RA, Haddix ML, Parton WJ, Paul EA, et al. (2008) Sensitivity of organic matter decomposition to warming varies with its quality. Global Change Biology 14: 1–10.

25. Hartley IP, Ineson P (2008) Substrate quality and the temperature sensitivity of soil organic matter decomposition. Soil Biology & Biochemistry 40: 1567–1574.

26. Peng S, Piao S, Wang T, Sun J, Shen Z (2009) Temperature sensitivity of soil respiration in different ecosystems in China. Soil Biology and Biochemistry 41: 1008–1014.

27. Dalias P, Anderson JM, Bottner P, Couteaux M-M (2001) Temperature responses of carbon mineralization in conifer forest soils from different regional climates incubated under standard laboratory conditions. Global Change Biology 7: 181–192.

28. Mahecha MD, Reichstein M, Carvalhais N, Lasslop G, Lange H, et al. (2010) Global Convergence in the Temperature Sensitivity of Respiration at Ecosystem Level. Science 329: 838–840.

29. Ohashi M, Gyokusen K (2007) Temporal change in spatial variability of soil respiration on a slope of Japanese cedar (Cryptomeria japonica D. Don) forest. Soil Biology & Biochemistry 39: 1130–1138.

30. Khomik M, Arain MA, McCaughey JH (2006) Temporal and spatial variability of soil respiration in a boreal mixedwood forest. Agricultural and Forest Meteorology 140: 244–256.

31. Metcalfe D, Meir P, Aragão LEOC, da Costa A, Almeida S, et al. (2008) Sample sizes for estimating key ecosystem characteristics in a tropical terra firme rainforest. Forest Ecology and Management 255: 558–566.

32. Tang J, Baldocchi DD (2005) Spatial-temporal variation in soil respiration in an oak-grass savanna ecosystem in California and its partitioning into autotrophic and heterotrophic components. Biogeochemistry 73: 183–207.

33. Liu S, Jiang Y, Shi Z (1998) Overview of the Baotianman Nature Reserve. A study on the Biological Diversity in Warm Temperate Forest in China. Beijing: China Science and Technology Press. 45.

34. Luan J, Liu S, Zhu X, Wang J, Liu K (2012) Roles of biotic and abiotic variables in determining spatial variation of soil respiration in secondary oak and planted pine forests. Soil Biology & Biochemistry 44: 143–150.

35. Hammond R, McCullagh PS (1978) Quantitative Techniques in Geography. UK: Clarendon Press.

36. Liu X, Zhang G, Heathman GC, Wang Y, Huang C-h (2009) Fractal features of soil particle-size distribution as affected by plant communities in the forested region of Mountain Yimeng, China. Geoderma 154: 123–130.

37. Six J, Elliott ET, Paustian K, Doran JW (1998) Aggregation and soil organic matter accumulation in cultivated and native grassland soils. Soil Science Society of American Journal 62: 1367–1377.

38. Garten CT, Post WM, Hanson PJ, Cooper LW (1999) Forest soil carbon inventories and dynamics along an elevation gradient in the southern Appalachian Mountains. Biogeochemistry 45: 115–145.

39. Lu R (2000) Soil and Agricultural Chemistry Analysis Methods (In Chinese). Beijing: Chinese Agricultural Scientific and Technology Press

40. Blair GJ, Lefroy RDB, Lisle L (1995) Soil carbon fractions based on their degree of oxidation and the development of a carbon management index. Australian Journal of Agricultural Research 46: 1459–1466.

41. Luan J, Xiang C, Liu S, Luo Z, Gong Y, et al. (2010) Assessments of the impacts of Chinese fir plantation and natural regenerated forest on soil organic matter quality at Longmen mountain, Sichuan, China. Geoderma 156: 228–236.

42. Kirschbaum MUF (1995) The temperature dependence of soil organic matter decomposition, and the effect of global warming on soil organic C storage. Soil Biology & Biochemistry 27: 753–760.

43. Davidson EA, Belk E, Boone RD (1998) Soil water content and temperature as independent or confounded factors controlling soil respiration in a temperate mixed hardwood forest. Global Change Biology 4: 217–227.

44. Knohl A, Søe AB, Kutsch W, Göckede M, Buchmann N (2008) Representative estimates of soil and ecosystem respiration in an old beech forest. Plant and Soil 302: 189–202.

45. Saiz G, Green C, Butterbach-Bahl K, Kiese R, Avitabile V, et al. (2006) Seasonal and spatial variability of soil respiration in four Sitka spruce stands. Plant and Soil 287: 161–176.

46. Kosugi Y, Mitani T, Itoh M, Noguchi S, Tani M, et al. (2007) Spatial and temporal variation in soil respiration in a Southeast Asian tropical rainforest. Agricultural and Forest Meteorology 147: 35–47.

47. Søe ARB, Buchmann N (2005) Spatial and temporal variations in soil respiration in relation to stand structure and soil parameters in an unmanaged beech forest. Tree Physiology 25: 1427–1436.

48. Jackson RB, Mooney HA, Schulze ED (1997) A global budget for fine root biomass, surface area, and nutrient contents. Proceedings of the National Academy of Sciences 94: 7362–7366.

49. Gill RA, Jackson RB (2000) Global patterns of root turnover for terrestrial ecosystems. New Phytologist 147: 13–31.

50. Matamala R, Gonzàlez-Meler MA, Jastrow JD, Norby RJ, Schlesinger WH (2003) Impacts of Fine Root Turnover on Forest NPP and Soil C Sequestration Potential. Science 302: 1385–1387.

51. Henry HAL, Juarez JD, Field CB, Vitousek PM (2005) Interactive effects of elevated CO2, N deposition and climate change on extracellular enzyme activity and soil density fractionation in a California annual grassland. Global Change Biology 11: 1808–1815.

52. Erika Marin-Spiotta WLSCWSRO (2009) Soil organic matter dynamics during 80 years of reforestation of tropical pastures. Global Change Biology 15: 1584–1597.

53. Janzen HH, Campbell CA, Brandt SA, Lafond GP, Townley-Smith L (1992) Light fraction organic matter in soils from long term crop rotations. Soil Science Society of America Journal 56: 1799–1806.

54. Laik R, Kumar K, Das DK, Chaturvedi OP (2009) Labile soil organic matter pools in a calciorthent after 18 years of afforestation by different plantations. Applied Soil Ecology 42: 71–78.

55. Luan J, Liu S, Zhu X, Wang J (2011) Soil carbon stocks and fluxes in a warm-temperate oak chronosequence in China. Plant and Soil 347: 243–253.

56. Powlson D (2005) Climatology: Will soil amplify climate change? Nature 433: 204–205.

57. Jones C, McConnell C, Coleman K, Cox P, Falloon P, et al. (2005) Global climate change and soil carbon stocks; predictions from two contrasting models for the turnover of organic carbon in soil. Global Change Biology 11: 154–166.

58. Ågren GI, Bosatta E (1998) Theoretical Ecosystem Ecology-Understanding Element Cycles. Cambridge: Cambridge Univ. Press.

59. Wang C, Yang J, Zhang Q (2006) Soil respiration in six temperate forests in China. Global Change Biology 12: 2103–2114.

6

Diversity and Communities of Foliar Endophytic Fungi from Different Agroecosystems of *Coffea arabica* L. in Two Regions of Veracruz, Mexico

Aurora Saucedo-García[1,2], Ana Luisa Anaya[2]*, Francisco J. Espinosa-García[3], María C. González[4]

1 Posgrado en Ciencias Biológicas, Instituto de Ecología, Universidad Nacional Autónoma de México, Distrito Federal, México, 2 Departamento de Ecología Funcional, Instituto de Ecología, Universidad Nacional Autónoma de México, Distrito Federal, México, 3 Laboratorio de Ecología Química, Centro de Investigaciones en Ecosistemas, Universidad Nacional Autónoma de México, Morelia, Michoacán, México, 4 Departamento de Botánica, Instituto de Biología, Universidad Nacional Autónoma de México, Distrito Federal, México

Abstract

Over the past 20 years, the biodiversity associated with shaded coffee plantations and the role of diverse agroforestry types in biodiversity conservation and environmental services have been topics of debate. Endophytic fungi, which are microorganisms that inhabit plant tissues in an asymptomatic manner, form a part of the biodiversity associated with coffee plants. Studies on the endophytic fungi communities of cultivable host plants have shown variability among farming regions; however, the variability in fungal endophytic communities of coffee plants among different coffee agroforestry systems is still poorly understood. As such, we analyzed the diversity and communities of foliar endophytic fungi inhabiting *Coffea arabica* plants growing in the rustic plantations and simple polycultures of two regions in the center of Veracruz, Mexico. The endophytic fungi isolates were identified by their morphological traits, and the majority of identified species correspond to species of fungi previously reported as endophytes of coffee leaves. We analyzed and compared the colonization rates, diversity, and communities of endophytes found in the different agroforestry systems and in the different regions. Although the endophytic diversity was not fully recovered, we found differences in the abundance and diversity of endophytes among the coffee regions and differences in richness between the two different agroforestry systems of each region. No consistent pattern of community similarity was found between the coffee agroforestry systems, but we found that rustic plantations shared the highest number of morphospecies. The results suggest that endophyte abundance, richness, diversity, and communities may be influenced predominantly by coffee region, and to a lesser extent, by the agroforestry system. Our results contribute to the knowledge of the relationships between agroforestry systems and biodiversity conservation and provide information regarding some endophytic fungi and their communities as potential management tools against coffee plant pests and pathogens.

Editor: Kathleen Treseder, UC Irvine, United States of America

Funding: The present research received the financial support from PAPIIT-DGAPA (UNAM) Project IN209611 given to ALA. The funders had no role in study design, data collection and analysis, decision to publish, or preparation of the manuscript.

Competing Interests: The authors have declared that no competing interests exist.

* E-mail: analanaya@ecologia.unam.mx

Introduction

Commercial coffee production relies mainly on the plant species *Coffea arabica* L. [1], a species native to the highlands of Ethiopia and Sudan [2]. In Mexico, this plant was introduced at the end of the eighteenth century and was incorporated into the local agrosystems, where coffee plants were cultivated under shade in diverse polycultures [3,4]. At the end of the 1970s, the Instituto Mexicano del Café promoted the transformation of traditional coffee polycultures into technified plantations [3], and at least five coffee production systems have been recognized in Mexico according to their vegetation structures, floristic composition, and management level [3,5]. Four of these five coffee agrosystems are shaded plantations: rustic or traditional, diverse polyculture, simple polyculture, and shaded monoculture. The fifth agrosystem is unshaded coffee monoculture [5]. These systems represent a gradient from the most traditional agroforestry system, with reduced management and a high proportion of native tree canopy,

to a lower proportion of native trees and a higher percentage of commercial shade trees. Unshaded coffee monoculture is at the end of the gradient of intensive management [3,5,6].

The relationship between coffee agroforestry types and their biodiversity has been attracting attention over the past two decades [6,7]. In general, studies on this topic have shown that shaded coffee plantations contain a higher level of associated biodiversity than unshaded coffee plantations [7–10]. In particular, it has been proposed that traditional shaded coffee systems can act as refuges for many species in regions where deforestation has drastically affected the original forests. This function could be decisive in biogeographically important areas where habitats have been severely transformed [3,6]. From a landscape perspective, the shaded plantations and their associated biodiversity preserve regional ecological processes and provide important ecosystem services [3,11], as opposed to the highly intensified unshaded plantations with reduced biodiversity. However, the conservation value of the different shaded coffee systems varies widely, depending on the taxonomic group of organisms that constitutes a part of the biotic community of coffee plantations [8,12].

Studies on the fungal diversity associated with Mexican coffee plantations, especially arbuscular mycorrhizal fungi (AMF) [13,14] and saprotrophic fungi [13], have reported no significant differences in fungal richness or fungal communities among different coffee plantations with a gradient of management intensity.

Other studies on the foliar endophytic fungi (EF) associated with coffee plants have been conducted in abandoned coffee plantations in Puerto Rico [15]; in various locations in Colombia, Hawai'i, Mexico, and Puerto Rico [16]; and in plantations in the center of Veracruz [17]. However, no studies have been conducted to investigate the effects of the different coffee agroforestry systems on the diversity of EF found on coffee leaves.

Endophytic fungi are microorganisms that live in plant tissues without causing apparent harm to the host [18,19]; therefore, plants with EF are asymptomatic [20]. These fungi have been found inhabiting healthy tissues in all plants in natural ecosystems [21–23], and they represent, individually and collectively, a continuum of variable associations with their host plants, from mutualism to latent pathogenicity [20]. In woody plants, such as *Coffea arabica*, the fungal endophytes are transmitted horizontally [23]; the growth of these fungi is highly localized within particular plant tissues [24,25]. With time, the EF accumulate in plant tissues, producing a heterogeneous mosaic of different species of endophytes in every plant organ [26,27].

The EF communities (EFC) may be influenced by the traits of a given host plant [26–31], distribution of the host plant [22,32,33], and characteristics of the locality in which the plant grows [27,34–36]. Studies on the EFC of cultivable host plants have shown variability among these communities based on the regions in which the plants are cultivated [37,38]; however, the variability in EFC among management systems is still poorly understood.

The aim of the present study was to analyze whether the coffee agroforestry system and the region where the coffee is cultivated could influence EF diversity and the EFC associated with *Coffea arabica* leaves. We analyzed and compared the colonization rates (CR), richness, diversity, and fungal communities of the EF inhabiting the leaves of *Coffea arabica* in different agroforestry systems in two different coffee regions, Huatusco and Coatepec, in the center of the state of Veracruz, Mexico. We selected coffee plants from a rustic plantation and from a simple polyculture in each of the coffee regions. We isolated 471 foliar endophytes, which we assigned to 31 morphospecies, from the four selected coffee plantations. We found differences in CR and EF diversity between the coffee regions, as well as differences in richness between the two different agroforestry systems of each region. The analysis of EFC similarity revealed that the endophytic communities in the two coffee plantations in Coatepec did not vary, while the EFC of the coffee plantations in Huatusco were different. There were similarities in the EFC of the rustic plantations and in the EFC of the simple plantations.

Studying the fungi associated with the phyllosphere of coffee plants enable us to recognize the presence of coffee pathogens and to evaluate, in future studies, the potentiality of some EF and EFC in controlling pathogens or pests of this important cultivated plant.

Materials and Methods

Ethics Statement

No specific permits were required for the described field studies. This work did not involve endangered or protected species.

Study sites

Four shaded coffee plantations from the central coffee-growing region of Veracruz, Mexico were selected as field study sites (Fig. 1). Two plantations were located in the Coatepec region and two were located in the Huatusco region. Although both of these regions are located in the tropical montane cloud forest, there are some differences between them. Both regions have a humid climate (A) C according to the classification of Köppen modified by García (1998) [39], but Coatepec experiences rainfall all year round (A) C (fm), whereas precipitation is seasonal in Huatusco (A) C (m). Another difference between the two regions is that the Coatepec region is located in the montane cloud forest, whereas the Huatusco region is located at the border of the montane cloud forest and the tropical deciduous forest [40]. In each region (Fig. 1), we selected one rustic plantation and one simple polyculture plantation according to a classification based on the vegetation structure of each plantation [5,41].

In rustic plantations, coffee is grown under the shade of native vegetation, such as *Heliocarpus* sp., *Quercus sartorii*, *Myrsine coriacea*, and *Trema micrantha*, with some selected shade trees, such as *Inga* spp. and *Citrus* spp. The canopy has a high degree of vertical stratification, and the vegetation contains abundant epiphytes [3,5,41]. These rustic plantations have low management intensity; the farmers predominantly use alternative management techniques, such as manual weed control and occasional pruning of coffee plants, with no addition of fertilizer.

In simple polyculture coffee plantations, the canopy of the native trees of the forest are removed and replaced with selected shade trees, such as legumes (*Inga* spp.), and crop species with commercial value, such as *Citrus* spp. and *Musa* sp. [5,41]. In this management system, the farmers apply agrochemicals, such as fertilizers and pesticides, to the coffee plants.

Sample collection and endophyte isolation

In November 2009, ten coffee plants, spaced at least 2 m apart, were chosen from each of the four plantations. To homogenize the sample collection, healthy, mature leaves from north-facing branches growing in the middle of the selected coffee plants were collected. Four coffee leaves from the middle part of the selected branches were collected, stored in a cool box above ice, and processed within 48 h.

Isolation of the EF was performed using the surface sterilization method for coffee leaves [15], washing them in running water and surface sterilizing them with sequential solutions of 70% ethanol (1 min), 2.6% sodium hypochlorite (3 min), and 70% ethanol (1 min). The leaves were then rinsed with sterilized, distilled water. Two round fragments (2 mm in diameter) were cut from the base, middle, and apex of the lamina of each leaf with a sterile leaf punch. Each surface-sterilized fragment was placed separately in a Petri dish with oatmeal agar (OA: 30 g oatmeal, 20 g agar, and 1 L distilled water) supplemented with chloramphenicol (50 mg/L). The leaf fragments were pressed briefly into the surface agar in the margins of the Petri dishes to create leaf prints and determine whether the superficial sterilization was successful. The Petri dishes were incubated at room temperature under natural light. The growth of the fungi and the leaf imprints were checked every three days for one month. The tips of hyphae from different fungi emerging from the same leaf fragment were subcultured on OA plates.

Morphological identification of endophytic fungi

All fungi isolates were examined after five and ten days and were grouped into morphotypes based on the following morphological traits: shape of the mycelium, texture of the mycelium

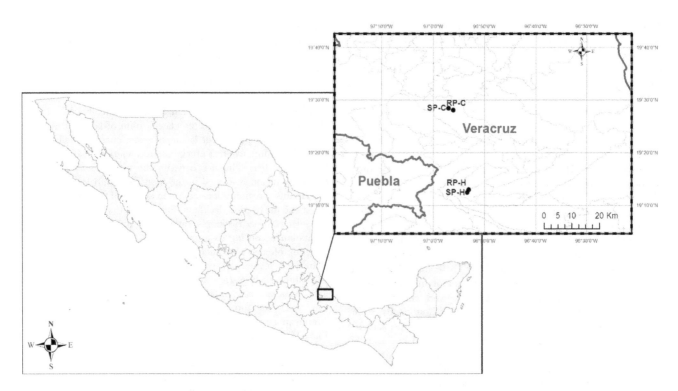

Figure 1. Map of survey area: geographical locations of the four coffee plantations, Veracruz, Mexico. RP-C: rustic plantation Coatepec, SP-C: simple polyculture Coatepec. RP-H: rustic plantation Huatusco, SP-H: simple polyculture Huatusco.

surface, color of the fungi, production of pigments and their diffusion into the medium, production of spores, and mycelium growth rates in the OA plates.

To enable morphological identification, 1 to 15 isolates were selected from each morphotype, depending on abundance of each morphotype. Selected isolates were cultured on OA and potato dextrose agar (PDA: 200 g scrubbed and diced potatoes, 15 g dextrose, 20 g agar, and 1 L distilled water) plates. The EF that did not sporulate on these media were transferred to PDA, OA, and malt extract agar (MEA 2%) plates and to plates with leaf extracts of healthy, mature coffee leaves (10% wt/vol) to activate sporulation.

We examined the above-mentioned traits of the mycelium morphology of the fungi isolates and their microscopic character-istics. The characteristics evaluated for the anamorph type were conidiomata, conidiogenous cells, conidiophores, and conidia morphology (e.g., size, color, shape, ornamentation), and the characteristics evaluated for the teleomorph type were sporomata, their associated structures, and spore morphology. The strains of isolated and identified endophytic fungi are part of a fungal collection of the Laboratorio de Alelopatía, Departamento de Ecología Funcional, Instituto de Ecología, Universidad Nacional Autónoma de México; the strains are freely available.

Following the method described in Waller et al. (1993) [42], the fungi belonging to the *Colletotrichum* genus were cultured in basal media supplemented with a different carbon source (i.e., ammonium tartrate or citric acid). Using that media enabled us to evaluate the substrate utilization of the isolated *Colletotrichum* strains and differentiate among species, as some species, such as the coffee pathogen *Colletotrichum kahawae*, cannot metabolize tartrate and citric acid [42,43].

The fungi were classified into morphospecies based on their growth rates and morphological and microscopic characteristics. Isolates that did not sporulate were identified as morphotypes

(Mycelia sterilia), based on their morphological characteristics. The relative frequency of EF for each coffee plantation was calculated as the abundance of a given species divided by the total number of fungi isolated from each coffee plantation.

Colonization and isolation rates

Colonization rate (CR) was calculated as the number of fragments from which one or more EF was isolated, divided by the total number of incubated fragments [44]. The isolation rate (IR) was defined as the number of EF isolated, divided by the total number of fragments incubated [45,46].

CR and IR were analyzed using a two-way ANOVA test, with regions (Huatusco and Coatepec) and agroforestry category (rustic and simple polycultures) as the factors. The difference between mean values was evaluated using Tukey's honestly significant differences (HSD) test. The statistical analysis was performed using Statistica software 8.0.

Diversity analysis

Because leaves serve as largely discrete, relatively uniform, bounded habitats to the microbes that inhabit them [47], we considered each leaf as a sampling unit, and the EF isolated from each leaf was considered to represent an EFC. To assess species richness and evaluate the sampling intensity, the observed and Jackknife 1 expected richness of EF for each coffee plantation were calculated with EstimateS software [48], using 1000 runs of bootstrapping with replacement. Jackknife 1 has proven to be a reliable estimator for various organisms [49], including fungi [35,50]. The observed and expected species accumulation curves were plotted using individual leaves as the unit for each plantation.

The richness and distribution of EF found in each coffee plantation were examined using the range diversity (RD) analysis developed by Arita et al. (2012) [51]. Using this method enabled us to analyze the richness and distribution of EF found in the coffee leaves of each plantation, as well as to evaluate the association and co-occurrence of morphospecies. The RD analysis was performed following the R script of Arita et al. (2012) [51], using the R program [52]. Analysis of variance was used to test for differences in endophyte richness between coffee plantations and coffee regions.

Following the R script of RD analysis [51], we computed the variance–covariance matrices among morphospecies and among leaves from each coffee plantation. Using the matrices enabled us to analyze the co-occurrence patterns of morphospecies in the coffee leaves from each plantation.

To test a possible association among species and a possible clustering of leaves in terms of shared species in each coffee plantation, we computed the ratio of variance for species (V_{sp}) and for leaves (V_{lv}) according to the method of Arita et al. (2012) [51]. If the ratio–variance value is higher than 1, there is a positive association among morphospecies or a greater similarity in morphospecies composition among leaves. If the ratio–variance value is less than 1, there is a negative association among morphospecies or no similarity in EFC among the leaves from each coffee plantation. We contrasted the values of ratio–variance obtained in the RD analysis (empirical values) with the values obtained with null models, which contrast real world assemblages against hypothetical patterns generated by randomizing some variables of a model [51]. The empirical values of V_{sp} were contrasted with null models in which we maintained the original frequency of EF richness found in every coffee plantation, but we assigned leaves to species randomly. The empirical values of V_{lv} were contrasted with null models in where we maintained the original frequency of number of species found in each coffee leaf, but we generate permutations to simulate the random assignment of species to leaves.

The Fisher's alpha and Shannon diversity (H') indexes of fungal endophyte species found in each coffee leaf were calculated with EstimateS software [48], using 1000 runs of bootstrapping with replacement to generate 95% confidence intervals for each diversity value. The statistical differences in foliar EF diversity between coffee plantations and regions were analyzed using an ANOVA test of two factors, and the differences between mean values were evaluated using Tukey's HSD test.

EFC similarities found among the coffee plants from the four plantations were analyzed using non-metric multidimensional scaling (NMDS) plots. The NMDS plots display the dissimilarities among EF communities graphically, and the distances between them on the plot represent their relative dissimilarity [53]. Three NMDS plots were constructed, each based on a different calculated ecological similarity index: Jaccard's index, based on the presence/absence of taxa among trees [54]; Bray–Curtis coefficient, based on the incidence and abundance of taxa found in the trees [54]; and Euclidean distance, a dissimilarity measure based in quantitative abundance data and the joint absences of taxa isolated among trees [54]. For each ecological similarity index calculated, a one-way analysis of similarity (ANOSIM) and a Bonferroni-corrected pair-wise comparison were performed to test for significant differences in the EFC of the coffee plants among plantations. The ANOSIM test uses a statistic (R) that ranges from 0 to 1. A zero R value represents a similarity between objects in different groups, and R values greater than zero indicate that objects are more dissimilar between groups than within groups [53]. The NMDS plots and ANOSIM analysis were performed using PAST software [55].

Results

Abundance and diversity of endophytic fungi

A total of 479 EF were isolated from the 80 Coffea arabica leaves collected. At least one EF was isolated in each the leaves examined, with one exception: no fungi were isolated from one coffee leaf from the simple polyculture plantation in Coatepec (SP-C). Therefore, that leaf was not included in the statistical analysis. The ANOVA test showed significant differences in CR and IR between the two regions (CR: F = 15.1249, p = 0.0002; IR: F = 14.523, p = 0.0003) and between the two agroforestry types (CR: F = 6.4181, p = 0.0134; IR: F = 8.3885, p = 0.0049). According to the post-hoc tests (Table 1), CR was significantly higher in Huatusco's rustic plantation (RP-H) and simple polyculture (SP-H) than in SP-C. In addition, significant differences in IR were found between the coffee leaves from RP-H and both Coatepec plantations, rustic (RP-C) and SP-C. There were no significant differences in CR or IR between the two agroforestry systems of each coffee region.

The 479 isolated EF were assigned to 31 morphospecies, including both identified genera and unidentified types. The relative frequencies of isolation of these fungi are shown in Table 2, and their descriptions are shown in Table S1. The particularly common EF genera were *Colletotrichum* and *Xylaria*.

The *Colletotrichum gloeosporioides* morphospecies was separated into two types (1 and 2) based on spore and conidiogenous cell size and mycelium traits. In the *Xylaria* genus, we recognized six morphotypes according to their morphological characteristics. Due to a lack of spore production, eight EF morphotypes, which represented only 5% of the isolated fungi, could not be identified; they were named Mycelia sterilia 1–8.

Interestingly, we observed changes in the appearance of the fresh leaf fragments cultured in the Petri plates according to the fungi isolated from them. The leaf segments in which the genus *Xylaria* was isolated withered after a few days of culture, while the leaf segments colonized by *Coniosporium* remained green in the Petri plates for almost one month.

Accumulation curves of observed species richness (gray triangles and circles) and Jackknife 1 estimated richness (open triangles and circles) are shown in Figure 2. Observed and estimated richness were higher in RP-H (circles in Fig. 2A) than in SP-H (triangles in Fig. 2A). Estimated richness (open circles and triangles) was higher than observed richness (gray circles and triangles) in all four coffee plantations.

The only coffee plantation in which the rarefaction curve reached the saturation point was the SP-C (Fig. 2B). Richness estimator analysis of this plantation showed that the estimated morphospecies richness was similar to the observed richness, indicating sufficient sampling work in SP-C.

The rarefaction curves of the other three plantations did not reach the saturation point, suggesting that endophyte richness was not fully recovered. The Jackknife 1 expected richness for those locations was higher than the observed richness, indicating that species richness was not exhaustively sampled.

The results of the RD analysis summarized in Table 3 show the total EF richness isolated from each coffee plantation and the number of leaves from which at least one EF was isolated. Analysis of variance showed that EF richness was higher in the coffee leaves from the Huatusco region than in the coffee leaves from Coatepec (F = 11.7512, p = 0.0009). In each region EF richness was higher in the rustic plantation than in the simple polyculture (F = 4.1639, p = 0.0448). There were coffee leaves from Coatepec from which only one EF morphospecies was isolated, and the maximum was six morphospecies per leaf. In contrast, the Huatusco coffee leaves

Table 1. Colonization rate (CR) and isolation rate (IR) of endophytic fungi in the four coffee plantations.

| | Huatusco | | Coatepec | |
	Rustic plantation	Simple polyculture	Rustic plantation	Simple polyculture
CR	0.88±0.04[a]	0.80±0.05[a]	0.73±0.05[ab]	0.57±0.05[b]
IR	1.30±0.09[a]	1.01±0.07[ab]	0.93±0.07[b]	0.75±0.09[b]

Mean of CR ± standard error and mean of IR ± standard error of fungal endophytes isolated from the four coffee plantations. Different letters indicate significant difference between coffee plantations at the p<0.05 level.

had a minimum of two or three morphospecies per leaf, with a maximum of nine different morphospecies per leaf.

On average, each EF morphospecies was isolated in five leaves from the Huatusco plantations and in four leaves from the Coatepec plantations. The minimum number of leaves in which a given morphospecies was isolated was one leaf, and the maximum was 14–20 leaves. The most widespread EF species in all four coffee plantations was *Colletotrichum gloeosporioides* 1.

The V_{sp} value, which indicates the degree of association among different species [56], was 1 in RP-H and SP-C (Table 3). Those values were similar to the V_{sp} of null models with 10 iterations (RP-H: $V_{sp} = 1.01\pm0.04$; SP-C: $V_{sp} = 1.02\pm0.28$). Empirical and null models showed no association among morphospecies in those

Table 2. Morphospecies of endophytic fungi from the four coffee plantations and their frequencies.

| Taxon | Huatusco | | Coatepec | |
	Rustic	Simple polyculture	Rustic	Simple polyculture
Alternaria citri				0.02
Beauveria brongniartii	0.01			
Colletotrichum aff *brassicicola*	0.03	0.06	0.03	0.03
Colletotrichum gloeosporioides 1	0.21	0.32	0.36	0.33
Colletotrichum gloeosporioides 2	0.13	0.01	0.03	0.07
Colletotrichum musae	0.04	0.02	0.01	
Colletotrichum sp. 1	0.03	0.07	0.02	0.04
Coniosporium sp.	0.03	0.02	0.10	0.08
Cryptopsoriopsis corticola	0.03		0.02	0.00
Cryptopsoriopsis sp. 1	0.04		0.02	0.07
Diplodia sp.	0.03	0.12	0.09	0.07
Glomerella cingulate	0.15	0.12	0.04	0.07
Guignardia mangiferae	0.03	0.06	0.09	0.01
Hyphomicete 1		0.01		
Hyphomicete 2				0.02
Mycelia esterilia 1	0.01	0.02	0.04	0.03
Mycelia esterilia 2	0.03			
Mycelia esterilia 3	0.01			
Mycelia esterilia 3		0.01		
Mycelia esterilia 4			0.01	
Mycelia esterilia 5			0.01	
Paecilomyces sp.	0.03	0.02	0.01	0.03
Phomopsis arnoldiae	0.02			0.01
Phomopsis sp.		0.02		
Xylaria 1.	0.08	0.05	0.03	0.04
Xylaria 2.	0.06	0.07	0.08	0.07
Xylaria 3	0.02		0.01	
Xylaria 4.	0.01			
Xylaria 5.			0.02	
Xylaria 6		0.01		

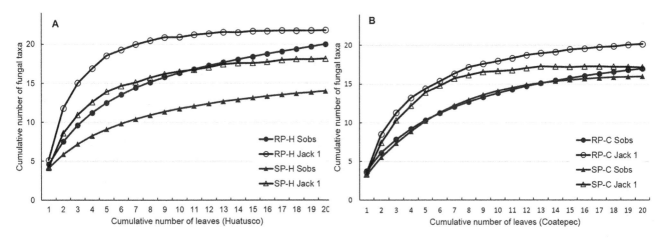

Figure 2. Rarefaction curves for fungal endophytes isolated from coffee leaves from the four coffee plantations. Species accumulation curves (gray circles and gray triangles) and Jackknife 1 estimated richness (open circles and open triangles) of rustic plantations (circles) and simple polycultures (triangles) of Huatusco region (A) and Coatepec (B). RP-H: rustic plantation Huatusco; RP-C: rustic plantation Coatepec; SP-H: simple polyculture Huatusco; SP-C: simple polyculture Coatepec.

plantations. In RP-C and SP-H, the empirical V_{sp} values were less than 1, while in the simulations values V_{sp} were 1 (RP-C: $V_{sp} = 1.00 \pm 0.16$; SP-H: $V_{sp} = 1.10 \pm 0.07$). Those results indicate negative associations among EF species in RP-C and in SP-H. Negative covariances were observed between *Colletotrichum gloeosporioides* and *Xylaria*, *Xylaria* and *Coniosporium* sp., and *Colletotrichum gloeosporioides* and *Coniosporium* sp. In general, positive covariances were found among fungi isolated in low frequencies, such as *Paecilomyces* sp., which was only isolated from leaves in the presence of one Coelomycete endophyte. The V_{lv} value, which indicates EFC similarity among leaves from the same coffee plantation, was higher than 1 in all four coffee plantations (Table 3). Those results contrast with the V_{lv} values obtained with the null models (RP-H: $V_{lv} = 0.87 \pm 0.18$; RP-C: $V_{lv} = 0.95 \pm 0.02$; SP-H: $V_{lv} = 1.29 \pm 0.16$; SP-C: $V_{lv} = 0.93 \pm 0.07$), and indicates considerable similarity in the shared species of the leaves of each coffee plantation. The highest EFC similarity was found in the coffee leaves from SP-H.

The ANOVA test of diversity indexes of EF isolated from the coffee leaves showed significant differences in diversity between agroforestry systems (Fisher's alpha: F = 40.4381, p<0.0001; Shannon's diversity: F = 22.8818, p<0.0001) and with the interaction between coffee region and agroforestry system (Fisher's

alpha: F = 43.9388, p<0.0001; Shannon's diversity: F = 40.3883, p<0.0001). The Shannon's diversity was different between coffee regions (F = 19.7207, p<0.0005), while the Fisher's alpha was not different between them (F = 0.4303, p = 0.5140) due to the variability in Fisher's alpha diversity in Huatusco region. Fisher's alpha EF diversity (Table 4) was significantly higher in the RP-H leaves than in the coffee leaves of the other three plantations; SP-H leaves had the lowest Fisher's alpha diversity value. Shannon's index of EF diversity was significantly higher in RP-H than in RP-C, SP-H, and SP-C; there were no significant differences in Shannon's index of EF diversity among RP-C, SP-H, and SP-C.

Analysis of EFC similarity among coffee plantations

The NMDS plots (Fig. 3) with Jaccard's index had a stress (S) value of 0.34, the NMDS plots with the Bray–Curtis coefficient had a 0.24 S value, and the Euclidean distance NMDS plots had a 0.16 S value. In NMDS plots, S values higher than 0.3 indicate a poor configuration of dissimilarities among the EFC in a multidimensional space, while S values lower than 0.2 indicate a better configuration of the dissimilarities [53]. Regardless of the ecological similarity index analyzed, there was not a clear

Table 3. Richness-distribution analysis of the endophytic fungi isolated from each coffee plantation.

	Huatusco		Coatepec	
	Rustic plantation	**Simple polyculture**	**Rustic plantation**	**Simple polyculture**
Total Richness	21	17	19	16
Leaves with endophytes (EF)	20	20	20	19
Mean of EF richness in coffee leaves	5.10±0.40	4.15±0.26	3.70±0.26	3.32±0.35
Min-Max of EF richness by leaf	3–9	2–6	1–5	1–6
Mean of distribution of EF	4.86±0.82	4.88±1.21	3.84±0.81	3.94±0.86
Min-Max distribution of EF	1–15 leaves	1–20 leaves	1–14 leaves	1–16 leaves
Ratio–variance for species (V_{sp})	0.949	0.621	0.641	1.311
Ratio–variance for leaves (V_{lv})	3.826	6.535	3.949	3.792

Values of total richness, media of richness ± standard error, distribution, media of distribution ± standard error, and ratio–variance of endophytic fungi in the four coffee plantations.

Table 4. Fisher's alpha and Shannon's diversity index of endophytic fungi in the four coffee plantations.

	Huatusco		Coatepec	
	Rustic	**Simple polyculture**	**Rustic**	**Simple polyculture**
Fisher's alpha	6.53±0.15 [a]	5.05±0.10 [c]	5.70±0.08 [b]	5.73±0.11 [b]
Shannon	2.40±0.03 [a]	2.07±0.03 [b]	2.08±0.03 [b]	2.13±0.03 [b]

Mean ± standard error. Different letters indicate significant difference between coffee plantations at the $p < 0.05$ level with ANOVA test of interaction between coffee region and agroforestry system.

clustering of EFC isolated from the coffee plants among the different plantations.

The results of the similarity analysis summarized in Table 5 indicate a significant difference in the EFC of RP-H and SP-H analyzed with Jaccard's index (R = 0.356), the Bray–Curtis coefficient (R = 0.324), and Euclidean distance (R = 0.258). Those results are better observed in the NMDS plots based on the Bray–Curtis coefficient (Fig. 3B) and Euclidean distance (Fig. 3C). Those figures show that the EFC of the different SP-H coffee plants are very similar (red circles in Fig. 3) and that those communities are different from the EFC isolated from the RP-H coffee plants (blue circles in Fig. 3).

The ANOSIM analyzed with the Euclidean distances also showed significant differences between the RP-H EFC and the SP-C EFC (R = 0.186). The NMDS plot based on Euclidean distance shows that the circles representing EFC from SP-C (green circles in Fig. 3C) are different to EFC from RP-H (blue circles in Fig. 3C).

The NMDS plots and ANOSIM based on the three ecological similarity indexes indicate that the EFC of the trees inhabiting the Coatepec region were similar, independent of the coffee agroforestry system. The similarity analysis also showed that the EFC isolated from each type of agroforestry system, rustic and simple polyculture, were similar, independent of the coffee region. There were similarities in the EFC of the rustic plantations (RP-H and RP-C) and in the EFC of the simple plantations (SP-H and SP-C).

Discussion

In this study, we identified 31 morphotaxa from the 479 isolated endophytic fungi. The number of morphospecies found in our study is lower than the number found in the study by Santamaría and Bayman (2005) [15], who studied the EF of coffee leaves from Puerto Rico and collected the same number of coffee leaves as in the current study. This difference might be due to the number of sites sampled; we sampled four sites, while Santamaría and Bayman sampled six sites. Because the interpretation of EF richness is method-dependent [21], the difference in the number of sites could affect the number of morphospecies recovered. However, it is far more likely that the difference is due to the type of management practiced at the sites studied. We isolated EF from coffee leaves in producing plantations, while Santamaría and Bayman isolated foliar EF from coffee plants growing in a secondary forest and a botanical garden. The traits of the ecological environment of the site strongly influence EF diversity [57].

The species reported in the present study are ubiquitous taxa isolated mainly from plants inhabiting tropical regions [22,25,35], woody cultured plants [16,58,59], and plants growing outside their native distribution areas [32,33]. The most common genera found in our study were *Colletotrichum* and *Xylaria*, which have often been isolated from coffee plants in American coffee-growing regions [15,16]. Other EF genera reported in the present study have also been isolated from coffee plants growing in different countries; these genera include *Phomopsis* [60], *Beauveria* [61], *Alternaria* [62],

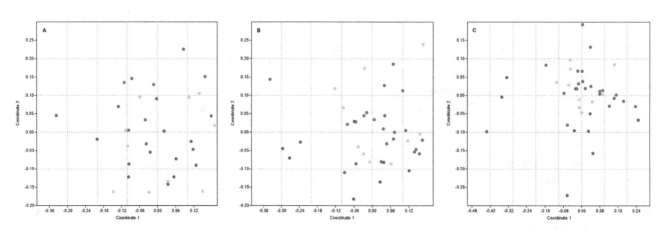

Figure 3. Non-metric multidimensional scaling (NMDS) plots of fungal endophytic communities from the four coffee plantations. NMDS plots based in Jaccard's index (A), Bray–Curtis coefficient (B), and Euclidean distance (C). Fungal endophytic communities of each coffee plantation are indicated by different colors: blue circles = rustic plantation Huatusco; pink circles = rustic plantation Coatepec; red circles = simple polyculture Huatusco; and green circles = simple polyculture Coatepec.

Table 5. ANOSIM pairwise test of EFC isolated from the four coffee plantations.

Pairwise test	R (Jaccard's index)	R (Bray-Curtis coefficient)	R (Euclidean distance)
RPH - RPC	0.211	0.164	0.145
RPH - SPH	0.356*	0.324*	0.258*
RPH - SPC	0.181	0.124	0.186*
RPC - SPH	0.003	0.111	0.110
RPC - SPC	0.060	0.004	0.111
SPH - SPC	0.183	0.141	0.169

The ANOSIM test was based on three different ecological similarity indexes: Jaccard's index, Bray–Curtis coefficient, and Euclidean distance. RPH: rustic plantation Huatusco; RPC: rustic plantation Coatepec; SPH: simple polyculture Huatusco; SPC: simple polyculture Coatepec; R = rank similarities; *significant difference p>0.05.

Coniosporium [17], *Guignardia* [15], *Paecilomyces* [16], and the teleomorph of *Diplodia*, *Botryosphaeria* [16].

Knowledge of the diversity of EF in coffee leaves is very important in identifying phytopathogenic fungi that might live as endophytes during part of their life cycle [63,64]. Of the EF isolated in the present study, the *Colletotrichum gloeosporioides* complex includes strongly aggressive pathogens, opportunistic pathogens [65], and endophytes [15,16]. In Latin America, *Colletotrichum gloeosporioides* is associated with blister spot, or "mancha mantecosa," in leaves [66]. The tartrate and citric acid biochemical tests conducted on isolates 1 and 2 of *Colletotrichum gloeosporioides* in this study, as well as the successful inoculation of these fungi in coffee plants without causing evident damage or disease (Velázquez-Bermudez, in preparation), show the endophytic nature of this species isolated from coffee leaves.

The genus *Xylaria* is recognized as a saprotrophic fungus [67,68] and as an endophytic fungus of many plants [69,70], included coffee plants [15,16]. We observed that the coffee leaf segments in which the genus *Xylaria* was isolated withered in a few days of Petri dish cultures, and some of them produced stromata only over the dead leaf segment. This phenomenon might be evidence of the saprobic phase of the xylariaceous fungus [71]. According to Vega et al. (2010) [16], the endophytes of this genus might play a saprotrophic role in coffee plants after their senescence.

Some genera, such as *Beauveria*, which was only isolated from RP-H leaves, and *Paecilomyces*, are recognized as entomopathogenic fungi and have been isolated and tested on some coffee pests [72]. Further studies on the entomopathogenic role of these genera will be very important in elucidating their possible ecological relation in each agrosystem.

The analysis of EF abundance in the coffee plantations in the two regions studied indicated higher CR and IR values in the Huatusco plantations than in the Coatepec plantations. This finding indicates that the geographical region (with its respective vegetation, climate, and soil characteristics) exerts an important influence on the colonization and abundance of foliar EF in coffee plants. Other studies have also shown differences in the abundance of EF between different localities in agrosystems [38] and natural systems [30,73].

On the other hand, we found no significant differences in CR or IR between the two agroforestry systems of each coffee region; however, we found higher CR and IR values in the rustic plantations than in the simple polycultures. Coffee plants in rustic plantations, as opposed to simple polycultures, are grown under a higher diversity of shade trees species, representing a more complex canopy. However, as shown in other studies, the canopy does not seem to influence the colonization and abundance of EF

in plants growing in the understory [27,74]. In agreement with our results, Arnold and Herre (2003) [27] found no differences in the CR of EF in plants growing under different canopy conditions (beneath the forest canopy and under cleared sites), although they found a higher number of fungal colony-forming units in the forest than in the cleared sites.

Besides the differences in canopy, the two coffee agroforestry systems also differ in their farming practices. However, the differences in management systems also do not seem to affect the colonization and abundance of EF. Similarly, Pancher et al. (2012) [37] found no differences in the number of EF isolated from grapevines (*Vitis vinifera*) cultivated in vineyards following different farming practices.

We found that the CR and IR of EF are not necessarily related to the diversity found in each sampling site. For example, Matsumura and Fukuda (2013) [75] reported higher frequencies of colonization of EF in trees inhabiting rural forests than in trees growing in a suburban forest in Japan. However, they did not find significant differences in Shannon's diversity index between these two forests. In contrast, we found similar colonization rates and abundance of EF in the Huatusco plantations, but the diversity values in the rustic plantation were higher than in the simple polyculture. On the other hand, we did not find significant differences in the EF diversity of the two Coatepec coffee plantations. Nevertheless, it is important to mention that the rarefaction curves for RP-C, RP-H, and SP-H did not reach saturation of morphospecies. This finding suggests that additional sampling efforts are needed in those plantations in order to obtain an accurate idea of the endophytic richness; the species that were not isolated might raise the diversity levels in those locations.

We found a higher total EF richness, a higher mean richness by leaf, and a higher expected richness in the rustic plantation than in the simple polyculture of each region. As mentioned previously, these results suggest that the complexity of the canopy and the alternative management techniques used in rustic plantations might influence the EF richness and diversity. Some studies have shown that the canopy might influence the richness and diversity of foliar EF growing in the understory [35,76,77]. A complex canopy could produce microclimatic conditions that promote a higher diversity of EF [76]. In rustic plantations, the complex canopy might also produce a higher diversity of litter types, and as those dead leaves appear to be a primary source of EF inoculum [36,71], the diversity of litter types might also produce a higher diversity of EF propagules.

Various studies have been conducted on biodiversity conservation in different coffee production systems in the central region of Veracruz, where Huatusco and Coatepec are located [78].

Contrary to our findings, studies regarding the fungi associated with different coffee agroforestry systems, particularly arbuscular mycorrhizal fungi (AMF) and saprotrophic fungi, showed no differences in richness or diversity among the different agrosystems [13,14]. It is possible that EF is more sensitive to agroforestry type than AMF and saprophytic fungi are.

In addition to the variability in richness and diversity of EF between coffee agrosystems, we found higher EF richness and diversity in RP-H than in the other three plantations, and a higher mean richness by leaf in the Huatusco plantations than in the Coatepec plantations. One reason for this finding is the difference in EF distribution in the two regions. Although we observed the typical distribution of EF reported in many studies on tropical and temperate forest trees [21,35,79], in which few EF morphospecies were frequent and the majority of morphospecies were found in low frequencies, we found some differences between regions. In general, the morphospecies in the Huatusco plantations had a broad distribution among the leaves, while the Coatepec morphospecies had a narrow distribution among the leaves. The differences in species distribution might influence the mean richness and diversity values of the coffee leaves. Furthermore, the Huatusco region (Fig. 1) is located at a lower longitude than Coatepec, and although it is immersed in a montane cloud forest, it runs adjacent to a tropical deciduous forest. These differences could influence the EF diversity of those plantations. Some studies have reported that Shannon's diversity index of EF increases when the longitude decreases [80]; this finding is consistent with our results for Huatusco and Coatepec. In the other hand, in Veracruz, there is also a higher diversity of other organisms, such as small- and medium-sized mammals [81] in the rustic Huatusco plantations than in the simple polycultures of Huatusco, and in the different agroforestry systems of Coatepec.

The Fisher's alpha and Shannon's index diversity analysis results differed in the present study. Both indexes consider the number of EF species (richness) and relative abundance of the individuals present in a given sample, but Fisher's alpha is not influenced by the sample size and is less affected by the abundance of the most common species than Shannon's index [82]. As such, Shannon's index indicated low variability among the four coffee plantations, whereas Fisher's alpha showed differences in diversity between the two regions and between agroforestry systems.

Although the total EF diversity found in the coffee plantations was not fully recovered, we found similar diversity values in other studies about EF diversity. In the present study, the mean of Fisher's alpha diversity of morphospecies (5.1–6.5) is similar to the mean of genotypic diversity (Fisher's alpha diversity of 5.2) of coffee plants in Mexico [16]. The similarity in diversity index values between the present study and other studies shows that morphotypes and morphospecies are valid taxonomic units [83] and that comparing the results of studies using different taxonomic units might be valid. The similitude in richness species estimators using morphotypes and genotypes has also been reported on foliar endophytes from beech trees (*Fagus sylvatica*) [79].

The Shannon's diversity values varied around 2 in the four coffee plantations. According to Gazis and Chaverri (2010) [84], the Shannon's index values in studies on EF are usually between 1.5 and 3.5; therefore, the Shannon's index for each coffee plantation in the present study is similar to those reported by other studies on EF.

In the analysis of association between species, the interactions between fungi are apparently different in each coffee plantation, but in general, the fungi isolated in high frequencies, such as *Colletotrichum gloeosporioides* 1 and *Xylaria* 1, have more negative co-occurrences than species isolated in low frequencies. The negative

covariances between *Colletotrichum gloeosporioides* and *Xylaria* spp. have been reported previously for endophytes inhabiting coffee leaves [15].

Species found in high frequencies co-occur with a low number of species, whereas the species isolated in low frequencies co-occur with a high number of species. Our results are consistent with those of Pan and May (2009) [85], who reported higher negative species co-occurrences between dominant EF and higher positive covariances between less common fungal species in maize plants. The positive covariances between EF could be attributed to a lack of competitive exclusion between those fungi or to a phenomenon known as facilitation. Fungal facilitation in endophytic communities can occur when infection of a plant host by one fungus species makes that host more vulnerable to infection by another fungus species [85]. The interspecific interactions of EF could influence the diversity and assemblage of EFC found in each coffee plantation.

The NMDS plots based on the different ecological indexes showed variation in EFC among the different plants of each coffee plantation. In addition, there was no EFC clustering among coffee plantations. When the EFC similarities of the coffee plantations were compared with an ANOSIM test, we found that the geographically closest plantations, the ones in the Coatepec region, were similar in their EFC, independent of their agroforestry systems. In agreement with this result, some studies have shown that EFC similarity is a function of the distance between sites, and thus, nearby sites have similar EFC [27,86,87]. The geographical condition of this region (microclimate and surrounding vegetation) could influence the assemblage of the EFC.

In contrast, we found dissimilarities in the EFC of the two Huatusco plantations, even though they were near each other. The ANOSIM test showed that the EFC of RP-H and SP-H were different according to the three ecological indexes of similarity that were analyzed. The EFC of RP-H and SP-C were also different as analyzed using Euclidean distance. Those results suggest that agroforestry system could influence the assemblage of EFC of coffee plants. Our results also showed that the EFC of the rustic plantations and the EFC of the simple polycultures were similar; in fact, the rustic plantations shared a higher number of morphospecies than the simple polycultures did.

Studies on the EF of herbaceous cultivable plants, such as cotton and maize, have shown no similarities in EFC based on farming practice [38,88]. However, in agreement with our results, Pancher et al. (2012) [37] reported dissimilarity in EFC isolated from *Vitis vinifera* plants under different vineyard management practices. These results indicate that the plantation management techniques in agroforestry systems, such as coffee plantations and vineyards, might influence the assemblage of EFC.

More sampling will help elucidate the factors that influence the EFC, but as shown in this study, some of those factors might be the geographical location and agroforestry system of the coffee plantations. Further studies evaluating the ecological roles of the different EFC found in the diverse coffee agroforestry systems will contribute to the knowledge of the role of the system of agroforestry management in the regulation of pest and pathogen populations [89,90].

The results of the present study show that the coffee agroforestry system produces variability in the colonization, richness, diversity, and composition of EFC in coffee plants. In addition, they demonstrate that the region in which coffee is cultivated is an important factor that influences these parameters. Future studies on the biodiversity conservation value of different coffee agroforestry systems will need to consider the coffee region as a determinant factor that affects biodiversity. In further studies, the use of molecular and physiological tools to identify, individually and collectively, the functional and ecological significance of EF in coffee plants under diverse ecological and geographical conditions will be equally significant. These studies will also provide an opportunity to understand the potential use of some EF as producers of relevant precursor substances in the regulation of different pests and pathogens, to discover new drugs, and to understand the potential role of EFC as potential controls of pest populations.

Supporting Information

Table S1 Description of morphospecies of foliar endophytic fungi isolated from the four coffee plantations.

References

1. Van Hilten HJ, Fisher PJ, Wheeler MA (2011) The Coffee Exporter's Guide. Geneva International Trade Centre. UNCTAD/GATT. 247 p.
2. Lashermes P, Combes MC, Robert J, Trouslot P, D' Hont A, et al. (1999) Molecular characterisation and origin of the *Coffea arabica* L. genome. Mol Gen Genet 261: 259–266.
3. Moguel P, Toledo VM (1999) Biodiversity conservation in traditional coffee systems of Mexico. Conserv Biol 13: 11–21.
4. Rice RA (1999) A place unbecoming: The coffee farm of Northern Latin America. Geogr Rev 89: 554–579.
5. Hernández-Martínez G, Manson RH, Contreras HA (2009) Quantitative classification of coffee agroecosystems spanning a range of production intensities in central Veracruz, Mexico. Agric Ecosyst Environ 134: 89–98.
6. Toledo VM, Moguel P (2012) Coffee and sustainability: the multiple values of traditional shaded coffee. J Sustain Agr 36: 353–377.
7. Gordon C, Manson R, Sundberg J, Cruz-Angón A (2007) Biodiversity, profitability, and vegetation structure in a Mexican coffee agroecosystem. Agric Ecosyst Environ 118: 256–266.
8. Perfecto I, Vandermeer J, Mas A, Soto-Pinto L (2005) Biodiversity, yield, and shade coffee certification. Ecol Econ 54: 435–446.
9. López-Gómez AM, Williams-Linera G, Manson RH (2008) Tree species diversity and vegetation structure in shade coffee farms in Veracruz, Mexico. Agric Ecosyst Environ 124: 160–172.
10. Perfecto I, Vandermeer J (2008) Biodiversity conservation in tropical agroecosystems: a new conservation paradigm. Ann N Y Acad Sci 1134: 173–200.
11. Ávalos-Sartorio B, Blackman A (2010) Agroforestry price supports as a conservation tool: Mexican shade coffee. Agroforest Syst 78: 169–183.
12. Perfecto I, Mas A, Dietsch T, Vandermeer J (2003) Conservation of biodiversity in coffee agroecosystems: a tri-taxa comparison in southern Mexico. Biodivers Conserv 12: 1239–1252.
13. Heredia AG, Arias RM (2008) Hongos saprobios y endomicorrizógenos en suelos. In: Manson RH, Hernández-Ortiz V, Gallina S, Mehltreter K, editors. Agroecosistemas cafetaleros de Veracruz: biodiversidad, manejo y conservación. México: INECOL INE-SEMARNAT. pp. 193–212.
14. Arias RM, Heredia-Abarca G, Sosa VJ, Fuentes-Ramírez LE (2012) Diversity and abundance of arbuscular mycorrhizal fungi spores under different coffee production systems and in a tropical montane cloud forest patch in Veracruz, Mexico. Agroforest Syst 85: 179–193.
15. Santamaría J, Bayman P (2005) Fungal epiphytes and endophytes of coffee leaves (*Coffea arabica*). Microb Ecol 50: 1–8.
16. Vega FE, Simpkins A, Aime C, Posada F, Peterson SW, et al. (2010) Fungal endophyte diversity in coffee plants from Colombia, Hawai'i, Mexico and Puerto Rico. Fungal Ecol 3: 122–138.
17. Carrión G (2006) La naturaleza de las interacciones entre la roya del cafeto y sus hongos hiperparásitos [PhD]. México: Universidad Nacional Autónoma de México. 110 p.
18. Petrini O (1991) Fungal endophytes of tree leaves. In: Andrews JH, Hirano SS, editors. Microbial ecology of leaves. New York: Springer-Verlag. pp. 179–197.
19. Stone JK, Bacon CW, White JF (2000) An overview of endophytic microbes: endophytism defined. In: Bacon CW, White JF, editors. Microbial endophytes. New York: Marcel Dekker. pp. 3–30.
20. Schulz B, Boyle C (2005) The endophytic continuum. Mycol Res 109: 661–687.
21. Arnold AE (2007) Understanding the diversity of foliar endophytic fungi: progress, challenges, and frontiers. Fungal Biol Rev 21: 51–66.
22. Arnold AE, Lutzoni F (2007) Diversity and host range of foliar fungal endophytes: are tropical leaves biodiversity hotspots? Ecology 88: 541–549.
23. Rodriguez RJ, White Jr JF, Arnold AE, Redman RS (2009) Fungal endophytes: diversity and functional roles. New Phytol 182: 314–330.
24. Lodge JD, Fisher PJ, Sutton BC (1996) Endophytic fungi of *Manilkara bidentata* leaves in Puerto Rico. Mycologia 88: 733–738.
25. Gamboa MA, Laureano S, Bayman P (2002) Measuring diversity of endophytic fungi in leaf fragments: does size matter? Mycopathologia 156: 41–45.
26. Espinosa-García FJ, Langenheim JH (1990) The endophytic fungal community in leaves of a coastal redwood population - diversity and spatial patterns. New Phytol 116: 89–97.
27. Arnold AE, Herre EA (2003) Canopy cover and leaf age affect colonization by tropical fungal endophytes: ecological pattern and process in *Theobroma cacao* (Malvaceae). Mycologia 95: 388–398.
28. Higgins KL, Coley PD, Kursar TA, Arnold AE (2011) Culturing and direct PCR suggest prevalent host generalism among diverse fungal endophytes of tropical forest grasses. Mycologia 103: 247–260.
29. Jiang H, Shi YT, Zhou ZX, Yang C, Chen YJ, et al. (2011) Leaf chemistry and co-occurring species interactions affecting the endophytic fungal composition of *Eupatorium adenophorum*. Ann Microbiol 61: 655–662.
30. Johnston PR, Johansen RB, Williams AF, Paula Wikie J, Park D (2012) Patterns of fungal diversity in New Zealand *Nothofagus* forests. Fungal Biol 116: 401–412.
31. Sanchez-Azofeifa A, Oki Y, Fernandes GW, Ball RA, Gamon J (2012) Relationships between endophyte diversity and leaf optical properties. Tress 26: 291–299.
32. Fisher PJ, Petrini O, Petrini LE, Sutton BC (1994) Fungal endophytes from the leaves and twigs of *Quercus ilex* L. from England, Majorca and Switzerland. New Phytol 127: 133–137.
33. Hoffman MT, Arnold AE (2008) Geographic locality and host identity shape fungal endophyte communities in cupressaceous trees. Mycol Res 112: 331–344.
34. Müller M, Hallaksela A-M (1998) Diversity of Norway spruce needle endophytes in various mixed and pure Norway spruce stands. Mycol Res 102: 1183–1189.
35. Gamboa MA, Bayman P (2001) Communities of endophytic fungi in leaves of a tropical timber tree (*Guarea guidonia*: Meliaceae). Biotropica 33: 352–360.
36. Herre EA, Mejía LC, Kyllo DA, Rojas EI, Maynard Z (2007) Ecological implications of anti-pathogen effects of tropical fungal endophytes and mycorrhizae. Ecology 88: 550–558.
37. Pancher M, Ceol M, Corneo PE, Longa CM, Yousaf S, et al. (2012) Fungal endophytic communities in grapevines (*Vitis vinifera* L.). Appl Environ Microbiol 78: 4308–4317.
38. Ek-Ramos MJ, Zhou W, Valencia CU, Antwi JB, Kalns LL, et al. (2013) Spatial and temporal variation in fungal endophyte communities isolated from cultivated cotton (*Gossypium hirsutum*). PLoS ONE 8: e66049. doi:10.1371/journal.pone.0066049
39. García E (1998) Carta Climas (Clasificación de Koppen, modificada por García). Scale 1:1000 000. Mexico: CONABIO.
40. Instituto Nacional de Estadística, Geografía e Informática (INEGI) (2008) Carta de uso de suelo y vegetación IV. Scale 1:250 000. Mexico: INEGI.
41. Hernández-Martínez G (2008) Clasificación Agroecológica In: Manson RH, Hernández-Ortiz V, Gallina S, Mehltreter K, editors. Agroecosistemas cafetaleros de Veracruz: biodiversidad, manejo y conservación. México: INECOL INE-SEMARNAT. pp. 15–34.
42. Waller JM, Bridge PD, Black R, Hakiza G (1993) Characterization of the coffee berry disease pathogen, *Colletotrichum kahawae* sp. nov. Mycol Res 97: 989–994.
43. Prihastuti H, Cai L, Chen H, McKenzie EHC, Hyde KD (2009) Characterization of *Colletotrichum* species associated with coffee berries in northern Thailand. Fungal Divers 39: 89–109.
44. Petrini O, Stone JK, Carroll FE (1982) Endophytic fungi in evergreen shrubs in western Oregon: a preliminary study. Can J Bot 60: 789–796.
45. Fröhlich J, Hyde KD, Petrini O (2000) Endophytic fungi associated with palms. Mycol Res 104: 1202–1212.

Acknowledgments

This paper constitutes a partial fulfillment of the Graduate Program in Biological Sciences of the Universidad Nacional Autónoma de México (UNAM), and a requisite to obtain the doctorate degree. A. Saucedo-García acknowledges the fellowship provided by Consejo Nacional de Ciencia y Tecnología (CONACyT) and PAPIIT-DGAPA UNAM. We are grateful to Q.A. Blanca E. Hernández-Bautista and Biol. Gabriel del Ángel García for their valuable technical assistance. We also thank Gloria Carrión and Brewster Philip John from Instituto Nacional de Ecología (INECOL) in Xalapa, Veracruz for their help with the location of the study sites. We thank the editor and two anonymous reviewers for their valuable comments, which helped us to improve the manuscript.

Author Contributions

Conceived and designed the experiments: ASG ALA FJEG. Performed the experiments: ASG. Analyzed the data: ASG ALA FJEG MCG. Contributed reagents/materials/analysis tools: ALA. Wrote the paper: ASG ALA.

46. Wang Y, Guo L-D (2007) A comparative study of endophytic fungi in needles, bark, and xylem of *Pinus tabulaeformis*. Can J Bot 85: 911–917.

47. Zimmerman NB, Vitousek PM (2012) Fungal endophyte communities reflect environmental structuring across a Hawaiian landscape. Proc Natl Acad Sci U S A 109: 13022–13027.

48. Colwell RK (2013) EstimateS: Statistical estimation of species richness and shared species from samples. Version 9.1. User's Guide and application. http://purl.oclc.org/estimates. Accessed 3 February.

49. Magurran AE (2004) Measuring biological diversity. Oxford. Blackwell Publising. 256 p.

50. Unterseher M, Schnittler M (2009) Dilution-to-extinction cultivation of leaf-inhabiting endophytic fungi in beech (*Fagus sylvatica* L.) – different cultivation techniques influence fungal biodiversity assessment. Mycol Res 113: 645–654.

51. Arita HT, Christen A, Rodríguez P, Soberón J (2012) The presence–absence matrix reloaded: the use and interpretation of range–diversity plots. Glob Ecol Biogeogr 21: 282–292.

52. R Core Team (2013) R: A language and environment for statistical computing. Vienna: R Foundation for Statistical Computing.

53. Quinn GP, Keough MJ (2002) Experimental design and data analysis for biologists. Cambridge. Cambridge University Press. 537 p.

54. Anderson MJ, Crist TO, Chase JM, Vellend M, Inouye BD, et al. (2011) Navigating the multiple meanings of β diversity: a roadmap for the practicing ecologist. Ecol Lett 14: 19–28.

55. Hammer Ø, Harper DAT, Ryan PD (2001) PAST: Paleontological statistics software package for education and data analysis. Palaeontol Electronica 4: 9. Available http://palaeo-electronica.org/2001_1/past/issue1_01.htm. Accessed 3 February 2014.

56. Arita HT, Christen JA, Rodriguez P, Soberón J (2008) Species diversity and distribution in presence-absence matrices: mathematical relationships and biological implications. Am Nat 172: 519–532.

57. Qi F, Jing T, Zhan Y (2012) Characterization of endophytic fungi from *Acer ginnala* Maxim. in an artificial plantation: media effect and tissue-dependent variation. PLoS ONE 7: e46785. doi:10.1371/journal.pone.0046785.

58. Arnold AE, Mejía LC, Kyllo D, Rojas EI, Maynard Z, et al. (2003) Fungal endophytes limit pathogen damage in a tropical tree. Proc Natl Acad Sci U S A 100: 15649–15654.

59. Douanla-Meli C, Langer E, Talontsi MF (2013) Fungal endophyte diversity and community patterns in healthy and yellowing leaves of *Citrus limon*. Fungal Ecol 6: 212–222.

60. Sette LD, Passarini MRZ, Delarmelina C, Salati F, Duarte MCT (2006) Molecular characterization and antimicrobial activity of endophytic fungi from coffee plants. World J Microbiol Biotechnol 22: 1185–1195.

61. Posada F, Vega FE (2006) Inoculation and colonization of coffee seedlings (*Coffea arabica* L.) with the fungal entomopathogen *Beauveria bassiana* (Ascomycota: Hypocreales). Mycoscience 47: 284–289.

62. Fernandes MDRV, Pfenning LH, Costa-Neto CMD, Heinrich TA, Alencar SMD, et al. (2009) Biological activities of the fermentation extract of the endophytic fungus *Alternaria alternata* isolated from *Coffea arabica* L. Braz J Pharmaceutical Sciences 45: 677–685.

63. Saikkonen K, Faeth SH, Helander M, Sullivan TJ (1998) Fungal endophytes: a continuum of interactions with host plants. Ann Rev Ecol Syst 29: 319–343.

64. Junker C, Draeger S, Schulz B (2012) A fine line—endophytes or pathogens in *Arabidopsis thaliana*. Fungal Ecol 5: 657–662.

65. Phoulivong S, Cai L, Chen H, McKenzie EHC, Abdelsalam K, et al. (2010) *Colletotrichum gloeosporioides* is not a common pathogen on tropical fruits. Fungal Divers 44: 33–43.

66. Waller JM, Bigger M, Hillocks RJ (2007) Coffee pests, diseases and their management. Wallingford. CABI. 434 p.

67. Petrini L, Petrini O (1985) Xylariaceous fungi as endophytes. Sydowia 38: 216–234.

68. Osono T, Takeda H (2002) Comparison of litter decomposing ability among diverse fungi in a cool temperate deciduous forest in Japan. Mycologia 94: 421–427.

69. Davis EC, Franklin JB, Shaw AJ, Vilgalys R (2003) Endophytic *Xylaria* (Xylariaceae) among liverworts and angiosperms: phylogenetics, distribution, and symbiosis. Am J Bot 90: 1661–1667.

70. Unterseher M, Peršoh D, Schnittler M (2013) Leaf-inhabiting endophytic fungi of European beech (*Fagus sylvatica* L.) co-occur in leaf litter but are rare on decaying wood of the same host. Fungal Divers 60: 43–54.

71. Osono T, Tateno O, Masuya H (2013) Diversity and ubiquity of xylariaceous endophytes in live and dead leaves of temperate forest trees. Mycoscience 54: 54–61.

72. Vega FE, Posada F, Catherine Aime M, Pava-Ripoll M, Infante F, et al. (2008) Entomopathogenic fungal endophytes. Biol Control, 46: 72–82.

73. Angelini P, Rubini A, Gigante D, Reale L, Pagiotti R, et al. (2012) The endophytic fungal communities associated with the leaves and roots of the common reed (*Phragmites australis*) in Lake Trasimeno (Perugia, Italy) in declining and healthy stands. Fungal Ecol 5: 683–693.

74. Hata K, Sone K (2008) Isolation of endophytes from leaves of *Neolitsea sericea* in broadleaf and conifer stands. Mycoscience 49: 229–232.

75. Matsumura E, Fukuda K (2013) A comparison of fungal endophytic community diversity in tree leaves of rural and urban temperate forests of Kanto district, eastern Japan. Fungal Biol 117: 191–201.

76. Unterseher M, Reiher A, Finstermeier K, Otto P, Morawetz W (2007) Species richness and distribution patterns of leaf-inhabiting endophytic fungi in a temperate forest canopy. Mycol Prog 6: 201–212.

77. Scholtysik A, Unterseher M, Otto P, Wirth C (2013) Spatio-temporal dynamics of endophyte diversity in the canopy of European ash (*Fraxinus excelsior*). Mycol Prog 12: 291–304.

78. Manson RH, Hernández-Ortiz V, Gallina S, Mehltreter K, editors (2008) In: Manson RH, Agroecosistemas cafetaleros de Veracruz: biodiversidad, manejo y conservación. México: INECOL INE-SEMARNAT. 330 p.

79. Unterseher M, Schnittler M (2010) Species richness analysis and ITS rDNA phylogeny revealed the majority of cultivable foliar endophytes from beech (*Fagus sylvatica*). Fungal Ecol 3: 366–378.

80. Wu L, Han T, Li W, Jia M, Xue L, et al. (2013) Geographic and tissue influences on endophytic fungal communities of *Taxus chinensis* var. *mairei* in China. Curr Microbiol 66: 40–48.

81. Gallina S, González-Romero A, Manson R (2008) Mamíferos pequeños y medianos. In: Manson RH, Hernández-Ortiz V, Gallina S, Mehltreter K, editors. Agroecosistemas cafetaleros de Veracruz: biodiversidad, manejo y conservación. México: INECOL INE-SEMARNAT. pp. 161–180.

82. Schulte RP, Lantinga EA, Hawkins MJ (2005) A new family of Fisher-curves estimates Fisher's alpha more accurately. J Theor Biol 232: 305–313.

83. Lacap DC, Hyde KD, Liew ECY (2003) An evaluation of the fungal 'morphotype' concept based on ribosomal DNA sequences. Fungal Divers 12: 53–66.

84. Gazis R, Chaverri P (2010) Diversity of fungal endophytes in leaves and stems of wild rubber trees (*Hevea brasiliensis*) in Peru. Fungal Ecol 3: 240–254.

85. Pan JJ, May G (2009) Fungal-fungal associations affect the assembly of endophyte communities in maize (*Zea mays*). Microb Ecol 58: 668–678.

86. Gange AC, Dey S, Currie AF, Sutton BC (2007) Site- and species-specific differences in endophyte occurrence in two herbaceous plants. J Ecol 95: 614–622.

87. Joshee S, Paulus BC, Park D, Johnston PR (2009) Diversity and distribution of fungal foliar endophytes in New Zealand Podocarpaceae. Mycol Res 113: 1003–1015.

88. Seghers D, Wittebolle L, Top EM, Verstraete W, Siciliano SD (2004) Impact of agricultural practices on the *Zea mays* L. endophytic community. Appl Environ Microbiol 70: 1475–1482.

89. Teodoro A, Klein AM, Tscharntke T (2008) Environmentally mediated coffee pest densities in relation to agroforestry management, using hierarchical partitioning analyses. Agric Ecosyst Environ 125: 120–126.

90. Vandermeer J, Perfecto I, Philpott S (2010) Ecological complexity and pest control in organic coffee production: uncovering an autonomous ecosystem service. BioScience 60: 527–537

Assessing Conservation Values: Biodiversity and Endemicity in Tropical Land Use Systems

Matthias Waltert[1]*, Kadiri Serge Bobo[2], Stefanie Kaupa[3], Marcela Leija Montoya[4], Moses Sainge Nsanyi[5], Heleen Fermon[1]

1 Department of Conservation Biology, Georg-August-Universität Göttingen, Göttingen, Germany, 2 Department of Forestry, University of Dschang, Dschang, Cameroon, 3 Göttingen, Germany, 4 Facultad de Ciencias Biológicas de la Universidad Autónoma de Nuevo León, Nuevo León, México, 5 Botany Programme, Center for Tropical Forest Science (CTFS)/Korup Forest Dynamics Plot (KFDP), Department of Plant and Animal Sciences, Faculty of Science, University of Buea, Buea, Cameroon

Abstract

Despite an increasing amount of data on the effects of tropical land use on continental forest fauna and flora, it is debatable whether the choice of the indicator variables allows for a proper evaluation of the role of modified habitats in mitigating the global biodiversity crisis. While many single-taxon studies have highlighted that species with narrow geographic ranges especially suffer from habitat modification, there is no multi-taxa study available which consistently focuses on geographic range composition of the studied indicator groups. We compiled geographic range data for 180 bird, 119 butterfly, 204 tree and 219 understorey plant species sampled along a gradient of habitat modification ranging from near-primary forest through young secondary forest and agroforestry systems to annual crops in the southwestern lowlands of Cameroon. We found very similar patterns of declining species richness with increasing habitat modification between taxon-specific groups of similar geographic range categories. At the 8 km^2 spatial level, estimated richness of endemic species declined in all groups by 21% (birds) to 91% (trees) from forests to annual crops, while estimated richness of widespread species increased by +101% (trees) to +275% (understorey plants), or remained stable (- 2%, butterflies). Even traditional agroforestry systems lost estimated endemic species richness by - 18% (birds) to - 90% (understorey plants). Endemic species richness of one taxon explained between 37% and 57% of others (positive correlations) and taxon-specific richness in widespread species explained up to 76% of variation in richness of endemic species (negative correlations). The key implication of this study is that the range size aspect is fundamental in assessments of conservation value via species inventory data from modified habitats. The study also suggests that even ecologically friendly agricultural matrices may be of much lower value for tropical conservation than indicated by mere biodiversity value.

Editor: Jon Moen, Umea University, Sweden

Funding: This study was supported financially by the German Academic Exchange Service (DAAD), the German Society for Tropical Ornithology (GTO), African Nature e.V., and the Volkswagen Foundation, Hanover, Germany. The funders had no role in study design, data collection and analysis, decision to publish, or preparation of the manuscript.

Competing Interests: The authors have declared that no competing interests exist.

* E-mail: mwalter@gwdg.de

Introduction

Since the seminal paper by Lawton *et al.* [1], numerous studies have dealt with biodiversity patterns of tropical land use gradients and analysed indicator properties of different taxa [2–5]. Looking more closely into these studies, it appears that patterns of alpha (point) diversity can be highly taxon-specific but that a general pattern of high beta turnover across habitats is visible in all taxa: most altered habitats usually also contain most altered biotic communities which may - or may not - be as diverse as primary forest but in any case composed of different species [6]. Although such inventory-based studies are being published and referred to in conservation journals, they more often allow conclusions rather on the 'biodiversity value' of modified landscapes rather than on their 'conservation value' [7]. E.g., while traditional agroforestry systems are often regarded as a potential quality matrix for the maintenance of diverse tropical forest biota [4,8] and may thus appear as effectively contributing to global conservation, their biodiversity may in cases have little to do with the original rainforest biota. A number of single-taxon studies have however

addressed more critically the comparative conservation value of agro-biodiversity by including degree of endemism/geographic range size [9–18] in the analyses, but we were unable to find a single multi-taxa land use gradient study which does so consistently for all taxa. We consider this lack of focus on conservation values as problematic since replacement of original assemblages by biota of different geographic range composition may have global conservation implications. We therefore advocate to more clearly separate 'biodiversity value' from 'conservation value' and to emphasise the latter issue much more in future studies. This should be especially important for multi-taxa studies which often receive a high number of citations, and which are most likely to influence landscape management. Here, using an existing dataset from southwestern Cameroon published elsewhere [13,15,16], we test the hypotheses that tropical deforestation and land use especially affect species of smaller geographic ranges and that agroecosystems favour richness of species of wider geographic ranges. We also hypothesized that the abundance of trees is a good predictor for the richness of biota of smaller geographic range categories. The dataset covers trees, understorey plants, fruit-feeding butterflies, and birds, sampled at 6 stations in each of four 8 km^2 areas belonging to different habitat types (near-primary forest, young secondary forest, cocoa-agroforestry systems and annual crops). The taxa were chosen because they are frequently used indicators given the relatively moderate skills and sampling efforts needed for assessment [1–4], and because their geographic ranges are relatively well established and accessible [19–25 and Appendix S1].

Results

Birds

Estimates of endemic bird species richness at sampling station (point) level were highest in, and did not differ significantly between (Tukey's Honest Significant Difference test, p>0.05), near-primary (NF, 64.2±4.1 S.D.) and secondary forest (SF, 64.9±5.3). They were however 27% lower in annual crops (AC: 47.1±12.9) and 21% lower in agroforestry systems (AF: 50.6±10.7) compared to NF (ANOVA, $F_{(3;20)} = 6.28$; p = 0.004 for estimated species richness). This pattern was very similar at habitat level (8 km^2 scale), but with a slightly lower decrease (by 21%) between NF (100±3.7) and AC (79.3±4.0) (Figure 1a).

Point level estimates of widespread bird species richness showed an opposite trend in having lowest species richness in NF (19.7±4.2) and SF (15.7±1.8), but increasing by 88% to AC (37.1±7.5), and having intermediate richness (increase by 42%) in AF (27.9±5.3)($F_{(3;20)} = 20.48$; p = 0.000). Estimates of widespread bird species richness did not differ significantly between NF and SF (Tukey's Honest Significant Difference test, p>0.05). Again, this pattern was similar at habitat level, where the increase from NF (30.7±3.8) to AC (66.5±5.6) amounted to 117%, and 48% to AF (45.3±2.8) (Figure 1b).

Butterflies

Point level estimates of endemic butterfly species richness were similar in NF (19.5±6.3) and SF (21.0±3.6) but declined by 79% from NF to AC (4.1±2.2), and by 46% to AF (10.5±4.6) ($F_{(3;20)} = 24.78$; p = 0.000 for observed, $F_{(3;20)} = 20,0527$; p = 0.000 for estimated richness). This pattern was almost identical at habitat level, with an 80% decrease from NF (47.2±3.8) to AC (9.5±1.1) and a 36% decrease from NF to AF (30.0±3.7) (Figure 1c).

In contrast, point estimates of widespread butterfly species richness increased from NF (8.1±3.8) towards AF (17.0±5.2) by 110%, but were again lower in AC (10.7±3.0), at similar levels to NF ($F_{(3;20)} = 6.82$; p = 0.002). Again, this pattern was similar at habitat level, with a 62% increase only between NF (21.7±3.1) and AF (35.2±3.0), and values of AC (21.2±2.7) very similar to NF (Figure 1d).

Understory plants

Point level estimates of endemic understorey plant species richness dropped by 91% from NF (18.2±5.1) to AF (1.6±1.4) and were similarly low in AC (4.1±3.5), representing a decrease of 77% compared to NF ($F_{(3;20)} = 22.18$; p = 0.000). Estimates at habitat level followed an identical pattern, with a decrease of 88% from NF (69.0±5.32) to AF (6.5±1.1) and of 71% to AC (15.0±3.4) (Figure 1e).

In contrast, widespread understorey plant species richness increased at point level from 0.0 (±0.0) species in NF to 15.4 (±4.3) species in AC, with an average of 10.8 (±0.8) species in AF. The percentage of increase from SF (4.1±3.3) to AC (15.8±2.7 species) amounted to 275%, but was only 43% to from SF to AF (7.2±1.6) ($F_{(3;20)} = 32.68$; p = 0.000). At habitat level, a similar pattern was found but with relatively higher values for SF

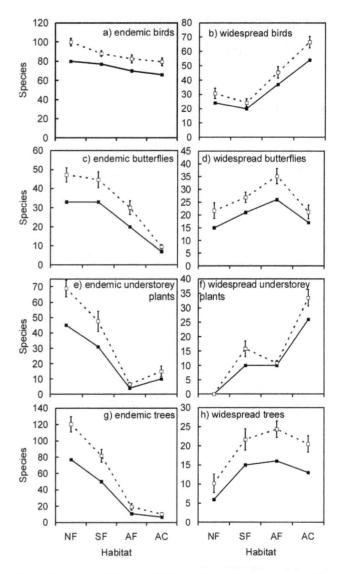

Figure 1. Species richness of different taxa, for two geographic range categories 'endemic' and 'widespread', at the 8 km^2 spatial scale, for different habitats. Continuous lines represent observed (Sobs), dashed lines estimated species richness (Jackknife 1). Whiskers indicate ±1 SD. Habitats: NF, near primary forest; SF, secondary forest, AF, agroforestry systems; AC, annual crops.

(15.8±2.7) and a proportional increase from SF to AC (33.5±2.8) of 112% and of 32% to AF (10.8±0.8) (Figure 1f).

Trees

Point estimates of endemic tree species richness dropped steadily from an estimated 31.1±9.5 species in NF, by 90% to 3.2±1.5 species in AC ($F_{(3;20)} = 23.34$; p = 0.000). In SF (19.6±9.6), the decrease from NF amounted to 37%. AF sampling stations (3.2±2.7) had similar values to AC, loosing 84% of species compared to SF. At habitat level, the pattern was again identical: the loss of estimated endemic tree species richness amounted to 91% between NF (120.3±9.3) and AC (10.3±1.7), the drop being still 32% from NF to SF (81.7±7.9), and still 76% from SF to AF (19.3±3.8) (Figure 1g).

Point estimates of widespread tree species richness were very low in NF (1.6±1.8), higher at SF and AC (6.7±3.9, 5.4±2.0, respectively), reaching 8.2±3.7 species in AF sampling stations ($F_{(3;20)} = 5.28$; p = 0.008). At habitat level, estimates of widespread tree species richness doubled from NF (10.2±2.4) to SF (21.7±2.8) and AC (20.5±2.1), representing an increase by 101%, and was even higher in AF (24.3±2.1) (Figure 1h).

Correlations

There were strong positive correlations between endemic species richness of one taxon/group and endemic species richness of others, explaining between 37 and 56% (R^2) of the variation (Spearman Rank correlation coefficients R amounted to between 0.61 and 0.75, $P \leq 0.001$). We also found strong positive correlations between widespread bird species richness and that of widespread understorey plants, as well as between widespread butterfly species richness and widespread tree species richness (Table 1).

Likewise, in most groups/taxa there were strong negative correlations between widespread species richness and endemic species richness (Spearman Rank corrleation coefficients R between -0.41 and – 0.87, $P < 0.05$).

There were also strong correlations between overall tree abundance and endemic species richness of birds (Spearman $R = 0.65$, $P < 0.001$), butterflies ($R = 0.69$, $P < 0.001$) and trees ($R = 0.78$, $P < 0.001$) and a moderately strong correlation between overall tree abundance and endemic understorey plant species richness ($R = 0.60$, $P = 0.002$).

Discussion

Our study, based on assessments at the 8 km^2-level, revealed that endemics of major indicator taxa show a steady decline in richness with increasing forest conversion, i.e. from near-primary forest to annual crops. In contrast, richness in widespread birds, understorey plants and trees increased along this gradient. Richness of widespread butterfly species also increased in secondary forest and agroforestry systems but reached again near-primary forest levels in annual crops.

If analyses at these spatial scales provide an indication of biodiversity patterns at the regional level, there may be important implications for the design and analysis of environmental impact assessments, as well as for global conservation strategies. The key conclusion is that endemic species richness is a potentially powerful indicator for conservation evaluation of modified habitats. One example: agroforestry systems have been found to maintain substantial levels of biodiversity and have therefore been largely appraised as an all too perfect fusion of economic yield and nature conservation [26,27]. Based on the results of our own approach, however, we may argue that this seemingly ideal land use form is prone to being overrated regarding its conservation value. Some of the potential reasons for these euphemistic appraisals are study biases towards investigating mere species

Table 1. Spearman Rank correlation coefficients R for relationships between estimated (first-order Jackknife) species richness of geographic range groups of different taxa.

Endemics vs Endemics (n = 24 sampling stations in all cases)

	Butterflies	Understorey Plants	Trees
Birds	**0.68**	**0.62**	**0.61**
	P<0.001	*P* = 0.001	*P* = 0.001
Butterflies		**0.61**	**0.71**
		P = 0.001	*P*<0.001
Understorey Plants			**0.75**
			P<0.001

Widespread vs Widespread (n = 24 in all cases)

	Butterflies	Understorey Plants	Trees
Birds	−0.02	**0.73**	0.22
	P = 0.93	*P*<0.001	*P* = 0.30
Butterflies		0.21	**0.65**
		P = 0.34	*P*<0.001
Understorey Plants			0.37
			P = 0.08

Endemics vs Widespread (n = 24 sampling stations in all cases)

	Endemic Birds	Endemic Butterflies	Endemic Understorey Plants	Endemic Trees
Widespread Birds	**−0.67**	**−0.87**	**−0.56**	**−0.71**
	P<0.001	*P*<0.001	*P* = 0.004	*P*<0.001
Widespread Butterflies		−0.05	**−0.51**	−0.25
		P = 0.82	*P* = 0.01	*P* = 0.24
Widespread Understorey Plants			**−0.68**	**−0.77**
			P<0.001	*P*<0.001
Widespread Trees				**−0.41**
				P = 0.04

Significant (*P*<0.05) values in bold.

richness and abundance, and only occasionally community composition [28] and beta diversity between primary- and agroforests [6]. Furthermore, agroforestry study sites are often situated close to primary or secondary forests, facilitating influx of mobile organisms into these systems [29] and thereby leading to overestimation of both biodiversity and conservation value of these modified systems.

Reviewing studies which compare biodiversity value of primary forests and agroforestry systems, Scales & Marsden [30] report reduced species richness in modified agroforests in 34 out of 43 studies. In some cases, declines along land use gradients may be so gradual that there seems to be no problem to rank agroforestry systems close to natural and secondary forests. Among the studies listed [30], there are also several which indicate higher or similar species richness in agroforestry systems compared to forest, even for vertebrates [31]. However, while such studies tell us something about functional diversity and ecology of modified systems, they are of little help when they are put into the context of the global biodiversity crisis. Our study differs from such work in that it explicitly addresses conservation values of land use systems based on several indicator groups. It largely confirms what earlier single taxon studies [13,15] indicated: namely that forest modification and land use affect endemic species of different indicator taxa in a very similar way, reflecting that species turnover from forest to farmland is to a large extent a replacement of endemic by widespread species. While ecological requirements of species of narrow geographic ranges are often little known, our results suggest this reflects to a large extent the change in tree abundance and species richness, the two parameters which were both correlated moderately to strongly with endemic species richness. However, there is still much to be learned about the ecology associated with species of narrow geographic ranges, and future studies should aim at exploring the causal relationships between endemic species declines and associated biotic and abiotic environment.

We advocate a pre-cautionary approach when putting biodiversity data from the tropical agricultural matrix into the context of conservation evaluation. We note that the current knowledge on the conservation benefits of tropical land use systems is still limited and that research at different spatial scales is still urgently needed. Given this state of knowledge, we suggest it is preferable to invest the limited funds for conservation of wet tropical forest region biodiversity into the proper protection and management of remaining natural forests.

Study area

The study was carried out within an appr. 40 km^2 section of the Support Zone (SZ) of Korup National Park, in the South-western region of Cameroon [16]. This region is part of the Guineo-Congolian forest [32] and also part of the Hygrophylous Coastal Evergreen Rainforest which occurs along the Gulf of Biafra within the Cross-Sanaga-Bioko Coastal Forest ecoregion [32,33]. This ecoregion is considered an important center of plant diversity because of its probable isolation during the Pleistocene [34] and holds an assemblage of endemic primates known as the Cameroon faunal group [35–36]. The region is also exceptionally rich in butterflies [37] and birds [38].

The studied sampling stations were all situated in the populated part of the SZ, where farming is restricted to the immediate surroundings of the villages, leaving most of the area forested. The land use types chosen represent different forms of common land use practice, and are situated along a gradient of human disturbance where near-primary forest (NF) serves as a reference. They basically differ in two important characteristics of habitats:

Habitat complexity referring to the vertical structure of vegetation and habitat heterogeneity expressed in the horizontal variation of the habitat's features. All sites outside the near-primary forest, i.e. secondary forest (SF), agroforestry systems (AF) and annual crop farms (AC), are located at the vicinity of the forest edge. The main characteristics of the chosen habitats are as follows [15,16]:

(1) NF: wet evergreen forest with high tree species richness. Closed canopy averages 35–45 m. The dominant trees are *Oubangia alata* and *Gilbertiodendron demonstrans*.

(2) SF: moist evergreen forest which has been cleared for farming along roads about 15 years ago. These forests have a relatively closed canopy. Canopy height averages 25–30 m. Characteristic trees are *Elaeis guineensis*, *Barteria fistulosa*, *Rauvolfia vomitoria* and *Pycnanthus angolensis*.

(3) CF: cocoa/coffee plantations shaded by natural forest trees of up to 25 m height. Apart from *Theobroma cacao* (Cocoa) and *Coffea robusta* (Coffee) trees, remnant *Elaeis guineensis* (Oil palm) and *Dacryodes edulis* (Plum) trees are characteristic.

(4) AC: open monoculture of manioc, remnant forest trees, remnant oil palms, no planted shade trees, dead wood, *Chromolaena odorata* and farm bush thickets; it is a dynamic habitat, due to the short cycles of the cultivated plants and associated human activities.

Methods

Data collection

Six sampling stations were selected in each of the above mentioned habitats, adding up to a total of 24 stations located at least 500 m apart from each other and covering an approximate area of ca. 8 km^2 in each habitat. Topographically, all study sites were situated at an altitude of about 250 m above sea level. For vegetation (tree and understorey plant) sampling, centred on each sampling station, plots of 50 m ×50 m were established. Each plot was divided into nine subplots of 10 m ×10 m (one subplot in the centre and eight others at the borders) so as to have 10 m in between subplots and spreading over 2,500 m^2 in total at each study site.

In each 10 m ×10 m subplot, a 1 m ×1 m small plot, established in its centre, was used to collect data for understorey plants. Understorey plants were defined as all vascular plants of less than 1.3 m height, and (overstorey) trees as all trees of more than 10 cm in diameter at 1.3 m height (DBH). In total, data were organised in 216 understorey plant samples (9 small plots ×24 sampling stations), as well as 216 tree samples (9 subplots ×24 sampling stations). In total, 1,230 understorey plant and 856 tree individuals were recorded and identified at least to morphospecies level. Species of uncertain identity were not assigned to geographic range categories. Therefore, of the original dataset containing 350 understorey plant and 226 tree morphospecies, we used 219 (63%) and 204 (90%) for our analyses, respectively.

Fruit-feeding butterfly data were collected within 50 m from the centre of each sampling station during the dry season, between 27 December 2003 and 10 March 2004. We used three cylindrical gauze-traps [39–41] baited with rotten bananas. These three traps were installed at about 1.5 m above the soil surface and controlled daily for nine days at each sampling station. Specimens collected on one sampling day in the three traps were pooled per sampling station, resulting in a total of 216 butterfly samples (24 sampling stations ×9 days). A total of 1,167 butterfly individuals of 119 species were collected, labelled and later identified using D'Abrera [19,20], Hecq [21] and Larsen [22]. All individuals were identified to species level.

Table 2. Geographic range descriptions for range size codes of endemic species, those with medium-sized ranges and widespread species, for birds, butterflies and plants, separately.

		Birds			Butterflies			Understorey Plants/Trees	
		Range	# spp.		Range	# spp.		Range	#/# spp.
Endemic	1	Cameroon-Gabon lowlands restricted	1	1	endemic to eastern Nigeria and southwestern Cameroon	2	1	Endemic to SW Cameroon/SE Nigeria border	14/24
				2	from Nigeria to the Cameroon-Gabon-Congo zone	20	2	From SW Cameroon/SE Nigeria border to W Benin/Togo border **OR** from SW Cameroon/SE Nigeria border to Gabon, Congo, Equatorial Guinea, including Fernando Po and Sao Tome and Principe islands.	44/54
				3	from Nigeria to D.R. Congo or DRC-Uganda border	26	3	From SW Cameroon/SE Nigeria, Gabon, Congo, Equatorial Guinea, including Fernando Po and Sao Tome and Principe islands to W Ivory Coast **OR** from SW Cameroon/SE Nigeria, Gabon, Congo, Equatorial Guinea to DRC/Uganda border and Angola	22/33
	2	Guineo-Congolian Forest Biome restricted	115	4	from Nigeria to east of Rift Valley	9			
Medium distributional range				5	from western West Africa to the Nigeria-Cameroon border	6	4	From Benin to DRC/Uganda border and Angola OR from SW Cameroon/SE Nigeria, Gabon, Congo, Equatorial Guinea to Sierra Leone OR from SW Cameroon/SE Nigeria, Gabon, Congo, Equatorial Guinea to Kenya and Angola	26/25
				6	from western West Africa to the Cameroon-Gabon-Congo zone	10	5	From Sierra Leone to D.R. Congo/Uganda, and Angola OR from SW Cameroon/SE Nigeria, Gabon, Congo, Equatorial Guinea to Senegambia OR from SW Cameroon/SE Nigeria, Gabon, Congo, Equatorial Guinea to E Rift Valley, and Angola	13/26
							6	From Senegambia to D.R. Congo/Uganda, and Angola	11/12
Widespread	3	African Rainforest	2	7	from western West Africa to D.R. Congo or DRC-Uganda border	14			
				8	from western West Africa to Uganda or Western Kenya	19	7	From Senegambia to Tanzania and Angola OR From Ivory Coast to Sudan, East Africa/Mozambique	5/3
				9	from western West Africa to east of Rift Valley	3	8	From Senegambia to Sudan, East Africa/Mozambique, and Angola	12/11
	4	Ubiquitous in Africa	62	10	found throughout Africa in suitable habitats	10	9	Throughout tropical Africa in suitable habitats	14/16

Also given are numbers of species in each category.

Bird surveys were carried out between 23 December 2003 and 5 March 2004 using point counts of birds within a range of 50 m from the centre of the sampling station. As most land use systems (AF, AC) were only of small size (<2 ha), small-scale point counting was the only possible method, regardless of the fact that bird point diversity in tropical forests may only reflect a small proportion of the overall alpha-diversity [42]. All visits were conducted between 6:00 and 9:00 am for 20 min, and both visual and acoustical detections were recorded. Fieldwork was done by the same observer throughout the survey and sites were visited nine times, respectively. A total of 4,530 records of 180 species were obtained and identified mainly with Borrow and Demey [23], but also Brown *et al.* [24] and Keith *et al.* [25].

Data analysis

We reviewed available geographic range information for all taxa and clustered species accordingly, resulting in groups of 'endemic'' and 'widespread'' species (Table 2). A third category of 'medium distributional range'' was introduced for butterflies and plants in order to obtain comparable percentages of endemic species across the four taxa. Hence, for butterflies 16 species, for understorey

plants 50 spp. and for trees 61 species were omitted from the analyses.

For bird distribution data, we used Fishpool and Evans [43] who classified 116 of our study species as being restricted to the Guinea-Congolian forest biome [33], spanning from east Guinea to west D.R. Congo and southwards to Congo and Gabon. This group of 116 bird species was categorised as 'endemic'. The remaining 64 species can be classified as non-biome-restricted [43], and were thus categorised as 'widespread''. This categorisation is also followed in an earlier publication [16].

For butterfly distribution data, we followed Larsen [22] to obtain geographic range information for butterflies, dividing them into ten range size categories, from 1 - most range-restricted to 10 - most widespread; 57 of the 119 species belonged to categories 1 to 4, with geographic ranges roughly situated within the Guinea-Congolian forest biome, and were categorised as 'endemic''. Those with categories 7 to 10 were grouped into the 'widespread'' category, which included 46 species.

We obtained understorey plant and tree geographic ranges from the Global Biodiversity Information Facility, Aubréville *et al.*, Hutchinson & Dalziel, and several volumes of the Flora of

Tropical East Africa (Appendix S1). As in butterflies and birds, we categorised geographic ranges ranging from 1 to 9, with 'endemic" species spanning at the most from southwest Cameroon to West Ivory Coast or eastwards to the D.R. Congo- Uganda border (categories 1–3). "Widespread" species were defined as those of the categories 7 to 9 with a geographic range size of at least the magnitude of the area from Senegambia to Tanzania and Angola. Eighty of the 219 understorey plant species and 111 of 204 tree species were thus defined as 'endemic", whereas 31 understorey plants and 30 tree species were classified as 'widespread". For 48 understorey plant species geographic range size data could not be procured and they were thus excluded from further analysis.

As biodiversity field studies in the tropics usually fail to record all of the present species [44] we calculated an 'estimated' species richness in order to gain a more accurate picture of the actual species richness. Such calculations of estimated species richness take into account the frequency patterns of the 'observed', species. We used the first-order jackknife method initially designed to estimate population size from capture to recapture data, allowing capture probabilities to vary by individuals [45]. This model can equally be applied to estimations of species richness [46–51]. Calculations of estimated species richness were made using EstimateS Win 7.5.0 and 8.2.0 by Colwell [52] using 200 randomizations.

Calculations of observed and estimated species richness were carried out for both point species richness, based on data of nine spatial subplots per point, spread over 2,500 m^2 (trees, understorey

plants) and nine temporal subsamples which covered a similar circular area (butterflies, birds), but also at the 8 km^2 habitat level, based on data from the six sampling stations in each habitat. One-way ANOVA was done to detect responses to habitat variation for each group of geographic range category. Spearman rank correlation coefficients were established with STATISTICA V.9 (StatSoft) to illustrate the relationships between the estimated species richness of taxa/geographic range groups and vegetation parameters.

Acknowledgments

We thank all our field assistants. We are especially indebted to the late Mr. John Njokagbo, Nguti, and the traditional councils of the villages Abat, Bajo, Basu, and Mgbegati for their co-operation. We also thank the Wildlife Conservation Society, Nguti, for assistance with vegetation data collection, as well as Yann Clough and two anonymous reviewers, for their help with the manuscript.

Author Contributions

Analyzed the data: MW SKB SK MLM HF. Wrote the paper: MW. Conceived and designed the study: MW SKB. Performed field work: SKB MSN. Compiled geographic range information: SK MLM.

References

1. Lawton JH, Bignell DE, Bolton B, Bloemers GF, Eggleton P, èt al. (1998) Biodiversity inventories, indicator taxa and effects of habitat modification in tropical forest. Nature 391: 72–76.
2. Schulze CH, Waltert M, Kessler P, Pitopang R, Shahabuddin, et al. (2004) Biodiversity indicator groups of tropical land use systems: comparing plants, birds, and insects. Ecol Appl 14: 1321–1333.
3. Perfecto I, Mas A, Dietsch T, Vandermeer J (2003) Conservation of biodiversity in coffee agroecosystems: a tri-taxa comparison in southern Mexico. Biodivers Conserv 12: 1239–1252.
4. Pardini R, Faria D, Accacio GM, Laps RR, Mariano-neto E, et al. (2009) The challenge of maintaining Atlantic forest biodiversity: A multi-taxa conservation assessment of specialist and generalist species in an agro-forestry mosaic in southern Bahia. Biol Conserv 142: 1178–1190. doi: 10.1016/j.biocon.2009.02.010.
5. Barlow J, Gardner TA, Araujo IS, Avila-Pires TC, Bonaldo AB, et al. (2007) Quantifying the biodiversity value of tropical primary, secondary, and plantation forests. Proc Nat Acad Sci USA 104: 18555–18560.
6. Kessler M, Abrahamczyk S, Bos M, Buchori D, Putra DD, et al. (2009) Alpha and beta diversity of plants and animals along a tropical land-use gradient. Ecol Appl 19: 2142–2156.
7. Barlow J, Gardner TA, Louzada J, Peres C (2010) Measuring the conservation value of tropical primary forests: the effect of occasional species on estimates of biodiversity uniqueness. Plos One 5: e9609. doi: 10.1371/journal.pone.0009609.
8. Bhagwat S, Willis KJ, Birks HJB, Whittaker R (2010) Agroforestry: a refuge for tropical biodiversity. Trends Ecol Evol 23: 261–267. doi: 10.1016/j.tree.2008.01.005.
9. Thomas CD (1991) Habitat use and the geographic ranges of butterflies from the wet lowlands of Costa Rica. Biol Conserv 55: 269–281.
10. Lewis OT, Wilson RJ, Harper MC (1998) Endemic butterflies on Grande Comore: habitat preferences and conservation priorities. Biol Conser 85: 1993–1996.
11. Fermon H, Waltert M, Larsen TB, Dall'asta U, Mühlenberg M (2000) Effects of forest management on diversity and abundance of fruit-feeding nymphalid butterflies in south-eastern Côte d'Ivoire. J Insect Conserv 4: 173–189.
12. Dunn RR, Romdal TS (2005) Mean latitudinal range sizes of bird assemblages in six Neotropical forest chronosequences. Global Ecol Biogeogr 14: 359–366. doi: 10.1111/j.1466-822x.2005.00155.x.
13. Bobo KS, Waltert M, Fermon H, Njokagbor J, Mühlenberg M (2006) From forest to farmland: butterfly diversity and habitat associations along a gradient of forest conversion in southwestern Cameroon. J Insect Conserv 10: 29–42.
14. Umetsu F, Pardini R (2007) Small mammals in a mosaic of forest remnants and anthropogenic habitats — evaluating matrix quality in an Atlantic forest landscape. Landscape Ecol 22: 517–530. doi: 10.1007/s10980-006-9041-y.
15. Bobo KS, Waltert M, Sainge NM, Njokagbor J, Fermon H, et al. (2006) From forest to farmland: species richness patterns of trees and understorey plants along

a gradient of forest conversion in Southwestern Cameroon. Biodivers Conserv 15: 4097–4117. doi: 10.1007/s10531-005-3368-6.
16. Waltert M, Bobo KS, Sainge NM, Fermon H, Mühlenberg M (2005) From forest to farmland: habitat effects on afrotropical forest bird diversity. Ecol Appl 15: 1351–1366.
17. Fritz SA, Bininda-Emonds ORP, Purvis A (2009) Geographical variation in predictors of mammalian extinction risk: big is bad, but only in the tropics. Ecol Lett 12: 538–549. doi: 10.1111/j.1461-0248.2009.01307.x.
18. Weist M, Tscharntke T, Sinaga MH, Maryanto I, Clough Y (2010) Effect of distance to forest and habitat characteristics on endemic versus introduced rat species in agroforest landscapes of Central Sulawesi, Indonesia. Mamm Biol 75: 567–571. doi: 10.1016/j.mambio.2009.09.004.
19. D'Abrera B (1980) Butterflies of the Afrotropical Region. Melbourne: Lansdowne Editions. pp 593.
20. D'Abrera B (1997) Butterflies of the Afrotropical Region, Part I. Melbourne and London: Hill House Publishers. 265 p.
21. Hecq J (1997) Euphaedra. Lambillionea. TervurenBelgium: Union des Entomoligistes Belges. 115 p.
22. Larson TB (2005) The butterflies of West Africa. Vol. 1-2. Stenstrup: Apollo Books. 865 p.
23. Borrow N, Demey R (2001) Birds of Western Africa. London: Christopher Helm. 512 p.
24. Brown LH, Urban EK, Newman K (1982) The Birds of Africa. Volume 1. London: Academic Press. 521 p.
25. Keith S, Urban EK, Fry CH (1992) The Birds of Africa. Volume 4. London: Academic Press. 632 p.
26. Klein A, Steffan-Dewenter I (2002) Effects of Land-Use Intensity in Tropical Agroforestry Systems on Coffee Flower-Visiting and Trap-Nesting. Conserv Biol 16: 1003–1014.
27. Schroth G, da Fonseca GAB, Harvey CA (2004) Agroforestry and biodiversity conservation in tropical landscapes. Washington, DC: Island Press. 523 p.
28. Su JC, Debinski DM, Jakubauskas ME, Kindscher K (2004) Beyond species richness: community similarity as a measure of cross-taxon congruence for coarse-filter conservation. Conserv Biol 18: 167–173.
29. Waltert M, Mardiastuti A, Mühlenberg M (2004) Effects of land use on bird species richness in Sulawesi, Indonesia. Conserv Biol 18: 1339–1346.
30. Scales BE, Marsden SJ (2008) Biodiversity in small-scale tropical agroforests: a review of species richness and abundance shifts and the factors influencing them. Environ Conserv 35: 160–172. doi: 10.1017/S0376892908004840.
31. Harvey CA, González Villalobos JA (2007) Agroforestry systems conserve species-rich but modified assemblages of tropical birds and bats. Biodivers Conserv 16: 2257–2292. doi: 10.1007/s10531-007-9194-2.
32. Olsen DME, Dinerstein ED, Wikramanayake ND, Burgess GVN, Powell EC, et al. (2001) Terrestrial ecoregions of the world: a new map of life on earth. BioScience 51: 933–938.

33. White F (1983) The Vegetation of Africa. Paris: UNESCO. 356 p.
34. Davis SD, Heywood VH, Hamilton AC (1994) Centres of Plant Diversity: A Guide and Strategy for their Conservation. Volume 1. Europe, Africa, South West Asia and the Middle East. Cambridge, UK: IUCN Publications Unit. 354 p.
35. Oates JF (1996) African Primates. Status survey and conservation action plan. Revised Edition. Gland, Switzerland: IUCN. 80 p.
36. Waltert M, Lien, Faber K, Mühlenberg M (2002) Further declines of threatened primates in the Korup Project Area, south-west Cameroon. Oryx 36: 257–265.
37. Larsen TB (1997) Korup butterflies - biodiversity writ large. Report on a butterfly study mission to Korup National Park in Cameroon. MundembaCameroon: WWF and Korup National Park. 65 p.
38. Rodewald PG, Dejaifve PA, Green AA (1994) The birds of Korup National Park and Korup Project Area, Southwest Province, Cameroon. Bird Conserv Int 4: 1–68.
39. DeVries PJ (1987) The Butterflies of Costa Rica and Their Natural History. Papilionidae, Pieridae, eds. Nymphalidae. New Jersey: Princeton University Press. 327 p.
40. DeVries PJ (1988) Stratification of fruit-feeding nymphalid butterflies in a Costa Rican rainforest. J Res Lepidoptera 26: 98–108.
41. Rydon AHB (1964) Notes on the use of butterfly traps in East Africa. J Lep Soc 18: 51–58.
42. Terborgh J, Robinson SK, Parker TA, Munn C, Pierpont N (1990) Structure and organization of an Amazonian forest bird community. Ecol Monogr 60: 213–238.
43. Fishpool LCD, Evans MI (2001) Important bird areas in Africa and associated islands: priority sites for conservation. Newbury and Cambridge: Pisces Publications and Birdlife International. 1144 p.
44. Nichols JD, Conroy MJ (1996) Estimation of species richness. In: Wilson DE, Cole FR, Nichols JD, Rudran R, Foster M, eds. Measuring and monitoring biological diversity. Standard methods for mammals. Washington: Smithsonian Institution Press. pp 226–234.
45. Burnham KP, Overton WS (1979) Robust estimation of the size of a closed population when capture probabilities vary among animals. Ecology 60: 927–936.
46. Heltshe JF, Forrester NE (1983) Estimating species richness using the Jackknife procedure. Biometrics 39: 1–11.
47. Colwell RK, Coddington JA (1994) Estimating terrestrial biodiversity through extrapolation. Philos Trans R Soc London Ser B 345: 101–118.
48. Boulinier R, Nichols JD, Sauer FR, Hines JE, Pollock KH (1998) Estimating species richness: the importance of heterogeneity in species detectability. Ecology 79: 1018–1028.
49. Chazdon RL, Colwell RK, Denslow JS, Guariguata MR (1998) Statistical methods for estimating species richness of woody regeneration in primary and secondary rain forests of northeastern Costa Rica. In: Dallmeier F, Comiskey JA, eds. Forest Biodiversity Research, Monitoring And Modeling: Conceptual Background And Old World Case Studies. Paris, France: Parthenon Publishing. pp 285–309.
50. Nichols JD, Boulinier T, Hines JE, Pollock KH, Sauer JR (1998) Inference methods for spatial variation in species richness and community composition when not all species are detected. Conserv Biol 12: 1390–1398.
51. Hughes JB, Daily GC, Ehrlich PR (2002) Conservation of tropical forest birds in countryside habitats. Ecol Lett 5: 121–129.
52. Colwell RK (2000) EstimateS – Statistical estimation of species richness and shared species from samples. Available: http://viceroy.eeb.uconn.edu/estimates. Accessed 2010 Dec 31.

8

Responses of Terrestrial Ecosystems' Net Primary Productivity to Future Regional Climate Change in China

Dongsheng Zhao, Shaohong Wu*, Yunhe Yin

Institute of Geographical Sciences and Natural Resources Research, Chinese Academy of Sciences, Anwai, Beijing, China

Abstract

The impact of regional climate change on net primary productivity (NPP) is an important aspect in the study of ecosystems' response to global climate change. China's ecosystems are very sensitive to climate change owing to the influence of the East Asian monsoon. The Lund–Potsdam–Jena Dynamic Global Vegetation Model for China (LPJ-CN), a global dynamical vegetation model developed for China's terrestrial ecosystems, was applied in this study to simulate the NPP changes affected by future climate change. As the LPJ-CN model is based on natural vegetation, the simulation in this study did not consider the influence of anthropogenic activities. Results suggest that future climate change would have adverse effects on natural ecosystems, with NPP tending to decrease in eastern China, particularly in the temperate and warm temperate regions. NPP would increase in western China, with a concentration in the Tibetan Plateau and the northwest arid regions. The increasing trend in NPP in western China and the decreasing trend in eastern China would be further enhanced by the warming climate. The spatial distribution of NPP, which declines from the southeast coast to the northwest inland, would have minimal variation under scenarios of climate change.

Editor: Ben Bond-Lamberty, DOE Pacific Northwest National Laboratory, United States of America

Funding: This study was supported by the National Scientific Technical Supporting Programs during the 12th Five-year Plan of China (2012BAC19B04, 2012BAC19B10). The funders had no role in study design, data collection and analysis, decision to publish, or preparation of the manuscript.

Competing Interests: The authors have declared that no competing interests exist.

* E-mail: wush@igsnrr.ac.cn

Introduction

The impact of climate change on ecosystems is an important topic that has elicited substantial interest across the world [1]. Analysis of recorded temperature in the past century shows a global surface average temperature rise of approximately 0.74°C. Based on the mean temperature recorded from 1980 to 1999, the global surface average temperature is projected to increase by about 1.1°C to 6.4°C by 2100 [2]. China's climate, dominated by the East Asian monsoon, is extremely sensitive to global change [3]. Following projections from numerous general circulation models (GCMs), China would experience obvious climate changes in the future, including increase in average temperature, frequent occurrences of extreme climatic events, spatial and temporal heterogeneity in enhancing precipitation, and enlargement of its arid [4]. These changes in the climate can induce substantial variations in the composition, structure, and function of terrestrial ecosystems, thus inducing changes in ecosystem services, which are closely associated with the living environment of human beings and socioeconomic sustainability.

The impact of climate change on the ecosystem is poorly quantified because long-term in situ measurements are very sparse, and remote sensing techniques are only partially effective [2,5]. Recent studies on the interaction between terrestrial ecosystems and climate change have primarily focused on enhancing the simulation of ecosystem models. Considering that ecosystem models can simulate not only the interaction among ecological processes but also the feedback between climate and ecosystems, these models are of paramount importance in understanding energy balance as well as the water and carbon cycles in an

ecosystem [6–9]. Net primary productivity (NPP) is the rate at which carbohydrates accumulate in a plant's tissues [10,11]. NPP is not only an important index to describe an ecosystem's structure and function but also a key element in describing carbon sequestration in an ecosystem during climate change [12].

Many models, including equilibrium biogeography and biogeochemical models, have been used in previous studies to simulate the impact of climate change on NPP in China. For example, Ni et al. employed an equilibrium biogeography model called BIOME3 to simulate changes in NPP under climate change scenarios [13]. The researchers found that NPP would increase in China from 2070 to 2099. Results from a process-based biogeochemical model (InTEC) indicated that the average forest NPP might be reduced from 2091 to 2100 under climate change, thereby inhibiting the CO_2 fertilizing effect in plants [14]. Ji et al. found that the NPP of terrestrial ecosystems in China would decrease from 1991 to 2100 based on simulations by an atmospheric–vegetation interaction model (AVIM2) [15]. These discrepancies in NPP trends can be partially attributed to ecosystem models because most of them disregard the role of vegetation dynamics in the carbon cycle during climate change. Dynamic vegetation models, including vegetation dynamics and biogeochemical processes, have supplied us with an effective approach to project transient responses of the ecosystem to rapid climate change [16,17]. Previous studies were almost entirely based on climate scenarios generated by GCMs with a resolution of 200 km to 300 km, which is too coarse for studies at regional or national scales. Regional climate models (RCMs) may be a better alternative in overcoming these shortcomings considering that these models provide greater spatial detail through dynamic downscaling of the GCM output [18].

Climate changes in China have significant regional differences; thus, the responses of terrestrial ecosystems to climate change vary in different regions. This study simulated the NPP of ecosystems in China under regional climate change scenarios (A2, B2, and A1B) based on a modified Lund–Potsdam–Jena Dynamic Global Vegetation Model (LPJ-DGVM). The temporal and spatial changes in the NPP of ecosystems in different regions were examined according to eco-regions. The purpose of the study was to elucidate the impacts of climate change on ecosystems at a regional scale and provide scientific basis for local adaptation and mitigating strategies.

Methods and Data

1. Methods

LPJ-DGVM [19] is an integrated dynamic biogeography–biogeochemistry model developed based on an earlier equilibrium model, BIOME3. LPJ-DGVM is constructed in a modular framework, which combines process-based representations of terrestrial vegetation dynamics and land–atmosphere carbon and water exchanges. LPJ-DGVM explicitly considers key ecosystem processes such as vegetation growth, mortality, carbon allocation, and resource competition. Vegetation structure and composition are described by ten plant functional types (PFTs) (Table 1), which are distinguished according to their phenology, physiology, physiognomy, disturbance response attributes, and bioclimatic constraints. Gross primary production is computed based on a coupled photosynthesis–water balance scheme established through canopy conductance. Net primary production is calculated by subtracting autotrophic respiration. The sequestrated carbon is stored in seven PFT-associated pools representing leaves, sapwood, heartwood, fine roots, fast and slow decomposition in the aboveground litter pool, and a below-ground litter pool. The decomposition rates of soil and litter organic carbon depend on soil temperature and moisture. Model input includes monthly mean air temperature, total precipitation, number of wet days and percentage of full sunshine, annual CO_2 concentration, and soil texture class. The full description of LPJ-DGVM was provided by Sitch et al. [19]; thus, only a short overview is provided in this paper. The LPJ-DGVM model is a typical dynamic vegetation model that has been widely utilized to study terrestrial ecosystem dynamics and interactions between climate change and ecosystems at global, regional, and site scales [8,9,16,17,20]. The version employed in the present study includes improved hydrology by Gerten et al. [20].

In our previous study [21], LPJ-DGVM was carefully modified by adding shrub and cold grass PFTs, which were parameterized based on various inventory and observational data, with respect to the characteristics of ecosystems in China. The results of the simulations by the modified LPJ-DGVM for China (LPJ-CN) were validated with data sets obtained from the sites [21]. The simulated NPP results from LPJ-CN matched the observed data ($R^2 = 0.64$, P<0.01), which was better compared with the original LPJ-DGVM data ($R^2 = 0.10$) employed by Ni [22]. Therefore, LPJ-CN is assumed to be appropriate for simulating NPP in China.

Considering that the East Asian monsoon caused by differences in the heat-absorbing capacity of the continent and the ocean dominates the climate in China, climatic types vary from tropical in the south to cold temperate in the north and from humid in the east to dry in the west. Diverse climates and complex topography result in high biodiversity in China. The macro-spatial distribution

of ecosystems forms different eco-regions with unique characteristics. Ecosystem responses to climate change in different eco-regions vary. According to the ecosystem regionalization scheme of China by Zheng et al. [23], China can be divided into eight eco-regions, namely, cold temperate humid region (I), temperate humid/sub-humid region (II), north semi-arid region (III), warm temperate humid/sub-humid region (IV), subtropical humid region (V), tropical humid region (VI), northwest arid region (VII), and Tibetan Plateau region (VIII). The impact of climate change on NPP was discussed in this paper according to the eco-region scheme.

According to the impact degree of future climate change on NPP, the change would be slight if the absolute change in NPP compared with the baseline term is <20%; the change would be moderate at 20% to 40%, the change would be severe at 40% to 60%, and the change would be extremely severe at over 60%. The slopes of linear regressions, obtained using the least squares method, were utilized to assess variation trends in the time series. Positive slopes indicate increasing trends, whereas negative slopes indicate decreasing trends. The significance levels of the trends were assessed by the non-parametric Mann-Kendall test.

2. Data

2.1 Climatic data. The data on climate scenarios utilized in this study were provided by the climate change research group of the Institute of Environment and Sustainable Development in Agriculture, Chinese Academy of Agricultural Sciences. Based on the greenhouse gas emission scenarios of the IPCC Special Report on Emission Scenarios (SRES) [24], the group produced high-resolution ($0.5° \times 0.5°$) climate data scenarios for late 21st century China through the Providing Regional Climate for Impacts Studies (PRECIS) system [25]. PRECIS was validated by applying reanalysis data derived from the European Centre for Medium-Range Weather Forecasts (ECMWF) as lateral boundary conditions; PRECIS was found to simulate terrestrial climate change in China effectively [26]. Projected climatic data, including three SRES emission scenarios (A2, B2, and A1B), for the period 1961 to 2100 were employed in the present study.

Compared with the baseline climate (1961–1990), the annual mean temperature of China (Fig. 1a) is projected to increase from 3.26°C (B2) and 4.64°C (A1B), with the highest increase in northwest and northeast China and the lowest increase in southeast China. The total annual precipitation across China (Fig. 1b) is projected to increase from 8% (B2) and 21% (A2), with the highest increase in northwestern China. A slight decrease in precipitation would occur in northeast China.

The periods were divided into baseline period (1961–1990), near term period (1991–2020), middle term period (2021–2050), and long term period (2051–2080). Each term was discussed according to the average of 30 years.

2.2 Soil data. This research adopted soil texture data from the map of soil texture types (1:14,000,000) [27], which contains information on the proportions of mineral grains in the top soil and geographical distribution of the different soil texture types. The soil textures were reclassified as clay, silt, sand, silty sand, sandy clay, silty clay, and clay with silt and sand according to the FAO classification standard for soil texture to address the data input requirements of LPJ-CN [28]. The soil data were then transformed into the ArcInfo grid format and resampled to the spatial resolution of $0.5° \times 0.5°$.

3. Modeling Protocol

LPJ was operated for a "spin up" period of 900 years to achieve equilibrium in stable vegetation structures and C pools. The model

Table 1. PFTs Bioclimatic limiting factors used by LPJ-CN.

PFTs	T_c min (°C)	T_c max (°C)	GDD min (°C)	T_w max (°C)
Tropical broad-leaved evergreen forest	*12.0* (15.5)			
Tropical broad-leaved raingreen forest	*12.0* (15.5)			
Temperate needle-leaved evergreen forest	−2.0	22.0	900	
Temperate broad-leaved evergreen forest	*0* (3.0)	*14.0* (18.8)	*1500* (1200)	
Temperate broad-leaved summergreen forest	−17.0	*.0* (15.5)	*1500* (1200)	
Boreal needle-leaved evergreen forest	−32.5	*−25.0* (2.0)	600	23.0
Boreal needle-leaved summergreen forest		*−2.0* (2.0)	350	23.0
Boreal broad-leaved summergreen forest		*−15.0* (2.0)	350	23.0
Shrub		*−5.0*	350	
Cold grass		−12.0		12
Temperate grass		*10.0* (15.5)		
Tropical grass	*15.0* (15.5)			

Note: The parameter values in round brackets given by Sitch et al. [19] are replaced by italic values in this study. Shrub and cold grass are newly added PFTs in LPJ. T_c min the minimum coldest monthly mean temperature for survival; T_c max the maximum coldest monthly mean temperature; GDD min the minimum growing degree days of over 5°C; T_w max the upper limit of temperature of the warmest month.

was thereafter subjected to transient climate. Observed atmospheric CO_2 concentration was applied to the ecosystem model simulation for the period of 1961 to 1990. In the projected simulation, the atmospheric CO_2 concentration would remain at the value of 1990. The scheme was designed to eliminate the fertilizing effects of atmospheric CO_2 concentration on the ecosystem.

Results

1. Variation Trend of NPP in Different Eco-regions

Figure 2 illustrates the inter-annual variations of average NPP in eight eco-regions under A2, B2, and A1B scenarios from 1991 to 2100. In the future, NPP may exhibit a significant decreasing trend in all eco-regions except in the Tibetan plateau and northwest arid regions. This projection is associated with the spatial heterogeneity of climate change and sensitivity of the ecosystem to climate change. In cold temperate humid region, the

NPP trend can be distinguished in two periods: before and after 2030. From 1991 to 2030, NPP would remain unchanged but would significantly decrease after 2030 (Fig. 2). NPP in the temperate humid/sub-humid region is projected to decline at a rate greater than 1.5 g C m^{-2} yr^{-1} and would demonstrate a large decadal fluctuation particularly in the B2 and A1B scenarios. The largest decrease is projected to occur in the warm, temperate humid/sub-humid region then in the tropical humid and sub-tropical humid regions (Table 2). The various climate trends in the three climate scenarios can produce slight differences in the NPP trend for each eco-region. Considering that the average NPP in the three eco-regions is higher than that in the other eco-regions, the rapid decrease could have a significant contribution to the decline of NPP in entire China. Increase in NPP may occur in the Tibetan plateau and northwest arid regions due to warming climate. The largest increase in NPP may be found in the Tibetan plateau region, with an average NPP of 1.7 g C m^{-2} yr^{-1} for the

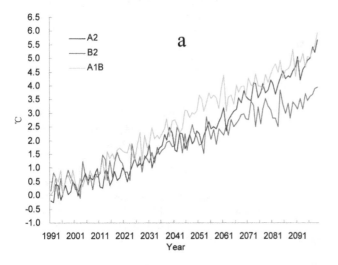

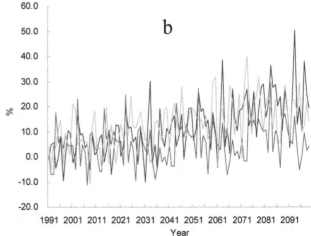

Figure 1. Inter-annual variations of temperature (a) and precipitation (b) anomalies averaged across China (relative to the reference averages from 1961 to 1990, projected by the PRECIS under SRES A2, B2, and A1B scenarios).

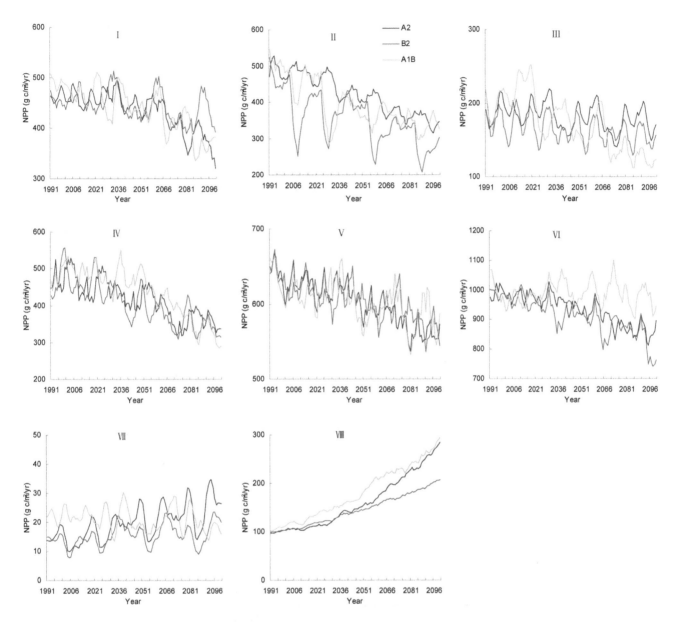

Figure 2. Inter-annual variation of NPP in different eco-regions (I refers to cold temperate humid region; II, temperate humid/sub-humid region; III, north semi-arid region; IV, warm temperate humid/sub-humid region; V, subtropical humid region; VI, tropical humid region; VII, northwest arid region; VIII, Tibetan Plateau region.) from 1991 to 2100 under SRES A2, B2 and A1B scenarios.

A2 and A1B scenarios. The NPP increment in the northwest arid region may be slight; approximately 0.13 g C m^{-2} yr^{-1} ($R^2 = 0.46$, P<0.001) for A2 and 0.14 g C m^{-2} yr^{-1} ($R^2 = 0.14$, P<0.001) for B2. The annual variations in NPP are projected to exhibit a mild decreasing trend in the A1B scenario, although this projection did not pass the statistical tests.

2. Spatial Variation of NPP

Figure 3 shows the spatial pattern of NPP in near term and its variation compared with the baseline term. The NPP in the majority of the ecosystems is projected to increase in near term in over 60% of the total land area of China under the A2, B2, and A1B scenarios. A slight increase in NPP is likely to be predominant. Over-moderate increment may exist only in a few ecosystems concentrated in the Tibetan Plateau and northwest arid region. A few ecosystems may experience a decrease in NPP

particularly under the A1B scenario, which shows the largest area of NPP decrement amounting to approximately 37% of the overall territorial ecosystems. A slight decrement dominates the largest areas of ecosystems with decreasing NPP, accounting for 62%, 64%, and 84% in the A2, B2, and A1B scenarios, respectively. Only a few ecosystems are likely to have a rate of decrement over moderate levels; these ecosystems are mainly distributed in the north semi-arid and northwest arid regions, particularly in the A2 scenario.

The trends of NPP, decreasing in east China and increasing in west China, are more obvious in the middle term compared with the near term (Fig. 4). A slight NPP decrement may dominate China's ecosystems in the A2, B2, and A1B scenarios, occupying 25% and 26% of the overall territorial ecosystem in the A2 and B2 scenarios, respectively, which mainly occurs in the sub-tropical humid and warm temperate humid/sub-humid regions. In

Table 2. NPP Slopes (g C m^{-2} yr^{-1}) in different eco-regions and its coefficient of determination under different climate scenarios.

Eco-regions	A2		B2		A1B	
	Slope	R^2	Slope	R^2	Slope	R^2
I	−0.95	0.61	−0.37	0.17	−1.11	0.64
II	−1.63	0.87	−1.53	0.50	−1.59	0.60
III	−0.19	0.11	−0.24	0.18	−0.94	0.62
IV	−1.36	0.66	−1.49	0.64	−1.76	0.71
V	−0.83	0.74	−0.60	0.55	−0.53	0.36
VI	−1.43	0.75	−1.65	0.70	−0.05	0.01
VII	0.13	0.46	0.04	0.14	−0.03*	0.01
VIII	1.66	0.97	1.00	0.99	1.69	0.98

Note: *means the significance level is P<0.5, or else P<0.001.

addition, ecosystems with over-moderate NPP decrement are limited to the temperate humid/sub-humid, north semi-arid, and northwest arid regions. Severe and extremely severe NPP decrements primarily exist in the central temperate humid/sub-humid region and the adjacent areas between the north semi-arid and northwest arid regions. The ecosystems with NPP increment are likely to decline, with approximately 53% of the total ecosystem in the B2 and A1B scenarios and 59% in the A2 scenario. In the B2 scenario, the area with a slight increment in NPP is reduced to 21% of the overall ecosystem, which is smaller than the value in the near term. Ecosystems with over-moderate NPP increment are still concentrated in the Tibetan Plateau and northwest arid region where the absolute NPP value is only 60 g C m^{-2}, a value that is considerably lower than those in the other regions.

Figure 5 illustrates the NPP distribution in the long term climate and change in NPP relative to the baseline term climate. In the A1B, A2, and B2 scenarios, the ecosystems with NPP decrement are expected to occupy 47%, 46%, and 52% of total area in China, respectively, which means that nearly half of the ecosystems in China may be exposed to the adverse effects of climate change. An NPP decrement may occur primarily in eastern China, which has an inherently high NPP. An NPP increment may occur in western China, which has low NPP. Ecosystems with slight NPP decrement would still predominate in China and would retain spatial patterns similar to that in the middle term. Areas with over-moderate NPP decrement are likely to expand further and are projected to become distributed in the temperate humid/sub-humid and warm temperate humid/sub-humid regions. Ecosystems may be severely impacted by climate change in the A1B scenario, and the area with the extremely severe NPP decrement would be greater than 7% of the overall territorial ecosystems. This projection is expected to occur in most parts of northern China. On the contrary, the NPP increment

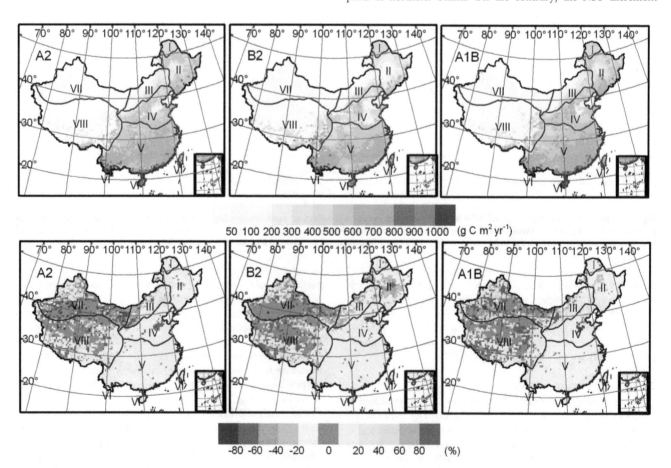

Figure 3. NPP spatial pattern (g C m^2 yr^{-1}) (upper panel) in the near term and its anomalies (%) (lower panel) with baseline over China modeled through climate change projections for SRES A2, B2, and A1B scenarios.

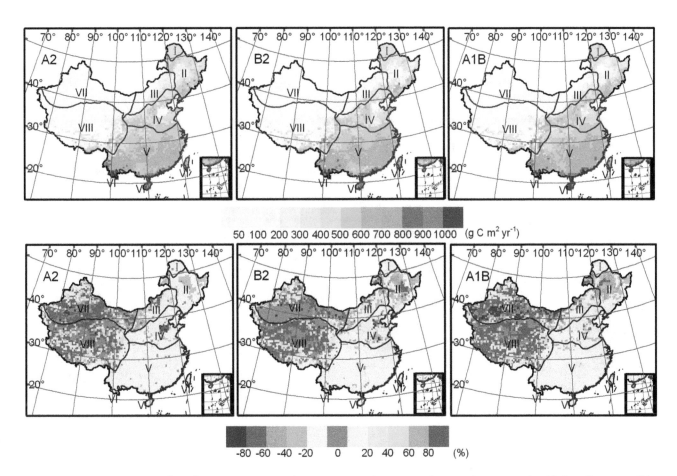

Figure 4. NPP pattern (g C m² yr⁻¹) (upper panel) in the midterm and its anomalies (%) (lower panel) with baseline over China modeled through climate change projections for SRES A2, B2, and A1B scenarios.

would be mainly concentrated in western China, particularly in the Tibetan Plateau region where the average NPP increment may reach approximately 60%. The NPP decrement in East China and increment in West China may be enhanced further in terms of spatial variation.

Discussion

1. Effects of Climate Change on NPP

The projected climate change characterized by warm and wet climate may cause variations in NPP in most areas of China according to our simulated results. Regional differences are mainly associated with the combinations of variations in temperature and precipitation in all three scenarios. The NPP decrease in cold temperate humid regions may be associated with the rising temperature because a slight increase in temperature can benefit the boreal forest growth, whereas a significant increase in temperature may reduce the water effectiveness through thawing frozen soil and enhancing evapotranspiration, which in turn can result in negative effects for the boreal forest growth [15,16,29]. The large annual variation in NPP in the temperate humid/sub-humid region can be attributed to frequent drought derived from the climate scenarios, particularly in B2. The decrement in precipitation can impair the photosynthesis rate of vegetation in the model because of reduced stomatal conductance [30,31].

The NPP decrement in warm temperate, sub-tropical, and tropical humid regions can be explained by increased water stress as a consequence of increased evapotranspiration that is not counterbalanced by an increase in rainfall. Piao et al. [29] also found that increase in temperature alone does not benefit vegetation NPP in temperate and tropical ecosystems because of water control. Studies on the historical tree ring records of tropical forests and temperature [32,33] suggest that water availability is a key limiting factor that controls vegetation NPP. Given that water limits photosynthesis in plants in the northwest arid region where plants are more sensitive to precipitation variation than temperature variation, increasing precipitation can result in an increase in carboxylation efficiency, which is the primary cause of enhanced productivity [30]. The NPP in the A1B scenario exhibits a different trend from that in A2 and B2. This different trend could have been caused by different precipitation trends. Before 2030, the A1B scenario has a greater amount of rainfall than the other two scenarios; however, the amount of rainfall would remain unchanged or even decrease after 2031. The relatively low temperature in the Tibetan Plateau limits plant growth, but warming may solve this limitation, thus leading to a general increase in NPP [13,34].

The results of this research are generally consistent with the results of previous studies [15,14,28,35]. However, some discrepancies exist. Our study suggests that NPP in eastern China, which is covered with forests, is likely to decrease even with the influence of climate change. This is similar to the result obtained by Ju et al. [14]. The spatial change in NPP, namely, a decrease in eastern China and an increase in western China under the climate change scenarios, is consistent with Wu et al. [35] and Ji et al.'s [15] results. The NPP changes in several sensitive areas located at the

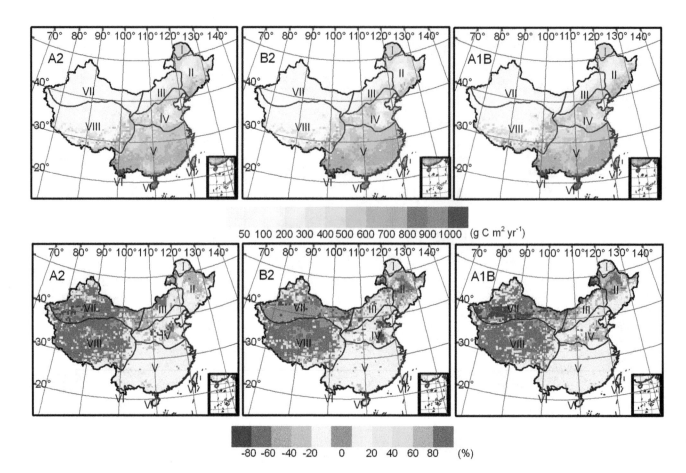

Figure 5. NPP pattern (g C m² yr⁻¹) (upper panel) in the long term and its anomalies (%) (lower panel) with baseline over China modeled through climate change projections for SRES A2, B2, and A1B scenarios.

temperate humid/sub-humid, warm temperate humid/sub-humid, and north semi-arid regions are different from the results of previous studies [15,35]. These areas have an increasing incidence of extreme climatic events; therefore, biogeography and biogeochemical models cannot simulate the effects of extreme climatic events on ecosystems. Dynamic vegetation models can capture these effects, leading to discrepancies in results.

2. Uncertainties in the Study

LPJ-CN, a process-based ecosystem model, was applied in this study to simulate NPP changes in China under scenarios A2, B2, and A1B. The study examined the NPP variation in three projected periods, which improves the understanding of the impact of the different warming levels. However, several uncertainties underlying future climate change and ecosystem responses exist. The trends of climate change in China are different in various climatic scenarios. However, we only probed the possibilities for the IPCC SRES A2, B2, and A1B scenarios. Given that the climatic system is complicated and nonlinear and China's landform is very complex with altitudes varying from −50 m to 8843 m, GCMs are often unable to depict the accurate spatial patterns of the climatic factors employed in ecosystem modeling with low resolution. Although the simulation capability of PRECIS with boundary condition from the general circulation model (GCMs)-HadAM3 [25] has been improved in this study by enhancing spatial resolution, the aforementioned problems still cannot be addressed entirely to eliminate uncertainties. Moreover,

some uncertainties remain with regard to the parameterization and process representations in the LPJ-DGVM model [36]. For example, the model does not provide an explicit representation of the nitrogen cycle; some scientific assumptions are unreasonable, and some lagging responses and self-adaptive mechanisms of the ecosystem to climate change are omitted in the LPJ-DGVM model. These issues have attracted the attention of researchers around the world, but there are insufficient studies to support the quantification of these processes in the model.

The LPJ-CN model predicts the temporal and spatial change of ecosystem NPP according to potential natural vegetation, which is free from the influence of anthropogenic activities. However, the great expansion in human activity can definitely have a major impact on ecosystem dynamics [37]. Liu et al. reported that approximately 5.2 Mha of grassland and woodland in China were converted to cropland in the 1990s because of reclamation activities [38]. Piao et al. found that the large-scale reforestation and afforestation programs since the 1980s in China have resulted in the increase of forest biomass carbon stocks [39]. Considering the lack of information on the amount and spatial patterns of land use in the future, the accurate estimation of the magnitude of NPP change associated with land use remains a challenge [40]. These uncertainties can be addressed by integrating human disturbances in future studies.

Conclusions

We utilized a dynamic vegetation model, LPJ-CN, to simulate the temporal and spatial responses of ecosystem NPP to climate change in 21st century China under A2, B2, and A1B scenarios. Our results indicate that rising temperature and slightly changing precipitation would lead to a serious impact on the natural ecosystem's NPP in general. The near-term, mid-term, and long-term impact of climate change may become worse gradually. In the near term, the impact of climate change on the ecosystem may be minimal; more than half of the ecosystems may benefit from the change. However, an adverse impact may occur in the middle term. In the long term, more than half of ecosystems may be exposed to adverse effects. NPP enhancement may appear mainly in the Tibetan Plateau and northwest arid region. The other ecoregions may be dominated by NPP increase. Although NPP may increase by a relatively high percentage in the Tibetan Plateau and northwest arid region, it cannot generate a significant influence on the overall NPP distribution in China because of its low initial productivity level. Therefore, the spatial distribution of NPP, which decreases from the southeast coast to the northwest inland, would not be altered under the climate change scenarios. The response of China's terrestrial ecosystems to climate change will contribute to our understanding of the vulnerability and adaptability of ecosystems to climate change in a regional scale. The result obtained in this study can provide a basis for environmental policy making.

Acknowledgments

We thank Prof. Yinlong Xu from Institute of Environment and Sustainable Development in Agriculture, Chinese Academy of Agriculture Sciences, for providing climate scenarios data.

Author Contributions

Conceived and designed the experiments: SW. Performed the experiments: DZ. Analyzed the data: DZ. Wrote the paper: DZ YY.

References

1. Heimann M, Reichstein M (2008) Terrestrial ecosystem carbon dynamics and climate feedbacks. Nature 451, 289–292.
2. IPCC (2007) Summary for Policymakers. In: Climate Change 2007: Impacts, Adaptation and Vulnerability. Contribution of Working Group II to the Fourth Assessment Report of the Intergovernmental Panel on Climate Change, M.L. Parry, O.F. Canziani, J.P. Palutikof, P.J. Van der Linden and C.E. Hanson, editors, Cambridge, Cambridge University Press.
3. Ding YH, Johnny CLC (2005) The East Asian summer monsoon: an overview. Meteorol Atmos Phys 89, 117–142.
4. Taskforce on China's National Assessment Report on Climate Change (2007) China's National Assessment Report on Climate Change. Science Press: Beijing.
5. Valentini R, Matteucci G, Dolman AJ, Schulze ED, Rebmann C, et al. (2000) Respiration as the main determinant of carbon balance in European forests. Nature 404: 861–865.
6. Tian H, Melillo JM, Kicklighter DW, Pan S, Liu J, et al. (2003) Regional carbon dynamics in monsoon Asia and its implications for the global carbon cycle. Global Planet Change 37: 201–207.
7. Koca D, Smith B, Sykes MT (2006) Modeling regional climate change effects on potential natural ecosystems in Sweden. Climatic Change 78: 381–401.
8. Morales P, Hick T, Rowell DP, Smith B, Sykes MT (2007) Changes in European ecosystem productivity and carbon balance driven by regional climate model output. Global Change Biol 13: 108–122.
9. Doherty RM, Sitch S, Smith B, Lewis SL, Thornton PK (2010) Implication of future climate and atmospheric CO_2 content for regional biogeochemistry, biogeography and ecosystem services across East Africa, Global Change Biol 16: 617–640.
10. Odum EP (1959) Fundamentals of ecology. Philadelphia: W.B. Saunders Co.
11. Lieth H, Whittaker RH (1975) Primary Productivity of the biosphere. New York: Springer-Verlag Press.
12. Nemani RR, Keeling CD, Hashimoto H (2003) Climate-driven increases in global terrestrial net primary production from 1982–1999. Science 300: 1560–1563.
13. Ni J (2000) A simulation of biomes on the Tibetan Plateau and their responses to global climate change. Mt Res Dev 20: 80–89.
14. Ju WM, Chen JM, Harvey D, Wang S (2007) Future carbon balance of China's forests under climate change and increasing CO_2. J Environ Manage 85: 538–562.
15. Ji JJ, Huang M, Li KR (2009) Prediction of carbon exchanges between China terrestrial ecosystem and atmosphere in 21st century. Sci China Ser D 51(6): 885–898.
16. Cramer W, Bondeau A, Woodward FI, Prentice IC, Betts RA, et al. (2001) Global response of terrestrial ecosystem structure and function to CO_2 and climate change: Results from six dynamic global vegetation models. Global Change Biol 7: 357–373.
17. Bachelet D, Neilson RP, Hickler T, Drapek RJ, Lenihan JM, et al. (2003) Simulating past and future dynamics of natural ecosystems in the United States. Global Biogeochem Cy 17(2), 1045, doi:10.1029/2001GB001508.
18. McGregor JL (1997) Regional climate modeling. Meteorol Atmos Phys 63: 105–117.
19. Sitch S, Smith B, Prentice IC. Arneth A, Bondeau A, et al. (2003) Evaluation of ecosystem dynamics, plant geography and terrestrial carbon cycling in the LPJ dynamic global vegetation model. Global Vegetation Model. Global Change Biol 9: 161–185.
20. Gerten D, Schaphoff S, Haberlandt U, Lucht W, Sitch S (2004) Terrestrial vegetation and water balance-Hydrological evaluation of a dynamic global vegetation model. J Hydrol 286: 249–270.
21. Zhao DS, Wu SH, Yin YH (2011) Variation trends of natural vegetation net primary productivity in China under climate change scenario. Chinese J Appl Ecol 22(4): 897–904.
22. Ni J (2003) Net primary productivity in forests of China: scaling-up of national inventory data and comparison with model predictions. Forest Eco Manag 176: 485–495.
23. Zheng D (1996) The system of physico-geographical regions of the Qinghai-Xizang (Tibet) Plateau. Sci China Ser D 39: 410–417.
24. Nakicenovic et al (2000) Special Report on Emissions Scenarios (SRES). Cambridge: Cambridge University Press.
25. Jones R, Noguer M, Hassell D, Hudson D, Wilson S, et al. (2004) Generating high resolution climate change scenarios using PRECIS. Met Office Hadley Centre, Exeter UK.
26. Xu YL, Jones R (2004) Validating PRECIS with ECMWF reanalysis data over China. Chinese J Agrometeorol 25(1): 5–9.
27. Zhang SH, Peng GB, Huang JY (2004) The feature extraction and data fusion of regional soil textures based on GIS techniques. Clim Environ Res 6: 65–79.
28. Ni J, Sykes MT, Prentice IC, Cramer W (2001) Modeling the vegetation of China using the process-based equilibrium terrestrial biosphere model BIOME3. Global Ecol Biogeogr 9: 463–479.
29. Piao SL, Ciais P, Friedlingstein P, de Noblet-Duconudre N, Cadule P, et al. (2009) Spatiotemporal patterns of terrestrial carbon cycle during the 20th century. Global Biogeochem Cy 23, GB4026, doi:10.1029/2008GB003339.
30. Hou YY, Liu QH, Yan H, Tian GL (2007) Variation trend of China terrestrial vegetation net primary productivity and its responses to climate factors in 1982–2000. Chinese J Appl Ecol 18(7): 1546–1553.
31. Yang YH, Fang JY, Ma WH, Guo DL, Mohammat A (2010) Large-scale pattern of biomass partitioning across China's grassland. Global Ecol Biogeogr 19: 268–277.
32. Clark DA, Piper SC, Keeling DC, Clark DB (2003) Tropical rain forest tree growth and atmospheric carbon dynamics linked to interannual temperature variation during 1984–2000. Pro Natl Acad Sci U S A 100: 5852–5857.
33. Liang EY, Shao XM, Kong ZC, Lin JX (2003) The extreme drought in the 1920s and its effect on tree growth deduced from tree ring analysis: A case study in north China. Ann Forest Sci 60: 145–152.
34. Zhao DS, Wu SH, Yin YH, Yin ZY (2011) Vegetation distribution on Tibetan Plateau under climate change scenario. Reg Environ Change 11: 905–915.
35. Wu SH, Yin YH, Zhao DS, Huang M, Shao XM, et al. (2010) Impact of future climate change on terrestrial ecosystems in China. Int J Climatol 30: 866–873.
36. Zaehle S, Sitch S, Smith B, Hatterman F (2005) Effects of parameter uncertainties on the modeling of terrestrial biosphere dynamics. Global Biogeochem Cy 19, GB3020, doi:10.1029/2004GB002395.
37. Bondeau A, Smith PC, Zaehle S, Schaphoff S, Lucht W, et al. (2007) Modelling the role of agriculture for the 20th century global terrestrial carbon balance. Global Change Biol 13: 679–706.
38. Liu JY, Liu ML, Tian HQ, Zhuang DF, Zhang ZX, et al. (2005) Spatial and temporal patterns of China's cropland during 1990–2000: An analysis based on Landsat TM data. Remote Sens Environ 98: 442–456.
39. Piao SL, Fang JY, Ciais P, Peylin P, Huang Y, et al. (2009) The Carbon balance of terrestrial ecosystems in China. Nature 458: 1009–1013.
40. Houghton RA (2007) Balancing the global carbon budget. Ann Rev Earth Pl Sc 35: 313–347.

9

Soil Organic Carbon and Total Nitrogen Gains in an Old Growth Deciduous Forest in Germany

Marion Schrumpf[1]*, Klaus Kaiser[2], Ernst-Detlef Schulze[1]

1 Max Planck Institute for Biogeochemistry, Hans-Knöll-Straße 10, 07745 Jena, Germany, **2** Soil Sciences, Martin Luther University Halle-Wittenberg, 06120 Halle (Saale), Germany

Abstract

Temperate forests are assumed to be organic carbon (OC) sinks, either because of biomass increases upon elevated CO_2 in the atmosphere and large nitrogen deposition, or due to their age structure. Respective changes in soil OC and total nitrogen (TN) storage have rarely been proven. We analysed OC, TN, and bulk densities of 100 soil cores sampled along a regular grid in an old-growth deciduous forest at the Hainich National Park, Germany, in 2004 and again in 2009. Concentrations of OC and TN increased significantly from 2004 to 2009, mostly in the upper 0–20 cm of the mineral soil. Changes in the fine earth masses per soil volume impeded the detection of OC changes based on fixed soil volumes. When calculated on average fine earth masses, OC stocks increased by 323 ± 146 g m^{-2} and TN stocks by 39 ± 10 g m^{-2} at 0–20 cm soil depth from 2004 to 2009, giving average annual accumulation rates of 65 ± 29 g OC m^{-2} yr^{-1} and 7.8 ± 2 g N m^{-2} yr^{-1}. Accumulation rates were largest in the upper part of the B horizon. Regional increases in forest biomass, either due to recovery of forest biomass from previous forest management or to fertilization by elevated CO_2 and N deposition, are likely causes for the gains in soil OC and TN. As TN increased stronger (1.3% yr^{-1} of existing stocks) than OC (0.9% yr^{-1}), the OC-to-TN ratios declined significantly. Results of regression analyses between changes in OC and TN stocks suggest that at no change in OC, still 3.8 g TN m^{-2} yr^{-1} accumulated. Potential causes for the increase in TN in excess to OC are fixation of inorganic N by the clay-rich soil or changes in microbial communities. The increase in soil OC corresponded on average to 6–13% of the estimated increase in net biome productivity.

Editor: Jose Luis Balcazar, Catalan Institute for Water Research (ICRA), Spain

Funding: This study was funded by the EU (CarboEurope-IP, Project No. GOGC-CT-2003n505572) and the Max-Planck Society. The funders had no role in study design, data collection and analysis, decision to publish, or preparation of the manuscript.

Competing Interests: The authors have declared that no competing interests exist.

* E-mail: mschrumpf@bgc-jena.mpg.de

Introduction

Terrestrial ecosystems are facing a number of environmental shifts, such as rising CO_2 in the atmosphere, increasing temperature, changes in precipitation, increasing nitrogen deposition, or increased frequency of extreme weather events. These factors can affect ecosystem carbon storage and thereby feed back to climate change [1,2]. A number of ecosystem manipulation experiments have been carried out to determine effects of changing environmental conditions on ecosystem processes [3]. It is, however, not possible to simulate environmental changes of entire forest ecosystems, and often only one factor is modified in controlled experiments while many changes occur simultaneously under real-world conditions. Determination of net responses of ecosystems to environmental changes, therefore, requires monitoring of ecosystem processes under natural field conditions.

Ballentyne et al. [4] showed that the net CO_2 uptake by terrestrial and aquatic ecosystems doubled during the last 50 years but it is still unresolved where the carbon goes [4,5]. Forests seem to be important terrestrial carbon sinks [6–8]. Besides environmental factors, long-term changes in forest cover and management need to be considered as possible causes for forest organic carbon (OC) accumulation or depletion [9]. Only few results on repeated forest soil OC inventories exist to date

[10]. These predominantly report increases in forest soil OC (Table 1), rarely declines (i.e., Belgium region Wallonia, [11] England and Wales, [12]). Many repeated soil inventories did not determine bulk densities directly. Instead, these were derived from pedotransfer functions, an approach that increases the uncertainty, especially if bulk density is not constant over time [13–15]. Accordingly, additional repeated inventories of forest soil OC stocks that include determination of all contributing variables are needed.

Following stoichiometric principles, gains in soil OC will go along with increases in other elements including nitrogen (N) [16]. The combined changes in soil OC and N can give hints on sources of accumulating organic matter as input of new plant litter, for example, has larger OC-to-N ratios than bulk soil organic matter. Another factor potentially affecting the soil OC-to-N ratio is N deposition. It is assumed to contribute to enhanced plant growth under elevated CO_2, leading to increased organic matter input and storage in soil [17]. Only part of the N added to ecosystems is taken up by plants, and soils often play the more important role in ecosystem N retention [18,19]. Thus, long-term N deposition should lead to N accumulation in soil if no N is lost due to leaching or denitrification. Although it is well known that C and N cycles are closely linked and the fate of deposited N is still not resolved, the focus of repeated soil

Table 1. Measured and modelled changes in soil C stocks in temperate forests.

Location		Soil depth	Change	Years	ref
		cm	g C m^{-2} yr^{-1}		
Europe	Region	0–100	60	1990–1999	[8]
			46	2000–2007	
Belgium	Region	0–30	68	1960–2000	[50]
Germany	Region	0–30	50	1990–2006	[51]
Sweden	Region	humus layer	25	1961–2002	[52]
Belgium	Region	0–30	−23	1950–2006	[11]
China	Plot	0–20	61	1979–2003	[14]
Germany	Plot	0–30	21 to 40	1974–2004	[9]
USA	Plot	0–10	−11 to +6	1976–2006	[30]
Germany	Plot	0–60	27–65	2004–2009	This study

inventories is often on OC only, while changes in N are not reported.

For the old-growth deciduous forest in the Hainich National Park, Germany, eddy covariance analyses indicated the site as a strong carbon sink of almost 500 g m^{-2} yr^{-1} [20], but Kutsch et al. [21] showed that some of that was erroneous due to advection. Based on soil carbon budgets calculated as difference between input and heterotrophic respiration, Kutsch et al. [22] concluded that the soil can only be a small net carbon sink of 1 to 35 g m^{-2} yr^{-1}. Tefs and Gleixner [23], on the other hand, observed soil gains of 164 g OC m^{-2} yr^{-1} and 32 g TN m^{-2} yr^{-1} between 2000 and 2004, using a repeated soil inventory. However, the Pürckhauer soil corer they used for collecting samples is highly prone to compaction due to its small diameter, which hampers sampling by depth increments. Average bulk densities obtained in another soil inventory had to be used to convert concentration differences into fluxes, which added much uncertainty to the results. The present study aimed at verifying the previous results by using an improved approach with combined determination of OC, TN, and soil masses at 100 paired soil cores down to 60 cm soil depth, sampled in 2004 and again in 2009. We hypothesize that (1) increased forest litter input leads to detectable gains in soil OC stocks after 5 years, and (2) continued N deposition results in soil N accumulation in excess to what is required according to stoichiometric principles to support OC gains.

Materials and Methods

Site description

The study site is located in the "Hainich National Park" in central Germany (51°04′46″N, 10°27′08″E, 440 m a.s.l.). Permission for soil sampling was granted by the headquarters of the park represented by Manfred Großmann. For detailed description of the study site, see Knohl et al. [20]. Mean annual rainfall at the eddy covariance tower site is 765 mm (range between 540 to 1050 mm from the year 2000 till 2012), and mean annual air temperature is 8.3°C (6.9 to 8.8°C in the same period) (Olaf Kolle, personal communication). The site is under an old-growth mixed beech forest (65% *Fagus sylvatica*, 25% *Fraxinus excelsior*, 7% *Acer*

pseudoplatanus and *A. platanoides*). Trees have a wide range of ages, with a maximum of 250 years. Maximum tree height varies between 30 and 35 m, and the leaf area index is 4.8 m^2 m^{-2}. Large amounts of standing dead wood and coarse woody debris are characteristic of this semi-natural forest, which was taken out of management about 60 years ago. The ground vegetation is dominated by geophytes and hemikryptophytes and was classified as a *Hordelymo-Fagetum* [24]. The substrate for soil formation is Triassic limestone, overlain by a Pleistocene loess layer of varying thickness (10–50 cm). Soils are classified as Eutric Cambisols [25]. The litter layer is mull-type and indicates high biological activity and intensive bioturbation.

Soil sampling

A regular grid with 30 m spacing was projected over the main footprint area (24 ha) of the eddy covariance tower, and 100 sampling points were randomly selected at grid points, marked with wooden poles having metal labels on top, and sampled in March 2004. In March 2009, new samples were taken 1.5 m north of the original sampling point. At each sampling, the litter layer was collected using a metal frame of 25 cm side length. Corers with an inner diameter of 8.7 (2004) and 8.3 cm (2009, both from Eijkelkamp Agrisearch Equipment BV, Giesbeek, The Netherlands) were used for mineral soil sampling. Closed corers were driven into soil with a motor hammer (Cobra Combi, Atlas Copco AB, Nacka, Sweden) and opened afterwards to obtain the intact soil cores. The depth of the borehole and the length of the extracted core were measured, and compared for estimation of soil compaction during coring. The average deviation between core length and hole was 0.3 cm. Soil cores were cut into 7 segments to obtain samples from 0–5, 5–10, 10–20, 20–30, 30–40, 40–50, and 50–60 cm depth. At some places, it was not possible to reach a depth of 60 cm because of stones. In such cases, a second attempt was made at 1 m distance. If necessary, the procedure was repeated a third time and the longest of the three cores used.

Sample treatment and analyses

Soil samples were stored at 4°C prior to processing. Coarse stones of a diameter >4 mm and roots of a diameter >1 mm were removed from the samples prior to drying at 40°C. Stone and root samples were air-dried separately. Then, soil samples were sieved to <2 mm. Particles >2 mm were combined with the coarse stones. Dry weights of roots and combined stone fractions were determined. Total C and N concentrations in <2 mm soil separates were determined using dry combustion at 1100°C (VarioMax, Elementar Analysensysteme GmbH, Hanau, Germany). Carbonate C was determined after dry combustion of the samples heated in a muffle furnace at 450°C for 16 h. Organic C was calculated as the difference between total and carbonate C. Litter layer samples were dried at 70°C, shredded, and a subsample further homogenized using a ball mill. Total C and N concentrations were determined using dry combustion (Vario EL II, Elementar Analysensysteme GmbH, Hanau, Germany). Nitrogen concentrations reflect total (organic plus inorganic) N (TN). Bulk density was determined based on the dry mass of total soil material of each depth increment. Accordingly, bulk density, fine earth material (<2 mm), and OC concentrations were measured on the same samples.

Statistics and calculations

Results are presented as means ± standard error unless indicated differently. Changes in C stocks can be calculated either for a fixed soil volume or a fixed mass of soil. Organic C and TN stocks for fixed soil volumes of soil layers were calculated based on bulk density, the relative contribution of fine earth material (soil <2 mm) to total soil mass, layer thickness, and element concentration [10]. Calculation of stocks for equivalent soil masses per area was carried out as described by Ellert and Bettany [26]. It is based on cumulative fine earth masses per area of the soil layers of each soil core (kg m^{-2}). We used the average fine earth masses of each depth increment of the 2004 sampling as reference soil mass for OC stocks in both years [10]. For the calculation of OC stocks, soil material from the uppermost and underlying soil layers is cumulatively added until the desired reference soil mass is reached. Organic C stocks are then calculated by multiplying the soil mass of each included layer with the corresponding OC concentration and adding the products up to achieve total amounts of OC per area for the respective reference soil masses. We did not determine the OC concentration of the stones. Therefore, we used the fine earth mass instead of the total soil mass as a reference.

Kolmogorov-Smirnov tests were used to test the datasets for normal distribution. To test for significant changes between the sampling years, paired-sample t-tests were conducted when differences between pairs were normally distributed. Otherwise, U-tests (Wilcoxon) were performed. To determine relations between variables, Pearson correlation coefficients were calculated and tested for significance with a two-tailed t-test ($p<0.05$, or $p<0.01$, as given in figures). Statistical analyses were performed using the software package SPSS 16.0 for Windows.

Results

Concentrations and stocks of OC and TN in 2004 [10] were significantly correlated with those in 2009 for all depth increments, indicating soil organic matter concentrations to be spatially dependent (Table 2). The closer spatial correlation of OC and TN stocks in deeper than in upper soil layers can be explained by the closer spatial dependence of the stones, and accordingly, the fine earth contents in deeper than in upper soil layers (Table 2). At 0–60 cm soil depth, the 2004 OC stocks explained 33% of the variance in stocks observed in 2009, whereas at 30–60 cm, the explained variance was 68%. The location dependence of soil OC stocks demonstrates the validity of the used paired sampling approach.

A number of variables differed between 2004 and 2009 (Fig. 1, Table S1 in Information S1). The litter layer held significantly

more OC in 2009 than in 2004 (121±24 g m^{-2} difference) and also the OC-to-TN ratio of the litter was larger in 2009 than in 2004 (by 6.4±1.8, data not shown). Concentrations of OC in the mineral soil were 7 to 9% larger in 2009 than in 2004, with differences being significant for the 0–5, 5–10, 10–20, and 30–40 cm depth layers. Increases in TN concentrations were even larger (8 to 12%) and significant throughout the profile down to 50 cm depth. The OC-to-TN ratio decreased significantly for the 0–5, 10–20, and 40–50 cm layers. Bulk densities and fine earth contents were greater in 2004 than in 2009 and negatively correlated with OC concentrations and with water contents (Fig. 1, Table in in Information S1). With OC and TN concentrations increasing and the fine earth amounts decreasing, there was no significant net change in OC stocks for individual layers. Only stocks of TN were significantly larger in the 5–10 and 40–50 cm layers.

We used the average soil mass per depth increment in 2004 as reference mass for each layer when calculating OC stocks based on equal soil masses. OC stocks of equal soil masses differed significantly for the layer corresponding to 10–20 cm soil depth in 2004 (133±62 g OC m^{-2}) and for the cumulative stock of the soil mass in 0–20 cm (323±146 g OC m^{-2}) (Fig. 2, Table S3 in Information S1). Cumulative OC stocks did not change significantly from 2004 to 2009 when soil masses from deeper soil layers were included (Fig. 2). The largest differences in TN were also within 0–20 cm depth (39±10 g TN m^{-2}), though, unlike OC, differences were significant for all individual layers to 20 cm depth when normalized to average soil masses obtained in 2004 (Fig. 2). Cumulative N stocks increased slightly below 20 cm depth from 39 to 47 g TN m^{-2}. Accordingly, the OC-to-TN ratio calculated for OC and TN in fixed soil masses was significantly lower in 2009 than in 2004 below the uppermost soil layer. The average OC-to-TN ratios of the changes were 11 for the 0–5 cm, and ~7 for the 5–10 cm, and 10–20 cm equivalents, and smaller than those of the volume-based changes calculated for the same soil layers (12, 10, and 9 for the 0–5, 5–10, and 10–20 cm layers, respectively).

Concentrations as well as stocks of OC and TN were strongly correlated in both years and at all soil depths (Fig. 3A, Table S2 in in Information S1). The OC concentrations and stocks were positively correlated to OC-to-TN ratios in both years and at all soil depths (Fig. 3 B, Table in in Information S1). When plotting changes in stocks of OC and TN between 2004 and 2009 (Figure 3C+D) for 205 kg fine earth m^{-2} (corresponding to the

Table 2. Correlation coefficients for significant ($p<0.05$) relations among bulk density (BD), fine earth mass per m^{-2} (FE), stone content (stone), water content (WC), OC concentration (OC), total nitrogen concentration (TN), OC-to-TN ratio (CN), OC stock (OCst), and TN stock (TNst) for paired samples taken in 2004 and 2009.

Soil depth	BD	FE	Stone	WC	OC	TN	CN	OCst	TNst
cm									
0–5	ns	0.21	ns	ns	0.45	0.52	0.38	0.28	0.23
5–10	0.21	0.28	0.23	0.25	0.39	0.46	0.40	0.30	0.35
10–20	0.31	ns	0.35	0.44	0.34	0.39	0.26	0.33	0.33
20–30	0.38	0.42	0.33	0.35	0.51	0.58	ns	0.44	0.44
30–40	ns	0.46	0.46	0.42	0.48	0.46	0.22	0.54	0.57
40–50	ns	0.44	0.55	ns	0.65	0.58	0.39	0.72	0.65
50–60	ns	0.41	0.38	ns	0.65	0.50	ns	0.71	0.65

ns: not significant.

Differences 2009-2004

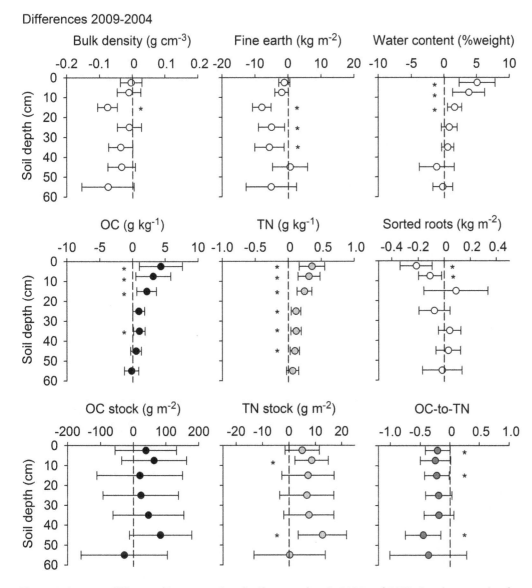

Figure 1. Average difference between pairs of soil cores taken in 2004 and 2009. Samples were taken for seven depth increments. Positive values indicate gains or increases with time. Error bars show 5–95% confidence intervals; asterisks indicate changes significantly different from zero.

average fine earth amount per m² in 0 to 20 cm depth in 2004), the regression indicates that with no change in OC, TN still increased by 19±4 g m^{-2}. This is gives an increase in N of 3.8±0.8 g m^{-2} yr^{-1} in excess of OC.

Discussion

In accordance with our hypothesis, it was possible to detect statistically significant increases in soil OC and TN concentrations within 5 years. However, the more relevant detection of changes in stocks was more difficult as similar to previous studies [13,14], our results showed that bulk densities were not constant over time. This could be due to variations in OC and water contents, or because of systematic sampling errors (e.g., compaction during coring). That highlights the necessity of determining OC and bulk density in the same inventory and to calculate stock changes based on equivalent soil masses rather than for defined soil depths [27].

Potential errors in the determination of bulk densities might have been smaller with samples taken from soil pits instead by a

soil core. However, using average bulk densities from soil pits and only OC concentrations from a corer would not necessarily result in more precise estimates of soil OC stocks and stock changes due to the small scale spatial variation of both variables and their negative correlation. Calculation of OC stocks based on equal soil masses alleviates some of the problems associated with variable bulk densities. It is, however, not possible to sample correctly by soil mass. Even with bulk density determined first and defining sampling depth for determination of comparable OC concentrations based on these results, small scale spatial variations of bulk densities still hamper correct sampling, especially in stone-rich soils.

Another potential problem of soil coring is the carry-over of soil material along the soil core while drilling. This cannot be fully avoided but the moist soil cores had a stable structure, and so contamination was restricted to just the immediate layer in contact with the corer wall. Given the large diameter of the soil core (8.5 cm), effects on total soil mass or mass of C per layer were minor. Differences in carry-over of material alongside the soil core

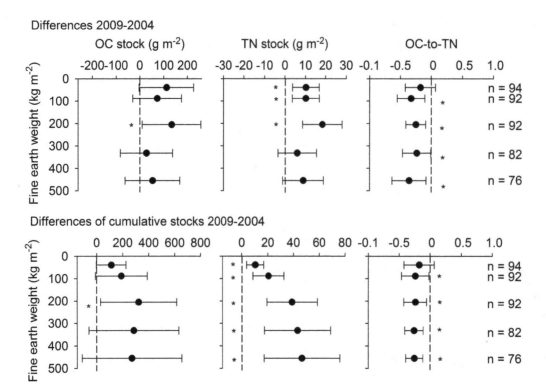

Figure 2. Differences in carbon and nitrogen stocks between 2004 and 2009, normalized to average soil masses in 2004. Therefore, the y-axis presents increasing cumulative fine earth masses for each depth increment of the 2004 sampling with soil depth. Upper graphs show the normalized differences of paired soil samples for each depth increment; lower graphs show the normalized differences of cumulative stocks. Positive values indicate gains or increases with time. Error bars show 5–95% confidence intervals; asterisks indicate changes significantly different from zero.

between sampling years would affect deeper layers stronger than upper soil layers, however, we observed largest increases in C concentrations in the upper horizons. There were no indications of more material being transported downwards in one year than another.

Larger amounts of roots sorted out in 2004 than in 2009 could have created an artificial increase in OC concentrations at 0–10 cm, but not at 10–20 cm depth (Fig. 1). Root picking was done following the same protocol in both years, thus, differences in roots likely are due to other causes. However, soil sampling in early spring 2004 was two weeks later and took two weeks longer than in 2009, so plants like wood garlic (*Allium ursinum*) and dog's mercury (*Mercurialis perennis*) already covered the ground, whereas they just started germination during the 2009 sampling. Accordingly, living root masses in 0–10 cm probably differed between sampling years.

Later sampling in 2004 could have also caused the smaller OC concentrations and OC-to-TN ratios of the litter layer, which indicate that litter in 2004 was already in a more advanced stage of decomposition than in 2009. The lack of an Oa horizon and the thin, discontinuous Oe horizon suggest no long-term OC accumulation in the litter layer at the study site. Therefore, we focussed the discussion on the mineral soil.

Increases in soil OC were most pronounced in the uppermost 0–20 cm of the mineral soil. Significant changes of stocks were observed for the equivalent soil mass in the 10–20 cm layer (27 ± 12 g OC m^{-2} yr^{-1}), and for the equivalent soil mass of in the entire 0–20 cm layer (65 ± 29 g OC m^{-2} yr^{-1}). These are in the same range of OC changes observed for other temperate forest sites (Table 1).

The 10–20 cm depth corresponds to the upper part of the B horizon. Gains of OC in deeper soil layers may reflect increases in

root litter, and/or downward OC transport by bioturbation or with percolating soil water. Turnover times of OC in subsoils are usually longer than in topsoils, possibly due to strong stabilization of OC by association with minerals [28,29]. Since the organic matter loading of minerals at 10–20 cm is less than that at the 0–5 cm [29], subsoils still offer more potential binding sites for OC at mineral surfaces than topsoil layers. So, long-term stabilization of additional OC in subsoil horizons seems possible.

Positive correlations between OC and the OC-to-TN ratio in the entire set of samples suggest that more input of new, litter-derived organic matter with large OC-to-TN ratios promotes larger OC concentrations at the study site. Similarly, increases in OC were positively correlated to increases of the OC-to-TN ratio, a relation that was also observed in the repeated forest soil inventory conducted by Kiser et al. [30].

The Hainich forest was intensively managed as coppice and forest grazing was practiced from the 16th to the 19th century [31,32]. By the end of the 19th century, the forest area was converted to selection management and the standing biomass increased from approximately 100 m^3 ha^{-1} to 250 m^3 ha^{-1} by the middle of the 20th century [32]. Currently, the timber volume at the study site is about 540 m^3 ha^{-1} [31]. This is still less than the observed maximum of 1000 m^3 ha^{-1} in the Hainich region [33]. Accordingly, forest growth as a result of recovery from former coppice management could contribute to the observed soil OC gains.

Nitrogen deposition or fertilization is assumed to enhance soil OC storage by causing reduced decomposition rates as well as increased litter input [34–40]. Estimated N deposition rates in the Hainich region were between 1.4 (M. Mund, personal communication) and 2.1 g m^{-2} yr^{-1} [41].

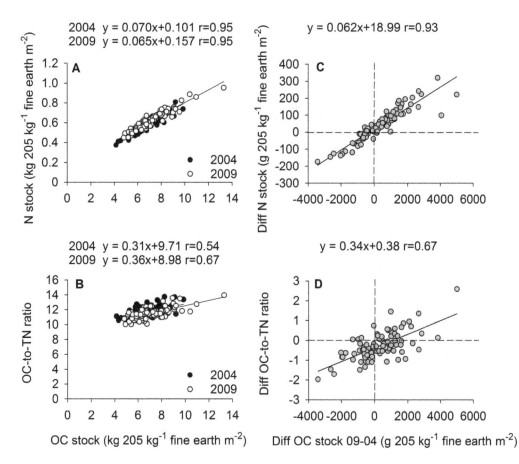

Figure 3. Relations between organic carbon (OC) and total nitrogen (TN). Graph A: correlations between cumulative OC and TN stocks for 205 kg fine earth m^{-2}, corresponding to the average amount of fine earth per m^2 in the 0–20 cm soil layer in 2004, for the years 2004 and 2009; Graph B: correlations between cumulative OC stocks and OC-to-TN ratios; Graph C: regression between differences between sampling times (2009–2004) for OC and TN stocks of sample pairs; Graph D: correlations between respective differences of paired samples between 2004 and 2009.

Forest fertilization experiments suggest an average N use efficiency of 20 to 25 kg C kg^{-1} N for trees and 10 to 25 kg C kg^{-1} N for forest soils [42]. Assuming an annual input between 1.4 and 3.8 g N m^{-2} yr^{-1} (Fig. 3C), N additions could explain a significant portion of the observed soil OC gains (14 to 95 g m^{-2} yr^{-1}).

Repeated forest inventories in the core area of the Hainich National Park revealed an increase in OC stocks from the year 2000 to 2007 with standing wood volume [33]. Observed changes correspond to an annual increase of approximately 450 g OC m^{-2} yr^{-1} or by 2.1% [33]. If this applies also to the study site, soil OC gains would be around 10% of aboveground increases. Litter trap studies at the Hainich tower site indicate an increase in annual litter fall of approximately 10 g OC m^{-2} yr^{-1} or by 2% from 2000 to 2007, further supporting the idea of increased forest production [22]. The observed change in soil OC corresponds on average to 0.9% yr^{-1} of existing stocks in the 0–20 cm soil layer, or 6 to 13% of the estimated sum of change in soil and biomass C (Net Biome Production, NBP). For the Hainich forest, the observed soil OC gain would therefore be less than half the accumulation rate of 29±15% of NBP assigned to forest soils by Luyssaert et al. [7].

Microbial biomass C (C_{mic}) usually increases with soil OC [43,44], thus, larger microbial biomass of low OC-to-TN ratios could compensate for some of the large OC-to-TN ratios of newly added litter in bulk samples. In order to assess the potential maximal impact of microbial biomass changes for changes in

stocks, we assumed C_{mic} in 2004 to be 2% of the OC at 0–10 cm depth [45], and 50% larger in 2009 than in 2004. Such an increase in microbial biomass would explain OC gains of 8.3 g m^{-2} yr^{-1} and N gains of 1.0 g m^{-2} yr^{-1} (assuming an average OC-to-TN ratio of 8.6 for soil microorganisms; [46]). Given that large changes occurred at 10–20 cm soil depth, where microbial biomass is typically smaller, and that soils were sampled later in spring in 2004, when microbial biomass was probably already larger, changes in microbial biomass alone cannot explain the observed increases in soil OC and TN. Also changes in the microbial community structure, e.g., a decrease in the fungi-to-bacteria ratio in response to N input [34], could have contributed to declining OC-to-TN ratios, but this cannot be tested with the available data.

The declining OC-to-TN ratios show that soils at the Hainich site are currently an even stronger sink of TN than of OC. Deposition and accumulation of inorganic N is one possible explanation. The observed increase in soil N exceeded estimated deposition rates for the study site. As the OC-to-TN ratio in the litter layer did not decline, deposited N accumulated preferentially in the mineral soil, which is in line with other studies showing forest soils to be more effective sinks of added N than the vegetation [47]. Also, mineralization of soil organic matter results in losses of OC as CO_2, while some of the released ammonium can be retained in the soil. The Hainich soil is rich in clay, and the clay mineral assemblage includes micas and illites, hydroxy-interlayered vermiculites and smectites, and kaolinite [48]. Due to its mineral assemblage, the site has a high capacity to fix ammonium [49]. Therefore, part of the increase in soil N could be due to ammonium fixation by clay minerals. Larger increases in TN than in OC in deeper soil layers could be due to vertical translocation of N and subsequent sorption, either of organic N or of ammonium, but this cannot be tested with the available data. Separation of organic and inorganic N would be helpful to disentangle the processes controlling soil N retention in future studies.

Summary and conclusive remarks

The present work supports earlier studies indicating that the old growth forest at the Hainich National park is still accumulating OC in soil. It adds to the growing number of studies showing that soils of temperate forests are currently C sinks. Soil OC increases are estimated to be ~10% of aboveground C gains. Soil OC and TN gains at the Hainich site seem to be due to increasing litter input, likely because the forest is accumulating biomass, either promoted by N input or due to recovery from historic use.

Nitrogen gains in excess of those expected from purely stoichiometric increases in OC indicate that the site is also a sink of deposited inorganic N.

Acknowledgments

We grateful acknowledge the assistance of Marco Pöhlmann, Alexander Sinz, Christian Wager, and Enrico Weber during soil sampling and with sample preparation. We are further indebted to the Routine Measurements and Analyses department (ROMA, Ines Hilke and Birgit Fröhlich) of Max-Planck Institute for Biogeochemistry.

Author Contributions

Conceived and designed the experiments: MS EDS MS. Performed the experiments: MS. Analyzed the data: MS. Contributed reagents/materials/analysis tools: MS EDS. Wrote the paper: MS KK EDS.

References

1. Reichstein M, Bahn M, Ciais P, Frank D, Mahecha MD, et al. (2013) Climate extremes and the carbon cycle. Nature 500: 287–295.
2. Heimann M, Reichstein M (2008) Terrestrial ecosystem carbon dynamics and climate feedbacks. Nature 451: 289–292.
3. Rustad LE (2008) The response of terrestrial ecosystems to global climate change: Towards an integrated approach. Science of the Total Environment 404: 222–235.
4. Ballantyne AP, Alden CB, Miller JB, Tans PP, White JWC (2012) Increase in observed net carbon dioxide uptake by land and oceans during the past 50 years. Nature 488: 70–72.
5. Levin I (2012) The balance of the carbon budget. Nature 488: 35–36.
6. Schulze ED, Luyssaert S, Ciais P, Freibauer A, Janssens IA, et al. (2009) Importance of methane and nitrous oxide for Europe's terrestrial greenhouse-gas balance. Nature Geoscience 2: 842–850.
7. Luyssaert S, Ciais P, Piao SL, Schulze ED, Jung M, et al. (2010) The European carbon balance. Part 3: forests. Global Change Biology 16: 1429–1450.
8. Pan YD, Birdsey RA, Fang JY, Houghton R, Kauppi PE, et al. (2011) A Large and Persistent Carbon Sink in the World's Forests. Science 333: 988–993.
9. Prietzel J, Stetter U, Klemmt HJ, Rehfuess KE (2006) Recent carbon and nitrogen accumulation and acidification in soils of two Scots pine ecosystems in Southern Germany. Plant and Soil 289: 153–170.
10. Schrumpf M, Schulze ED, Kaiser K, Schumacher J (2011) How accurately can soil organic carbon stocks and stock changes be quantified by soil inventories? Biogeosciences 8: 1193–1212.
11. Stevens A, van Wesemael B (2008) Soil organic carbon dynamics at the regional scale as influenced by land use history: a case study in forest soils from southern Belgium. Soil Use and Management 24: 69–79.
12. Bellamy PH, Loveland PJ, Bradley RI, Lark RM, Kirk GJD (2005) Carbon losses from all soils across England and Wales 1978–2003. Nature 437: 245–248.
13. Hopkins DW, Waite IS, McNicol JW, Poulton PR, Macdonald AJ, et al. (2009) Soil organic carbon contents in long-term experimental grassland plots in the UK (Palace Leas and Park Grass) have not changed consistently in recent decades. Global Change Biology 15: 1739–1754.
14. Zhou GY, Liu SG, Li Z, Zhang DQ, Tang XL, et al. (2006) Old-growth forests can accumulate carbon in soils. Science 314: 1417–1417.
15. Smith P, Chapman SJ, Scott WA, Black HIJ, Wattenbach M, et al. (2007) Climate change cannot be entirely responsible for soil carbon loss observed in England and Wales, 1978–2003. Global Change Biology 13: 2605–2609.
16. Hessen DO, Agren GI, Anderson TR, Elser JJ, De Ruiter PC (2004) Carbon, sequestration in ecosystems: The role of stoichiometry. Ecology 85: 1179–1192.
17. Schlesinger WH (2009) On the fate of anthropogenic nitrogen. Proceedings of the National Academy of Sciences of the United States of America 106: 203–208.
18. Phoenix GK, Emmett BA, Britton AJ, Caporn SJM, Dise NB, et al. (2012) Impacts of atmospheric nitrogen deposition: responses of multiple plant and soil parameters across contrasting ecosystems in long-term field experiments. Global Change Biology 18: 1197–1215.
19. Templer PH, Mack MC, Chapin FS, Christenson LM, Compton JE, et al. (2012) Sinks for nitrogen inputs in terrestrial ecosystems: a meta-analysis of N-15 tracer field studies. Ecology 93: 1816–1829.
20. Knohl A, Schulze ED, Kolle O, Buchmann N (2003) Large carbon uptake by an unmanaged 250-year-old deciduous forest in Central Germany. Agricultural and Forest Meteorology 118: 151–167.
21. Kutsch WL, Kolle O, Rebmann C, Knohl A, Ziegler W, et al. (2008) Advection and resulting CO2 exchange uncertainty in a tall forest in central Germany. Ecological Applications 18: 1391–1405.
22. Kutsch W, Persson T, Schrumpf M, Moyano FE, Mund M, et al. (2010) Heterotrophic soil respiration and soil carbon dynamics in the deciduous Hainich forest obtained by three approaches. Biogeochemistry 100: 167–183.
23. Tefs C, Gleixner G (2012) Importance of root derived carbon for soil organic matter storage in a temperate old-growth beech forest – Evidence from C, N and C-14 content. Forest Ecology and Management 263: 131–137.
24. Oberdorfer E (1994) Pflanzensoziologische Exkursionsflora. München: Verlag Eugen Ulmer.
25. WRB IWG (2007) World Reference Base for Soil Resources 2006, first update 2007. Rome: FAO.
26. Ellert BH, Bettany JR (1995) Calculation of organic matter and nutrients stored in soils under contrasting management regimes. Canadian Journal of Soil Science 75: 529–538.
27. Wuest SB (2009) Correction of Bulk Density and Sampling Method Biases Using Soil Mass per Unit Area. Soil Science Society of America Journal 73: 312–316.
28. Rumpel C, Kögel-Knabner I (2011) Deep soil organic matter-a key but poorly understood component of terrestrial C cycle. Plant and Soil 338: 143–158.
29. Schrumpf M, Kaiser K, Guggenberger G, Persson T, Kögel-Knabner I, et al. (2013) Storage and stability of organic carbon in soils as related to depth, occlusion within aggregates, and attachment to minerals. Biogeosciences 10: 1675–1691.
30. Kiser LC, Kelly JM, Mays PA (2009) Changes in Forest Soil Carbon and Nitrogen after a Thirty-Year Interval. Soil Science Society of America Journal 73: 647–653.
31. Mund M (2004) Carbon pools of European beech forests (fagus sylvatica) under different silvicultural management. 256 p.
32. Wäldchen J, Schulze ED, Mund M, Winkler B (2011) Der Einfluss politischer, rechtlicher und wirtschaftlicher Rahmenbedingungen des 19. Jahrhunderts auf die Bewirtschaftung der Wälder im Hainich-Dün-Gebiet (Nordthüringen). Forstarchiv 82: 35–47.
33. Hessenmöller D, Schulze ED, Großmann M (2008) Bestandesentwicklung und Kohlenstoffspeicherung des Naturwaldes "Schönstedter Holz" im Nationalpark Hainich. Allgemeine Forst- und Jagdzeitung 179: 209–219.
34. Frey SD, Knorr M, Parrent JL, Simpson RT (2004) Chronic nitrogen enrichment affects the structure and function of the soil microbial community in temperate hardwood and pine forests. Forest Ecology and Management 196: 159–171.
35. Knorr M, Frey SD, Curtis PS (2005) Nitrogen additions and litter decomposition: A meta-analysis. Ecology 86: 3252–3257.
36. Olsson P, Linder S, Giesler R, Hogberg P (2005) Fertilization of boreal forest reduces both autotrophic and heterotrophic soil respiration. Global Change Biology 11: 1745–1753.
37. Johnson DW, Curtis PS (2001) Effects of forest management on soil C and N storage: meta analysis. Forest Ecology and Management 140: 227–238.
38. Hyvonen R, Persson T, Andersson S, Olsson B, Agren GI, et al. (2008) Impact of long-term nitrogen addition on carbon stocks in trees and soils in northern Europe. Biogeochemistry 89: 121–137.
39. Bowden RD, Davidson E, Savage K, Arabia C, Steudler P (2004) Chronic nitrogen additions reduce total soil respiration and microbial respiration in temperate forest soils at the Harvard Forest. Forest Ecology and Management 196: 43–56.

40. Hagedorn F, Spinnler D, Siegwolf R (2003) Increased N deposition retards mineralization of old soil organic matter. Soil Biology & Biochemistry 35: 1683–1692.
41. Hahn V, Buchmann N (2004) A new model for soil organic carbon turnover using bomb carbon. Global Biogeochemical Cycles 18.
42. de Vries W, Solberg S, Dobbertin M, Sterba H, Laubhann D, et al. (2009) The impact of nitrogen deposition on carbon sequestration by European forests and heathlands. Forest Ecology and Management 258: 1814–1823.
43. Booth MS, Stark JM, Rastetter E (2005) Controls on nitrogen cycling in terrestrial ecosystems: A synthetic analysis of literature data. Ecological Monographs 75: 139–157.
44. Alvarez CR, Alvarez R, Grigera S, Lavado RS (1998) Associations between organic matter fractions and the active soil microbial biomass. Soil Biology & Biochemistry 30: 767–773.
45. Wardle DA (1992) A Comparative-Assessment of Factors Which Influence Microbial Biomass Carbon and Nitrogen Levels in Soil. Biological Reviews of the Cambridge Philosophical Society 67: 321–358.
46. Cleveland CC, Liptzin D (2007) C: N: P stoichiometry in soil: is there a "Redfield ratio" for the microbial biomass? Biogeochemistry 85: 235–252.
47. Nadelhoffer KJ, Colman BP, Currie WS, Magill A, Aber JD (2004) Decadal-scale fates of N-15 tracers added to oak and pine stands under ambient and elevated N inputs at the Harvard Forest (USA). Forest Ecology and Management 196: 89–107.
48. Wäldchen J, Schöning I, Mund M, Schrumpf M, Bock S, et al. (2012) Estimation of clay content from easily measurable water content of air-dried soil. Journal of Plant Nutrition and Soil Science 175: 367–376.
49. Nieder R, Benbi DK, Scherer HW (2011) Fixation and defixation of ammonium in soils: a review. Biology and Fertility of Soils 47: 1–14.
50. Lettens S, van Orshoven J, van Wesemael B, Muys B, Perrin D (2005) Soil organic carbon changes in landscape units of Belgium between 1960 and 2000 with reference to 1990. Global Change Biology 11: 2128–2140.
51. Oehmichen K, Demant B, Dunger K, Grüneberg E, Henning P, et al. (2011) Inventarstudie 2008 und Treibhausgasinventur Wald. Braunschweig: Johann Heinrich von Thünen-Intistut. 343 343. 141 p.
52. Berg B, Johansson MB, Nilsson A, Gundersen P, Norell L (2009) Sequestration of carbon in the humus layer of Swedish forests – direct measurements. Canadian Journal of Forest Research-Revue Canadienne De Recherche Forestiere 39: 962–975.

10

Shade Tree Diversity, Cocoa Pest Damage, Yield Compensating Inputs and Farmers' Net Returns in West Africa

Hervé Bertin Daghela Bisseleua[1,2,3,4*], **Daniel Fotio**[2], **Yede**[2], **Alain Didier Missoup**[2], **Stefan Vidal**[3]

1 MDG Centre West and Central Africa, Dakar, Senegal, 2 Laboratory of Entomology, IRAD, Yaoundé, Cameroon, 3 Georg-August-University Goettingen, Department of Crop Science, Entomological Section, Goettingen, Germany, 4 Centre for Environmental Research and Conservation, The Earth Institute of Columbia University, New York, New York, United States of America

Abstract

Cocoa agroforests can significantly support biodiversity, yet intensification of farming practices is degrading agroforestry habitats and compromising ecosystem services such as biological pest control. Effective conservation strategies depend on the type of relationship between agricultural matrix, biodiversity and ecosystem services, but to date the shape of this relationship is unknown. We linked shade index calculated from eight vegetation variables, with insect pests and beneficial insects (ants, wasps and spiders) in 20 cocoa agroforests differing in woody and herbaceous vegetation diversity. We measured herbivory and predatory rates, and quantified resulting increases in cocoa yield and net returns. We found that number of spider webs and wasp nests significantly decreased with increasing density of exotic shade tree species. Greater species richness of native shade tree species was associated with a higher number of wasp nests and spider webs while species richness of understory plants did not have a strong impact on these beneficial species. Species richness of ants, wasp nests and spider webs peaked at higher levels of plant species richness. The number of herbivore species (mirid bugs and cocoa pod borers) and the rate of herbivory on cocoa pods decreased with increasing shade index. Shade index was negatively related to yield, with yield significantly higher at shade and herb covers<50%. However, higher inputs in the cocoa farms do not necessarily result in a higher net return. In conclusion, our study shows the importance of a diverse shade canopy in reducing damage caused by cocoa pests. It also highlights the importance of conservation initiatives in tropical agroforestry landscapes.

Editor: Andrew Hector, University of Zurich, Switzerland

Funding: Financial support was provided by Georg-August University of Goettingen (Dept of Crop Sciences, Entomological section). The funders had no role in study design, data collection and analysis, decision to publish, or preparation of the manuscript.

Competing Interests: The authors have declared that no competing interests exist.

* E-mail: herve.bisseleua@mdgwca.org

Introduction

Human economy grows at the competitive exclusion of nonhuman species. Ecological changes due to agricultural intensification are known to increase anthropogenic biodiversity loss [1,2]. However, conservation biologists and economists increasingly acknowledge the need to incorporate environmentally sustainable cocoa production strategies into conservation strategies [3,4]. Recently, cocoa agro-ecosystems have received substantial attention because of their social, economic and ecological importance [5,6]. Cocoa is important for national macroeconomic balances and provides livelihoods to millions of people in developing and developed countries.

Shaded plantations facilitate dispersal of forest fauna between fragments. Plant and animal biodiversity found within shaded cocoa systems could augment ecosystem services like pest control, pollination, weed control, fungal disease limitation, and soil fertility [7,8,9]. However, increasing and widespread intensification of management practices, including removal of shade trees and frequent weeding, is resulting in different cocoa production systems ranging from forest-like environments to full-sun cocoa [10]. How these different cocoa habitats differ in their fauna and flora, and how this affects functionally important species groups and ecosystem functioning is largely unknown. However, species diversity of birds and insects has functional consequences and influences ecosystem processes and services such as natural pest control [11]. Additionally, the type of interactions among species in an agro-ecosystem and the sensitivity of each species to different types of environmental fluctuations predict the stability of that system.

A considerable number of ecologists have acknowledged the role of cocoa agroforests as a refuge for biodiversity, specifically for ants, spiders and wasps [12]. Several studies have emphasized the role of ants in biological control in cocoa plantations [13,14,15], or their influence on other predators in agro-ecosystems [16]. Moreover, as cocoa plantations get intensified (with the reduction or elimination of shade trees); it is likely that the response of ant diversity to unpredictable outbreaks may vary. However, the extent to which cocoa agroforests are managed, with respect to the shade tree cover, species richness of the shade trees and herbaceous vegetation, and whether they provide valuable habitat and improve ecosystem functioning has barely been investigated in the West African cocoa belt [17]. Studies from Mesoamerica and

Southeast Asia abound [18,19], but cannot simply be transferred to West Africa considering the differences in management, tree phenology and structure and composition of ground-living herbaceous plants [20,21].

The multi-strata cocoa agroforest in Cameroon harbor both rare and common species of aesthetic and cultural interest, and maintain valuable ecosystem services that are ensured by high species diversity [21]. Such wildlife-friendly farming approaches enable coexistence of agricultural activity and biodiversity in the same landscapes. Intensification may alter species diversity of relevance for conservation and ecosystem functioning [22,23,24]. We therefore need to predict consequences of agricultural intensification (specifically the reduction or elimination of shade) in order to develop pro-poor agroforestry strategies and incentives to conservation-friendly, ecologically complex agroforestry systems in West Africa. In addition, we must also strengthen the ecological knowledge of farmers to improve the farmer's ability to manage his/her local landscape [25].

This study focuses specifically on pest infestation and input use in cocoa agroforests with the aim of improving our understanding of how diversified and complex shade, in addition to biodiversity conservation, can provide ecosystem services such as biological pest control. We tested the hypotheses that, (i) shade tree removal may alter pest control and, (ii) reduction of inputs would enable coexistence of agricultural activity and biodiversity in the same landscape. The article highlights the contribution of complex shade agroforests in reducing pest infestation and input use. We also discuss recommendations derived from a different approach in conservation management of both cultivated forests such as traditional cocoa agroforests and the wider landscape of southern Cameroon (many of which are also applicable in other cocoa regions).

Materials and Methods

Study sites

In Cameroon, cocoa was originally grown by smallholders under a structurally and floristically diverse canopy of shade trees that provided a habitat for a high diversity of flora and fauna [26,27]. The typical production system involves clearing virgin forests to plant new trees, and later replacing old cocoa plantations with food crops [28,29]. Our study took place in five major cocoa-growing regions (Ngomedzap, Bakoa, Obala, Talba and Kedia) in the Central Region of Cameroon between 2°35' N and 4 °15' N and 11°48' and 11°15' E. The mean annual temperature is about 25 °C with a relatively small thermal variation. The mean annual rainfall is about 1600 mm per year. The five regions differed in land-use management ranging from less extensive (Ngomedzap), intermediate (Bakoa and Obala) to more intensive (Talba and Kedia) cocoa agroforests. Landscape characteristics are summarized in Table 1.

Four cocoa plantations ranging from 1 to 3 ha, and located at least 500 m from one another were selected in every region while ensuring that the plantations were managed exclusively by their owners using production techniques common to small landholders in the region [10]. The selected plantations differed in shade intensity, shade density, weed intensity, weed density and cocoa density. In each chosen plantation we assessed floristic (forest tree and herb species) and insect diversity. No specific permits were required for the described field studies and locations/activities. We received permissions from the cocoa growers associations from the selected regions to conduct the field studies. The locations and field studies are privately-owned by cocoa growers but not protected in any way. The field studies and locations did not involve endangered or protected species.

Vegetation survey

We collected data on the vegetation characteristics within four 20×30 m plots in each plantation. We recorded the number of shade trees species. Unknown trees were given a unique morphospecies number. We estimated canopy cover within a 30 m radius circle at 10 subpoints within the circle; the center and at approximately 15 m N, S, E and W of the center subpoint. To estimate canopy cover we took readings with a hand-held concave densiometer at each of the 10 subpoints. To estimate canopy structure (depth), at each of the 10 subpoints we recorded the height of the lowest and highest canopy vegetation immediately above the subpoint. We used a digital rangefinder to improve our estimates of canopy height. The differences in the highest and lowest vegetation heights were used to estimate canopy depth at the 10 subpoints within each circle. We also recorded all herb species in 15' quadrates of 2×1 m per plot. Scientific and vernacular names (the latter given by local stakeholders) were recorded. Species that could not be identified in the field were identified at the National Herbarium of Cameroon (Yaoundé).

To represent land-use intensity, we created a shade index based on eight variables: number of trees, number of tree species, tree density, number of herbs, number of herb species, average tree height, percent shade cover, and percent herb cover. The mean of each variable was divided by the highest value of the same variable recorded in the plantation. We then summed the resulting values for all variables in one plantation, and divided this by the number of variables (i.e. eight) to obtain a value between zero and one for each plantation, where zero would represent the least diverse and one the most diverse shade. In each plot rainfall was recorded per day. For the analyses we used the mean of the average monthly rainfall per plot during the study period.

Insect pests and natural enemy surveys

In each plantation we selected 30 cocoa trees at least 15 m apart, which we monitored weekly for pests and predators over two cocoa growing seasons, from March to December [30]. In each tree we quantified the total number of pods damaged by the cocoa pod borer (CPB; *Conopomorpha cramerella* Snellen), the total number of CPB holes on cocoa pods and CPB larvae per cocoa pod, the black pod rot (BPR; *Phytophthora megakarya* Brasier & M.J. Griffin), and the fresh feeding lesions caused by mirid bug *Sahlbergella singularis* Hagl. as well as the number of adult mirids.

On the same 30 selected trees, we sampled active ants between 9 AM and 1 PM [2] and the numbers of spider webs and social wasp nests. We also used 10 plastic observation plates (10 cm diameter) equipped with baits of about 4 g, composed of pieces of tinned tuna fish, honey, and cookie crumbs, to sample ground-foraging ants on 5 of the selected cocoa trees and 5 forest trees. Two persons monitored all plates on a subplot by observing each baited plate for 1 minute. For each ant species appearing on the plate, 5–10 specimens were caught with forceps and preserved in 70% ethanol for later identification. Ant species that occurred as singletons were sampled immediately to avoid missing them.

Household and village surveys

In each region, we also randomly selected and interviewed 200 farmers, including the farmers among whom we sampled the biological information. We investigated their economically motivated preference for shade tree removal in their cocoa agroforests. We also collected socio-economic data (age and size of cocoa production area, total cocoa yield and revenue from cocoa

Table 1. Landscape characteristics of the regions.

Region	Rainfall regime (mm)	Age of cocoa plantation (yrs)	Agricultural land	Forest land
Ngomedzap	>1900	>50 "rustic plantation"	20% cocoa fields	70% Pristine forest
			10% annual crop (cassava, plantain)	With Forest reserve
Bakoa	<1100	~30	50% cocoa fields	20% secondary forest No reserve
			25% annual field crops (maize, yams, citrus)	
			5% Patchy pasture fields	
Obala	>1300	~40	70% cocoa fields	5% secondary forest No forest reserve
			25% annual crop fields of mixed crops (homegardens: cassava, groundnuts, maize, tomatoes etc...), agroforestry trees (citrus, safou, avocado, etc...).	
Talba	~1200	15–20	70% Cocoa fields	25% pristine forest No reserve
			5% annual field crops (banana, plantain)	
Kedia	~1050	8–15	65% cocoa fields	5% secondary forest
			25% annual field crops (maize) 5% pasture lands	

production, costs for agro-chemicals including herbicides, pesticides and fertilizers, and agricultural technologies). We examined the differences in cocoa landholding, agrochemical costs, alternative forest products, cocoa yields and annual net returns per hectare using univariate analysis of variance.

Statistical analysis

We use our different surveys per agroforest to generate sample-based rarefaction curves (MaoTao estimates) with EstimateS Version 8.2.0 [31] to compare plant (tree and herb) richness. We rescaled samples based rarefaction curves to the number of individuals to best compare richness between regions [32,33]. Using the Chao-Jaccard Estimated Abundance Indices [34], we re-computed Chao1 for abundance distribution with a coefficient of variation higher than 0.5 and calculated first order jack-knife estimators of species richness.

A multiple linear regression analysis was used to describe the relationship between predators (ant/spiders/wasps) and vegetation variables (number of trees, number of tree species, tree density, number of herbs, number of herb species, average tree height, percent shade cover, and percent herb cover). We also used simple linear regression analysis to describe the relationship between predators (ant/spiders/wasps) and the presence of native or exotic shade tree species as well as the relationship between pod rot and shade index. Because of the likeliness that environmental gradients such as rainfall and the shade cover gradient confound each other we conducted general linear model and correlation analyses controlling for rainfall against biodiversity data. We also conducted non-parametric tests with all biodiversity data (predator richness, tree and herb richness) and shade index to provide alternatives to ANOVA.

Data on species richness of ant and the number of spiders/wasps were analyzed by multiple regressions against shade index and rainfall. Where needed we additionally perform kruskal Wallis test on biodiversity data (predator richness, tree and herb richness) and shade index. General linear model and correlation analyses conducted in Systat 11 [35] were also used to analyze data on yields. We used yield as the dependent variable and shade index, predators (ant/spiders/wasps) as independent variables in the

multivariate regression analysis to separate the influence of management strategy from confounding factors such as age. We also looked at the relationship between the input costs and the net return. We used log-transformed data on species count to meet the condition of normality.

Results

Species richness and similarity of plants

We recorded a total of 102 tree species and 260 herbaceous species belonging to 56 families of trees and 113 families of herbs, respectively. The shade index differed in each region with cocoa plantations near pristine forests (Ngomedzap) having the highest index, cocoa plantations in forest galleries (Bakoa) and in homegardens (Obala), having an intermediate index and cocoa plantations near secondary forests (Talba) and in artificial forests (Kedia) having the lowest shade index (Table 2). The uses for each tree species is detailed in Table 3. Species similarity between regions was low for trees and herbs. Of all tree and herb species recorded, 31% were shared between Bakoa and Ngomedzap, 27% were shared between Obala and Ngomedzap and only 16% were shared between Ngomedzap and Kedia; and between Talba and Ngomedzap. However, similarity of tree and herb species did not significantly differ between regions (ANOVA, Chao-Jaccard Estimated ($F_{4, 19} = 0.23$, $P = 0.87$) for tree species, Chao-Jaccard Estimated ($F_{4, 19} = 2.8$, $P = 0.13$) for herb species). The ANOVA of the shade indices revealed statistically significant differences among the five regions ($F_{4, 19} = 10.94$, $P < 0.001$). Tree and herbaceous species richness significantly decreased with decreasing shade index (Tree species ($F_{1, 19} = 14.7$, $P < 0.0001$, Fig. 1a); Herb species ($F_{1, 19} = 10.3$, $P < 0.0001$, Fig. 1b).

Species richness and similarity of natural enemies

We recorded 38 species of ants within the cocoa agroforests, which represent between 56% and 73% of the maximum number of species determined by commonly used estimators for species richness (Chao: 72.55 ± 12.90; First order jacknife: 55.58 ± 11.87). Species richness of ants significantly decreased ($y = 0.50 + 11.3x$, $r^2 = 0.68$, $F_{1, 19} = 37.9$, $P < 0.0001$) with decreasing shade index (Fig. 2b). Ant species similarity between cocoa agroforests was

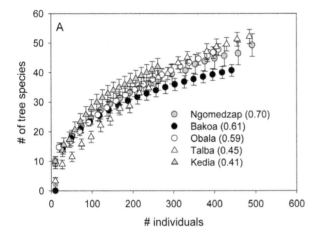

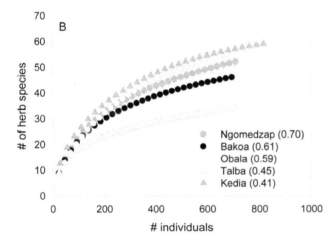

Figure 1. Species accumulation curves for trees (a) and herb species (b) in cocoa agroforests in relation to shade index. Error bars show 95% confidence intervals and non-overlapping bars show significant differences between shade indexes. Figures in parentheses are average values of the shade index for each region.

cocoa plantation location (ANOVA, Chao-Jaccard Estimated ($F_{1, 19}$ = 0.9, P = 0.52). The number of spider webs and wasp nests was significantly higher ($F_{2, 19}$ = 157.2, P<0.0001) at higher shade indices (Fig. 2a). We also noted that species richness of ants significantly increased ($F_{1, 19}$ = 38.0, P<0.0001) with increasing shade indices (Fig. 2b).

Factors affecting species richness and pest damage

The number of spider webs and wasp nests significantly increased with increasing density of native shade trees ($F_{1, 19}$ = 11.5, r^2 = 0.39, P<0.005) (Fig. 3a). This number also tends to decrease with the density of exotic shade trees (Fig. 3b). Density of native and exotic shade trees did not have significant effects on ant richness (Fig. 3c and 3d). Rainfall and shade cover were correlated to some extend (Pearson's correlation, $F_{2, 19}$ = 4.9, r^2 = 0.22, P<0.05). However, ant richness was positively related with shade cover ($F_{1, 19}$ = 12.2, r^2 = 0.40, P<0.01), and herbaceous cover ($F_{1, 19}$ = 12.7, r^2 = 0.40, P<0.001). In the multivariate regression analyses, predator (ant richness and number of spider webs/wasps nests) richness was significantly affected by the percentage of shade cover, herb cover, and regions, respectively ($F_{1, 19}$ = 14.8, P<0.0001, r^2 = 0.70). Predator richness (ant richness and number of spider webs/wasps nests), the number of herbivores and the rate of herbivory were not affected by rainfall in all our analyses. The number of herbivores (mirid bugs and cocoa pod borers) and the rate of herbivory on cocoa decreased with increasing shade index (number: y = 26.9–36.9x, r^2 = 0.76, $F_{1, 19}$ = 57.8, P<0.0001; herbivory: y = 85.1–97.5x, r^2 = 0.64, $F_{1, 19}$ = 32.4, P<0.0001). Pod rot caused by *Phytophthora megakarya* did not show any relationship with the shade index. The number of herbivores and the rate of herbivory showed a positive correlation with ant richness (number: $F_{1, 19}$ = 36.8, r^2 = 0.67, P<0.0001; rate of herbivory: $F_{1, 19}$ = 22.6, r^2 = 0.56, P<0.0001) and the number of spider webs and wasp nests (number: $F_{2, 19}$ = 69.1, r^2 = 0.78, P<0.0001; rate of herbivory: $F_{1, 19}$ = 30.6, r^2 = 0.63, P<0.0001).

In the multivariate regression analyses, cocoa yield was significantly affected by the percentage of shade index, predator richness (ant, spider, wasps) and the age of cocoa trees ($F_{1, 19}$ = 58.9, P<0.0001, r^2 = 0.95). Native shade trees negatively affected yield ($F_{1, 19}$ = 5.9, r^2 = 0.25, P<0.05) as compared to exotic shade trees (Fig 4). Yield was significantly higher at shade and herb cover <50% (Shade cover: $F_{1, 19}$ = 14.83, P<0.001; Herb cover: $F_{1, 19}$ = 34.77, P<0.0001).

Impact of shade index on annual return

When analyzing cocoa farmer survey we found that the management of shade trees significantly differed ($F_{4, 19}$ = 78.2,

higher than that of trees and herbs. An average of 51% of ant species was shared between agroforests. Again, similarity of ant assemblages in cocoa agroforests did not differ significantly with

Table 2. Variables used to calculate shade index in cocoa agroforests in Cameroon.

Variable name	Description	Minimum	Mean (SE)	Maximum
# Tree individuals	Number per hectare	17	88.9 (15.0)	220
# Tree species	Number of shade per hectare	4	8.0 (0.6)	13
Shade cover	In percent, measured above ground	25	73.3 (4.0)	95
Mean tree height	In meter, shade trees with dbh >5 cm	36	54.8 (2.4)	72.0
# Herb individuals	Number of herbs per hectare	72	103 (33.2)	216
# Herbaceous species	Number of herb species per hectare	12	25.1 (1.4)	36
Herbaceous cover	In percent, measured in quadrate	5	45.0 (7.5)	100
Cocoa tree density	Number per hectare	900	1230.5 (54.7)	2000

N.B. min and max were calculated over all 5 regions using all cocoa plantations.

Table 3. List of 43 common forest tree species recorded and used as explanatory variables to explain shade index in cocoa agroforests in Cameroon.

Species	Family	Local/common name	Conserva-tion star*	Economic importance/uses				
				Timber	Food/spice	Medicine	Fuel-wood	Other
Spondias lutea Linn.	Anacardiaceae	Cassimaga			X(fruit)			
Xylopia aethiopica (Dunal) A. Rich	Annonaceae	Akui		X	X(spice)	X		
Alstonia boonei De Wild.	Apocynaceae	Ekouk/Emien	Green			X		
Voacanga africana	- \\ -	Voacanga				X		
Elaeis guineensis Jacq.	Arecaceae		X		X(Oil)			Wine
Newbouldia laevis (P.Beauv.) Seem.	Bignoniaceae	Nouentchè/Mbikam				X		
Spathodea campanulata P. Beauv. Subsp.	- \\ -	Evovone/Tulipier	Green				X	
Ceiba pentandra Gaertn.	Bombacaceae	Doum/Fromager	Pink					Shade
Cordia platythyrsa Baker	Boraginaceae	Ebe/African cordia	Blue	X		X		
*Canarium schweinfurthii*Engl.	Burseraceae	Abel/Aiele	Red	X	X (fruit)			
Dacryodes edulis (G.Don) H.J. Lam	- \\ -	Plum/Safou	Green		X(fruit)			
*Monopetalanthus microphyllus*A. Chev.	Caesalpiniaceae	Ekop/Yellow ndoung		X				
Musanga cecropioides	Cecropiaceae	Asseng/Parasolier	X					
Terminalia superba Engl. & Diels	Combretaceae	Akom/Fraké	Pink	X				
Diospyros spp.	Ebenaceae	N'nom Elem		X				
Discoglypremna caloneura (Pax) Prain	Euphorbiaceae	Dambala						
Hevea brasiliensis Muell. Arg.	- \\ -							Latex
Ricinodendron heudelotii Mull. Arg.	- \\ -	Ezezang/Djansang	Green		X			
Guibourtia demeusei(Harms) J. Léonard	Fabaceae	Essingang/Bubinga		X				
Pterocarpus soyauxii Taub.	- \\ -	Mbel/Red Padauk	Red	X				
Hypodaphnis zenkeri Stapf.	Lauraceae	Ataag				X		
Petersianthus macrocarpus Liben	Lecythidaceae	Abing/Abale		X				
Entandrophragma cylindricum Sprague	Meliaceae	Assie/Sapelli	Red	X				
Khaya senegalensis	- \\ -	Mahogany		X				
Lovoa trichilioides Harms	- \\ -	Bibolo/Dibétou		X				
Albizia adianthifolia W.Wight	Mimosaceae	Sal'yeme/Bangbaye	Pink			X		
A. ferruginea (Guill. & Perr.) Benth.	- \\ -	Evouvous/Ossoto'o	Pink			X		
A. zygia (DC) J.F. Macbr.	- \\ -	Sal'yeme/Ketomb	Pink			X		
Piptadeniastrum africanum Brenan	- \\ -	Atui/Dabema	Red	X		X		
Tetrapleura tetraptera Taub.	- \\ -	Akpa				X		
Ficus exasperata Vahl.	Moraceae	Akol/Akole	X			X	X	
Ficus mucuso Welw. ex Ficalho	- \\ -	Toily/Figuier	X			X		
Micicia excelsa (Welw.) C.C. Berg.	- \\ -	Abang/Iroko	Scarlet	X				
Morus mesozygia Stapf.	- \\ -	Abang/Yellow iroko		X				
Morinda lucida Benth.	Rubiaceae	Akeng		X				
Cola acuminata	Sterculiaceae	Kola			X (fruit)			
Cola nitida (Vent.)Schott & Endl.	- \\ -	Kola			X (fruit)			
Cola lepidota K. Schum.	- \\ -	Kola	Gold		X (fruit)			
Mansonia altissima A. Chev/Chev.	- \\ -	Nkul/Bete	Gold	X				
Triplochiton scleroxylon K. Schum.	- \\ -	Ayous	X	X				
Duboscia macrocarpa Bocq.	Tiliaceae	Akak				X		
Eribroma oblongum (Mast) Pierre.	Ulmaceae	Eyong		X				

*In descending order of conservation importance: black, gold, blue, scarlet, red, pink and green [36].
Source: Household cocoa farmer survey and field survey.

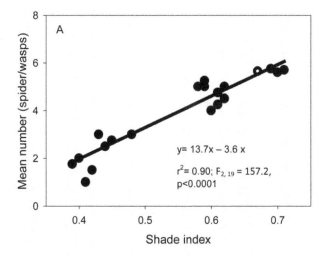

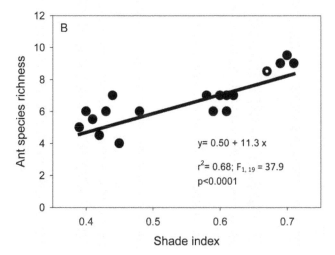

Figure 2. Mean number of spider webs/wasp nests (a), and ant species richness (b) in cocoa agroforests in relation to shade index.

Discussion

We linked a biodiversity estimate to a management indicator on cocoa agroforests, thereby covering the full range from extensive to extremely intensive land use pattern. When differences in environmental conditions had been accounted for, we found evidence that plant species richness declined with increasing land use intensity.

We found that shade cover and environmental gradient such as rainfall do not confound each other. From all analyses controlling for rainfall, we found that although there is a correlation between shade and rainfall ($r = 0.47$), both variable do not impact biodiversity data in a similar way. Rainfall in cocoa agroforests in southern Cameroon is not the predictor of diversity of predators (ant, spiders and wasps) and plants. Instead, the shade index per plot and the shade tree diversity were more suitable for predicting diversity of ants, spiders, and wasps, respectively. We also found that under same rainfall condition, shade management by farmers has a significant impact on predator and plant richness. Because shade is strongly correlated with all biodiversity data, we focused our analyses on shade impacts on biodiversity data and yields and we have downgraded rainfall effects. Plant species richness is often closely related to the diversity of other trophic levels [9]. We also found that land use changes are driven by well-known socio-economic factors and culturally mediated innovations [36]. These observations highlight synergies that emerge from diversified cocoa agroforests and the conditions necessary to move from an unsustainable syndrome of production to a sustainable one. To our knowledge, this is the first study examining the relationship between biodiversity, agricultural matrix and pest control in cocoa agroforests in tropical Africa.

Our results document that differences in management among regions, specifically shade and herb layer management between smallholders strongly impacts cocoa landscapes and ecosystem service, such as biological pest control. We observed that shading and choice of shade trees are separate variables in the management choices of the smallholders, and consequently, these factors are correlated only to some extent. Common management practices in cocoa agroforests tend to decrease tree diversity over time. These include the progressive thinning of shade canopies (partly motivated by the need to maximize yields; [2]) and official recommendations to substitute old forest trees by the often exotic faster growing leguminous species [36] in order to provide conditions for soil rejuvenation [37,38]. The mix of exotic and native species may not produce enough resources, such as fruits and breeding sites, needed for beneficial insects. Thus, the high proportion of exotic species in cocoa agroforests may contribute to the relatively low maintenance of a forest-based beneficial fauna [19,39]. For example, greater diversity of shade trees in cocoa plantations was positively related to ant and parasitoids richness, and thus supported more natural enemies [6,40]. Shade reduction may also increase the spread of invasive species, such as ants, in cocoa agroforests [41].

Our data on herbivores and herbivory supported the hypothesis that density of functionally monophagous herbivores will be reduced with increasing shade index [11]. Farms with greater vegetation heterogeneity and thus greater functional diversity of ants, spider and wasp species could exhibit stronger resilience of services after climatic disturbances or outbreaks through "insurance" species [42,1]. Moreover, our results showed a positive association between ant richness, wasp nests, spider webs and shade indices. It is known that in cocoa agroecosystems ants play important roles in biological control by chemically deterring pest feeding [13] or directly by preying upon them. A higher richness of

$P < 0.0001$) among regions. An average of 56% of farmers removed shade trees from their cocoa field. This figure included 4% of farmers in Ngomedzap (highest shade index), 58% in Bakoa, 69% in Obala and more than 72% of farmers in Talba and Kedia (lowest shade index). Reasons mentioned by farmers for shade removal were to reduce the incidence of pod rot and to increase yields. However, some shade trees were retained by farmers in their farms for fruits (70% of respondents), medicine (13% of respondents), timber (15% of respondents) and local spice (2% of respondents), such as the njangsang tree (*Ricinodendron heudelotii*) and the bush mango (*Irvingia gabonensis*).

Farming household surveys from the 5 regions also revealed that intensified cocoa production increased annual net returns from US$ 1194/ha on plots with 0.69 shade index to US$ 2349/ha with 0.59, and to US$ 3801/ha on plots with 0.41 shade index (for a less diverse farm) (Fig. 5). However, we observed that higher inputs in the cocoa farms did not necessarily result in a higher net return ($y = 1490.5 + 6.7x$, $r^2 = 0.18$, $F_{1, 19} = 3.87$, $P = 0.06$).

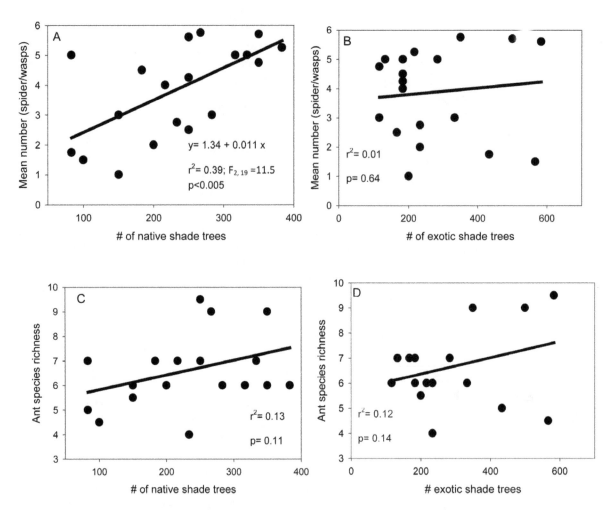

Figure 3. Relationship between the mean number of spider webs (*a* and *b*), ant species richness (*c* and *b*) and the type of shade trees (native and exotic) in cocoa agroforestry systems.

ants may enhance their ability to adapt and respond to changing conditions such as pest outbreaks or exploit new resources efficiently. However, the increased ecosystem function is not only due to diversity *per se* but rather the intraspecific differences in foraging or behaviour within beneficial insect communities that help to enhance the response to herbivory or to boost functionality under the insurance hypothesis [43].

The extent and diversity of the herbaceous layer only moderately affected spider and wasp numbers as compared to shade cover and tree richness, which suggests that canopy structure rather than herbs are the key variable for most parasitoids species and predators, such as spiders, in cocoa agroforest landscapes. Vegetation heterogeneity has been high-lighted several times as being a surrogate for habitat suitability for beneficial insects in human dominated landscapes [44,4], but this is the first time these variables were addressed at the scale of contrasting land-use types.

Our results showed that the matrix quality is important in the relationship between insect pests such as the cocoa pod borers and natural enemy control by wasps. Diverse cocoa agroforests represent a good quality matrix that promotes migration among fragments and maintains populations as meta-populations and therefore maintains biodiversity and ecosystem services at the landscape level [45]. Less diverse cocoa systems represent a low quality matrix that would hinder migration of beneficial insects

such as wasps [4]. The lack of migration thus may cause local (within fragment) extinctions to turn into regional extinctions. Consequently, the nature of the agroecosystems that make up that matrix is important, not only as a potential repository of biodiversity, but also as a habitat through which organisms can migrate from fragment to fragment (i.e. the matrix). Therefore, to optimize the attractiveness of cocoa agroforests to beneficial insect species, the nature of cocoa plantations as part of the landscape matrix should be considered in term of species composition of the planned and unplanned crop and noncrop biodiversity.

We showed for smallholder agroforests that higher inputs do not necessarily result in a higher net return. This finding is remarkable because it has identified win-win situations in biodiversity-yield relationship in species-rich agroforests. Conserving biodiversity in these systems is associated with maintaining a diversity of shade trees, rather than simply the number of trees per se, combined with moderate inputs of pesticides and labor per unit area that will enhance biological pest control [6]. This suggests the possibility of establishing premium prices to promote shade tree diversity and habitat complexity in tropical human-dominated landscapes with the purpose of conserving biodiversity. Therefore, conservation of highly sensitive taxa should take into account lower yields resulting from diverse shade. Furthermore, analyses of the relationship between yield, shade index and net return suggest that increasing premium values may generate a dramatic shift from a plantation

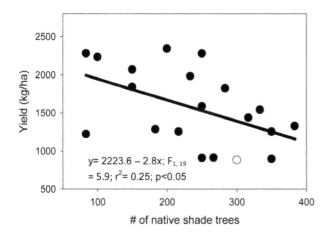

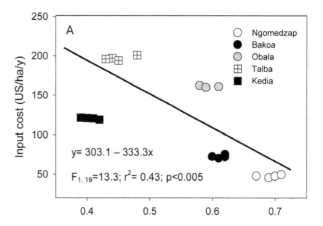

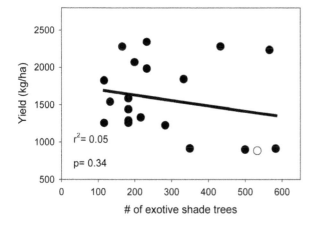

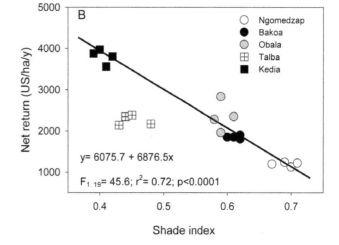

Figure 4. Yield-shade trees (native and exotic) relationship in cocoa agroforestry systems.

Figure 5. Cost of input (pesticides and labour) (a) and net returns (b) in cocoa agroforests in relation to shade index.

with high yield but low species richness to a plantation with low yield and high species richness. Nevertheless, high yields realized by intensification do not necessarily reduce functional biodiversity if a proper shade-vegetation structure is maintained. Policies and incentives aiming at helping cocoa farmers to overcome the costs of conversion from low-biodiversity systems to more diverse systems may, therefore, generate simultaneous increases in biodiversity and net income. Conservation programs of traditional land-use strategies must encourage cultural preferences for shade tree diversity and habitat complexity of tropical dominated-human landscape. Additionally, education of smallholders about unacknowledged ecosystem services provided by diversified and heterogeneous shade systems could further promote the implementation of certifications schemes. Such incentives will enhance the conservation value of traditional cocoa agroforests as an important refuge for tropical biodiversity and sources of valuable ecosystem services.

Conclusion

The results of this study provide a conceptual framework for conservation initiatives in cocoa agroforest landscapes. Initiatives could be most (cost-) effective if they are preferentially implemented in low-intensity cocoa agroforestry systems that still support high levels of biodiversity. Our models show no simple trade-off between biodiversity and net income. However, a

threshold in species richness at a 0.5 shade index in cocoa agroforests that is economically and ecologically profitable should be encouraged to balance economic and ecological needs.

This applies not only at the national level, but also at the international level, and highlights the importance of conservation initiatives on tropical human-dominated landscape of West Africa that host some of the most species rich farmlands, but are severely threatened by intensification [21]. Incentives from payment-for-ecosystem services and certification schemes should encourage farmers to keep heterogeneous shade tree cover. Conservationists and policy makers should nevertheless be aware that measures required to effectively conserving biodiversity and targeted species in these landscapes need to more drastically reduce land use intensity and will therefore be more costly. Participatory knowledge sharing between farmers, agronomists and ecologists will help to encourage heterogeneous shade systems that balance economic and ecological needs and provides a 'diversified food-and-cash crop' livelihood strategy.

Acknowledgments

We thank the interviewers and farmers who participated in the survey. We thank CIRAD and GTZ Cameroon for providing logistic support during field survey. We are also grateful to two anonymous reviewers for their valuable comments. We also thank Bruce Jaffee for editing the final version of the manuscript.

Author Contributions

Helped to analyze the relationship between shade cover and environmental factors and other confounding variables and to prepare responses to

reviewers' comments: DF. Conceived and designed the experiments: HBDB YY ADM SV. Performed the experiments: HBDB YY ADM. Analyzed the data: HBDB SV. Contributed reagents/materials/analysis tools: HBDB ADM SV. Wrote the paper: HBDB ADM SV.

References

1. Tsharntke T, Klein AM, Kreuss A, Steffan-Dewenter I, Thies C (2005) Landscape perspectives on agricultural intensification and biodiversity-ecosystem service management. Ecol Let 8: 857–874
2. Bisseleua DHB, Missoup AD, Vidal S (2009) Biodiversity conservation, Ecosystem functioning and economic incentives under cocoa agroforestry intensification. Conserv Biol 5: 1176–1184
3. Clough Y, Faust H, Tscharntke T (2009a) Cacao boom and bust: sustainability of agroforests and opportunities for biodiversity conservation. Conserv Let 2: 197–205
4. Tscharntke T, Clough Y, Bhagwat SA, Faust H, Herted D, et al. (2011) Ecological principles of multifunctional shade-tree management in cacao agroforestry landscapes. J Appl Ecol 48: 619–629.
5. Bos MM, Tylianakis JM, Steffan-Dewenter I, Tscharntke T (2008) The invasive Yellow Crazy ants and the decline of forest and diversity in the Indonesian cacao agroforests. Biol Inv 10: 1399–1409.
6. Clough Y, Putra DD, Pitropang R, Tscharntke T (2009b) Local and landscape factors determine functional bird diversity in Indonesian cacao agroforestry. Conserv Biol 142: 1032–1041.
7. Cassano RC, Schroth G, Faria D, Delabie JHC, Bede L (2009) Landscape and farm scale management to enhance biodiversity conservation in the cocoa producing region of southern Bahia, Brazil. Biodiv Conserv 18: 577–603.
8. Norgrove L, Csuzdi C, Forzi F, Canet M, Gounes J (2009) Shifts in soil faunal community structure in shaded cacao agroforests and consequences for ecosystem function in Central Africa. Trop Ecol 50(1): 71–78.
9. Clough Y, Barkmann J, Juhrbandt J, Kessler M, Wanger TC, et al. (2011) Combining high biodiversity with high yields in tropical agroforests. Proc Natl Acad Sci USA 108(20):8311–8316.
10. Jagoret P, Bouambi E, Menimo T, Domkam I, Batomen F (2009) Analyse de la diversité des systèmes de pratiques en cacaoculture. Cas du Centre Cameroun. Biotech Agro Soc Env 12 (4): 367–377.
11. Tscharntke T, Sekercioglu CH, Dietsch TV, Sodhi NS, Hoehn P, et al. (2008) Landscape constraints on functional diversity of birds and insects in tropical agroecosystems. Ecology 89: 944–951.
12. Greenberg R, Bichier P, Angón C (2000) The conservation value for birds of cacao plantations with diverse planted shade in Tabasca, Mexico. Anim Conserv 3: 105–112.
13. See YA, Khoo KC (1996) The influence of *Dolichoderus thoracicus* (Hymenoptera: Formicidae) on cocoa pod damage by *Conopomorpha cramerella* (Lepidoptera: Gracillariidae) in Malaysia. Bull Ento Res 86: 467–474.
14. Davidson DW, Cook SC, Snelling R, Chua TH (2003) Explaining the abundance of ants in lowland tropical rainforest canopies. Science 300: 969–972.
15. Philpott SM, Greenberg R, Bichier P, Perfecto I (2004) Impacts of major predators on tropical agroforest arthropods: comparisons within and across taxa. Oecologia 140: 140–149.
16. Philpott SM, Greenberg R, Bichier P (2005) The influence of ants on the foraging behavior of birds in an agroforest. Biotropica 37: 468–471.
17. Bisseleua DHB, Vidal S (2008) Plant biodiversity and vegetation structure in traditional cocoa forest gardens in southern Cameroon under different land-use management. Biodiv Conserv 17: 1821–1835.
18. Philpott SM, Bichier P, Rice RA, Greenberg R (2008) Biodiversity conservation, yield, and alternative products in coffee agroecosystems in Sumatra, Indonesia. Biodiv Conserv 17: 1805–1820
19. Steffan-Dewenter I, Kessler M, Barkmann J, Bos M, Buchori D, et al. (2007) Tradeoffs between income, biodiversity, and ecosystem functioning during tropical rainforest conversion and agroforestry intensification. PNAS 104: 4973–4978
20. Zapfack L, Engwald S, Sonke B, Achoundong G, Birang M (2002) The impact of land conversion on plant biodiversity in the forest zone of Cameroon. Biodiv Conserv 11: 2047–2061
21. Bobo SK, Waltert M, Sainge MN, Njokagbor J, Fermon H, et al. (2006) From forest to farmland: species richness patterns of trees and understorey plants along a gradient of forest conversion in Southwestern Cameroon. Biodiv Conserv 15: 4097–4117.
22. Rice RA, Greenberg R (2000) Cacao Cultivation and the Conservation of Biological Diversity. Ambio 29(3): 167–173.
23. Philpott SM, Armbrecht I (2006) Biodiversity in tropical agroforests and the ecological role of ants and ant diversity in predatory function. Ecol Entomol 31: 369–377.
24. Cicuzza D, Kessler M, Clough Y, Ramadhanil P, Leitner D, et al. (2011) Conservation Value of Cacao Agroforestry Systems for Terrestrial Herbaceous Species in Central Sulawesi, Indonesia. Biotropica, 1–8: DOI 10.1111/j.1744-7429.2010.00741x..
25. Laube I, Breitbach N, Bohning-Gaese K (2008) Avian diversity in a Kenyan agroecosystem: effects of habitat structure and proximity to forest. J Ornith 149: 181–191.
26. Schroth G, Da Fonseca GAB, Harvey CA, Gascon C, Vasconcelos HL, et al. (2004) Agroforestry and biodiversity conservation in tropical landscapes. Island Press, Washington, D.C.
27. Sonwa DJ, Nkongmeneck BA, Weise SF, Tchatat M, Adesina AA, et al. (2007) Diversity of plants in cocoa agroforests in the humid forest zone of Southern Cameroon. Biodiv Conserv 16: 2385–2400
28. Sunderlin WD, Ndoye O, Bikié H, Laporte N, Mertens B, et al. (2000) Economic Crisis, Small-Scale Agriculture, and Forest Cover Change in Southern Cameroon. Env Conserv 27(3): 284–290.
29. Ndoye O, Kaimowitz D (2000) Macro-economics, Markets and the Humid Forests of Cameroon, 1967–1997. J Mod Afr Stud 38(2): 225–253.
30. Bisseleua DHB, Yede, Vidal S (2011) Dispersion models and sampling of cacao mirid bug *Sahlbergella singularis* (Haglung) (Hemiptera: Miridae) on *Theobroma cacao* L. in southern Cameroon. Envir Ento 40(1): 111–119.
31. Colwell RK (2006) *Estimates S*: Statistical estimation of species richness and shared species from samples. Version 8. Persitent URL <purl.oclc.org/estimates>.
32. Gotelli NJ, Colwell RK (2001) Quantifying biodiversity: procedures and pitfalls in the measurement and comparison of species richness. Ecol Let 4: 379–391
33. Longino JT, Coddington J, Colwell RK (2002) The ant fauna of a tropical rain forest: estimating species richness three different ways. Ecology 83: 689–702.
34. Chao A, Chazdon RLM, Colwell RK, Colwell STJ (2005) A new statistical approach for assessing compositional similarity based on incidence and abundance data. Ecol Let 8: 148–159.
35. SPSS Inc. (2000) Systat 11 for Windows. SPSS Inc., Chicago
36. Laird SA, Awung LG, Lysinge RJ (2007) Cocoa farms in the Mount Cameroon region: biological and cultural diversity in local livelihoods. Biodiv Conserv 16: 2401–2427.
37. Duguma B, Gockwski J, Bakala J (2001) Smallholder cacao (*Theobroma cacao* Linn.) cultivation in agroforestry systems of West and Central Africa: Challenges and opportunities. Agrof Syst 51: 177–188.
38. Snoeck D, Abolo D, Jagoret P (2010) Temporal changes in VAM fungi in the cocoa agroforestry systems of central Cameroon. Agrof Syst DOI 10.1007/s10457-009-9254-6
39. Bos MM, Steffan-Dewenter I, Tscharnkte T (2007) The contribution of cacao agroforests to the conservation of lower canopy ant and beetle diversity in Indonesia. Biodiv Conserv 16: 2429–2444.
40. Klein AM, Steffan-Dewenter I, Tscharntke T (2006) Rainforest promotes trophic interactions and diversity of trap-nesting Hymenoptera in adjacent agroforestry. J Anim Ecol 75: 315–323.
41. Wielgoss A, Tscharntke T, Buchori D, Fiala B, Clough Y (2010) Temperature and a dominant dolichoderine ant species affect ant diversity in Indonesian cacao plantations. Agric Ecosyst Env 135: 253–259.
42. Loreau M, Mouquet N, Gonzalez A (2003) Biodiversity as spatial insurance in heterogeneous landscapes. PNAS 100: 12765–12770
43. Yachi S, Loreau M (1999) Biodiversity and ecosystem productivity in a fluctuating environment: The insurance hypothesis. PNAS 96: 1463–1468.
44. Sperber CF, Nakayama K, Valverde MJ, Neves FS (2004) Tree species richness and diversity affect parasitoid diversity in cacao agroforestry. Basic Appl Ecol 5: 241–151.
45. Perfecto I, Vandermeer J (2010) The agroecological matrix as alternative to the landsparing/agriculture intensification model. PNAS 107: 5786–5791

11

The Impact of Land Abandonment on Species Richness and Abundance in the Mediterranean Basin

Tobias Plieninger[1]*, Cang Hui[2,3], Mirijam Gaertner[2], Lynn Huntsinger[4]

1 Department of Geosciences and Natural Resource Management, University of Copenhagen, Frederiksberg, Denmark, **2** Centre for Invasion Biology, Department of Mathematical Sciences, Stellenbosch University, Matieland, South Africa, **3** Mathematical and Physical Biosciences, African Institute for Mathematical Sciences, Cape Town, South Africa, **4** Department of Environmental Science, Policy, and Management, University of California, Berkeley, California, United States of America

Abstract

Land abandonment is common in the Mediterranean Basin, a global biodiversity hotspot, but little is known about its impacts on biodiversity. To upscale existing case-study insights to the Pan-Mediterranean level, we conducted a meta-analysis of the effects of land abandonment on plant and animal species richness and abundance in agroforestry, arable land, pastures, and permanent crops of the Mediterranean Basin. In particular, we investigated (1) which taxonomic groups (arthropods, birds, lichen, vascular plants) are more affected by land abandonment; (2) at which spatial and temporal scales the effect of land abandonment on species richness and abundance is pronounced; (3) whether previous land use and current protected area status affect the magnitude of changes in the number and abundance of species; and (4) how prevailing landforms and climate modify the impacts of land abandonment. After identifying 1240 potential studies, 154 cases from 51 studies that offered comparisons of species richness and abundance and had results relevant to our four areas of investigation were selected for meta-analysis. Results are that land abandonment showed slightly increased (effect size $=0.2109$, $P<0.0001$) plant and animal species richness and abundance overall, though results were heterogeneous, with differences in effect size between taxa, spatial-temporal scales, land uses, landforms, and climate. In conclusion, there is no "one-size-fits-all" conservation approach that applies to the diverse contexts of land abandonment in the Mediterranean Basin. Instead, conservation policies should strive to increase awareness of this heterogeneity and the potential trade-offs after abandonment. The strong role of factors at the farm and landscape scales that was revealed by the analysis indicates that purposeful management at these scales can have a powerful impact on biodiversity.

Editor: Edward Webb, National University of Singapore, Singapore

Funding: T.P. has received funding from the German Ministry of Education and Research (FKZ 01UU0904A) and the European Community's Seventh Framework Programme under Grant Agreement No. 603447 (Project HERCULES). M.G. acknowledges support from the DST-NRF Centre of Excellence for Invasion Biology. The funding sources are not involved in study design, in the collection, analysis and interpretation of data, in the writing of the report and in the decision to submit the article for publication.

Competing Interests: The authors have declared that no competing interests exist.

* E-mail: tobias.plieninger@ign.ku.dk

Introduction

Increasing competition for land is one of the most significant processes of global environmental change [1,2]. Though obscured by the attention given to global land scarcity, the phenomenon of land abandonment – change towards termination of crop cultivation or livestock grazing [3] – is equally on the rise [4,5]. Cropland abandonment has affected an estimated 1.47 million km² worldwide from the 1700s to 1992 [6]. Agricultural abandonment has been substantial throughout the 20th century in North America, the former Soviet Union and Southern Asia, followed by Europe, South America and China since the 1960s [7]. A set of underlying and proximate ecological (e.g. declining soil fertility), social (rural depopulation) and economic (e.g. globalization of agro-commodity markets, declining farm profitability) drivers determine the patterns and processes of land abandonment, usually through interaction at various spatial and temporal scales [8]. Land abandonment has a range of consequences for the provision of ecosystem processes, including functions and services that are not well-understood and often context-specific, for example wildfire frequency and intensity, nutrient cycling, carbon sequestration, cultural landscape values and water balance [3]. Here we conduct a meta-analysis of the literature to examine the consequences of land abandonment in the Mediterranean Basin.

Consequences of Land Abandonment

Two fundamentally different biodiversity consequences are possible: On the one hand, land abandonment may contribute to "passive landscape restoration" [9] or "rewilding" [10], thus facilitating the restoration of natural ecosystem processes and reducing direct human influence on landscapes. Several studies confirm that, for example, woodland bird and large mammal populations benefit from large-scale land abandonment (see [11] and references therein). On the other hand, abandonment of agricultural landscapes may threaten farmland biodiversity, in particular functional diversity [12] associated with anthropogenic landscapes of high nature value. "High nature value farming" is a predominantly European concept that recognizes that the conservation of biodiversity in some settings depends on the continuation of low-intensity farming systems [13–15]. Processes induced by abandonment of agricultural uses that may threaten local biodiversity include habitat loss, decrease in habitat patchiness, competitive exclusion, invasions of non-native plants, litter accumulation, increased predation, and increased wildfires [3].

Put into a larger perspective, the dispute between "rewilding" and "high nature value farming" advocates reflects the ongoing scholarly debate of whether biodiversity conservation should pursue "land sparing" (embracing "rewilded" ecosystems) or "land sharing" (calling for "high nature value" farming) [16,17]. Trajectories of land abandonment are accompanied by considerable societal trade-offs, not only between the different kinds of biodiversity that are supported or degraded, but also between ecosystem functions and services such as aesthetic values, carbon sequestration, or wildfire regimes in landscapes [5]. Despite the implications of these diverging views for conservation, the biodiversity impacts of land abandonment have only started to be assessed beyond local-scale case studies [18,19].

Objective

The Mediterranean Basin is one of the original 25 global biodiversity hotspots [20], exhibiting high levels of plant and animal richness and endemism [11,21,22]. There are numerous case studies on the impacts of land abandonment. To upscale these local-scale case study insights, we performed a meta-analysis focusing on the effects of land abandonment on plant and animal species richness and abundance in agroforestry, arable land, pastures, and permanent crops of the Mediterranean Basin. Based on a previously developed review protocol [23], we investigated (1) which taxonomic groups (arthropods, birds, lichen, vascular plants) are more affected by land abandonment; (2) at which spatial and temporal scales the effect of land abandonment on species richness and abundance is pronounced; (3) whether previous land use and current protected area status affect the magnitude of changes in the number and abundance of species; and (4) how prevailing landforms (mountain vs. lowland areas) and climate modify the impacts of land abandonment. Mediterranean-type environments are particularly suitable for meta-analysis, as they vary less in climate, disturbance regimes, and further key aspects than other biome types [24]. Previous reviews have covered land abandonment [3,8,18], but did not perform formal meta-analyses and/or did not cover a particular biodiversity hotspot. Our intention is to identify knowledge gaps and to inform conservation policy.

Materials and Methods

Study Area: Mediterranean Basin

The Mediterranean Basin is one of the world's regions where land abandonment is prevalent [25,26], especially in upland areas [27]. Precise data on land abandonment are not available, but FAO forest statistics indicate that most of the abandoned Mediterranean farmland is in the European Union member countries, Israel, Turkey and Algeria [28]. Old fields have always been part of a dynamic equilibrium in Mediterranean landscapes, but permanent land abandonment has increased throughout the 20th century [29]. In most northern Mediterranean countries forest cover has increased by about 2% per annum [11].

Modernization of agricultural production in fertile lowland areas and a population exodus from rural areas to urban centers have been the most decisive drivers of Mediterranean land abandonment [30–33]. Agricultural land uses are generally given up when farming fails to adjust to changed economic conditions. The physical constraints of soils, topography, climate, and remoteness limit the options for adaptation to more intensive, mechanized, and profitable farming techniques on the marginal lands of the Mediterranean Basin. Agricultural policies have further accelerated the concentration of agricultural activities on more fertile and accessible land and the abandonment of marginal lands, though some more recent rural development policies have mitigated this trend [27,34,35].

The rich biodiversity of the Mediterranean Basin is the consequence of a particular biogeography, geological history, landscape ecology, and human history. Most notably, human land uses have shaped ecosystems for more than 10,000 years and have enhanced biological and landscape diversity [29,31]. Given that the Mediterranean biome has been predicted to experience the greatest proportional change in biodiversity by 2100, mainly through land use and climate change [36], questions about the impacts of land abandonment on biodiversity are critical.

Study Selection

Our methodology was derived from the guidelines of the Collaboration for Environmental Evidence [37]; following these standards, a sampling protocol was peer-reviewed and published a priori (online repository: [23]). A Preferred Reporting Items for Systematic Reviews and Meta-Analyses (PRISMA) checklist was applied (Table S1). The minimum requirement for inclusion of a case study in the meta-analysis was that it reported summary data on plant or animal species richness or abundance comparing managed versus abandoned farmland. While species richness is argued to be a limited indicator of biodiversity, it is by far the most commonly used proxy for biodiversity in the primary studies that we evaluated. Among the simplest and most robust diversity measures available [38,39], it underlies many ecological models and conservation strategies [40]. It is important to note that whether or not a given outcome, in terms of species richness or some other measure of biodiversity, is a "desirable" outcome is subjective, and will vary by region, landscape, and social factors and is beyond the scope of this study. The following definitions and study inclusion criteria were used:

- *Relevant populations*: Plant and animal populations that may change with the abandonment of agroforestry, arable land, pastures, and permanent crops (Figure 1). We based our definitions on the CORINE land cover nomenclature to delimit agroforestry, arable land, pastures, and permanent crops [41]. "Agroforestry" is defined as annual crops or pasture growing with forestry species such that the two interact; "arable land" refers to irrigated or non-irrigated lands used for annually cultivated and harvested crops; "pastures" are characterized by dense herbaceous cover, generally grazed or harvested for fodder; and "permanent crops" are crops that persist and are harvested over a longer than annual timeframe [41].

- *Relevant exposure*: The complete or partial abandonment of livestock grazing and/or crop cultivation. We understood abandonment as the ceasing of cultivation or grazing on farmland over a period of at least five years.

- *Types of comparators*: Comparisons between species richness and abundance before and after abandonment of particular sites and comparisons of abandoned land to adjacent reference farmland at the same moment in time ("space-for-time substitution") [42].

- *Relevant outcomes*: Quantitative measures of richness and/or abundance of terrestrial plant and animal species. Only taxonomic group, not individual species abundances were included.

- *Relevant types of study design*: Observational field studies and manipulative field experiments. Control plots that were not abandoned should be in similar ecological settings, ideally close to abandoned plots.

We searched the following databases for relevant documents: ISI Web of Science, BIOSIS Citation Index, CAB Abstracts, Scopus, ProQuest Agricola, and ProQuest Dissertations & Theses. To minimize publication bias, we additionally included grey literature by considering the first 50 pdf and word documents provided by each of the following sources: Google, Google Scholar, and Dogpile. We considered studies in English, French, Italian, Portuguese, and Spanish. Search terms referred to the defined population, intervention and outcomes. Terms were broad enough to capture all relevant papers. The following logical search string was used: (abandon* OR "old fields" OR fallow) AND (biodiversity OR richness OR abundance OR composition OR

Figure 1. Examples of active and abandoned farmlands. (a) arable land in Burgos province, Spain; (b) grassland in a North Adriatic pastoral landscape, Croatia; (c) permanent crops: Olive groves on Lesvos, Greece; (d) agroforestry: *Quercus pyrenaica* dehesa in León province, Spain (sources: (a) J. Arroyo; (b) I. Kosić; (c) T. Plieninger, T. Kizos; (d) A. Taboada).

assemblage) AND Mediterranean. Titles and abstracts were stored in a single Endnote database.

The search was performed in May 2013, yielding a total of 2012 studies. We obtained an additional 3 studies from colleagues. After removal of duplicates, 1240 studies remained. Study selection was a three-stage process. First, 632 studies with relevant titles were selected. Second, selection was made based on abstracts, after which 204 studies remained. To be considered in the meta-analysis, a study had to provide summary data (i.e. mean, standard deviation, and sample size) for species richness or abundance on managed vs. abandoned farmlands. When studies had collected relevant data but not published the required summary data, authors were contacted by e-mail. We contacted 33 authors, received information from 24 of them, and processed 19 of these datasets. Full paper content was assessed in the third stage, leaving 51 studies that provided all the information needed for the meta-analysis (means, standard deviations, sample sizes etc.) (Figure S1, see Table S2 for full references).

Repeatability of study inclusion was checked through a random subset of ca. 10% of references whose titles (150 papers) and abstracts (65 papers) were assessed by an independent reviewer. Inclusion consistency was calculated using kappa statistics [43]. Agreement between reviewers was good in both steps (k = 0.61 in the first stage and k = 0.72 in the second stage).

Study quality was assessed before data extraction. All articles that were finally selected met the requirements specified by our systematic review protocol.

Data Treatment and Analysis

Observations of multiple taxa, different study sites, and/or different measurement times within one study were included separately in the dataset and considered independently. For each observation, we extracted means, standard deviations, and sample sizes (see Table S3). When summary statistics were not presented numerically, they were extracted from tables and graphs, using image analysis software in some cases. If original data were provided but summary statistics were lacking, summary statistics were calculated on the basis of raw data. In cases of insufficient information, corresponding authors were contacted to gather the required data. Location data given in the study were used to obtain parameter estimates for explanatory variables from other data sources (GoogleEarth, European Environment Agency, WorldClim) in order to extract variables that were not provided in the studies (Table 1, Table S4). The spatial resolution of WorldClim data is 1 km^2. We considered a total of 8 independent variables (Table 1, plus 3 less important variables in Table S5).

Data were synthesized through meta-analysis to address the overall effects of land abandonment on plant and animal richness and abundance. In meta-analysis, effect size is a difference value relative to the standard deviation. Additional aspects were addressed through meta-regression and sub-group analyses. Specifically, we measured the effect size of each case by the standardized mean difference, $d = (\mu_1 - \mu_2)/sd_p$. μ_1 and μ_2 are means of a focal dependent variable (e.g. population density and diversity) after and before land abandonment, respectively. The pooled standard deviation is sd_p and equals the square root of $((n_1 - 1)sd_1^2 + (n_2 - 1)sd_2^2)/(n_1 + n_2 - 2)$, where n_1 and sd_1 are the sample size and standard deviation for experiments after land abandonment, and n_2 and sd_2 land without abandonment. The heterogeneity analysis of the data was examined using a Q-test, and the significance of the null hypothesis (i.e. the effect size equals zero) was examined by a Z-test as in [44]. We further conducted a meta-regression on temperature and precipitation as continuous

Table 1. Explanatory variables provided by primary studies and additional data sources that were included in the meta-analysis and percentage of observations for which these data could be gathered.

Explanatory variable	Description	Source	Plant species richness (%)	Fungi species richness (%)	Animal species richness (%)	Animal abundance (%)
Unit size	Sample unit size (m^2)	Primary studies	93	100	-	-
Extent	Extent of study area (km^2)	Primary studies, GoogleEarth	95	100	100	100
Time since abandonment	Time elapsed since land was abandoned (years)	Primary studies	98	100	86	85
Previous land use	Agroforestry/arable land/pastures/permanent crops	Primary studies	100	100	100	100
Landform	Situated in mountain/lowland area	European Environment Agency	100	100	100	100
Protected area status	Situated in NATURA 2000 network of protected areas	European Environment Agency	88%	100	100	100
Temperature	Mean annual temperature (°C)	WorldClim	100	100	100	100
Precipitation	Mean annual precipitation (mm)	WorldClim	100	100	100	100

moderators. All calculation was done by using the *metafor* package [45] in R software [46].

Results

Overall, our data set included 154 cases in 51 individual studies, published between 1974 and 2013 (Table S1). In particular, we found 89 cases in 38 primary studies for plant species richness (including fungi species richness), 24 cases in 14 studies for animal species richness, and 21 cases in 10 studies for animal abundance. Eighty-nine percent of cases of animal species richness and 76% of the cases of animal abundance referred to arthropods. Forty percent of cases considered agroforestry, 27% pastures, 20% permanent crops, and 12% arable land (Table 2). Iberian semi-sclerophyllous and semi-deciduous forests (44% of cases), North-eastern Spain and Southern France Mediterranean (11%), and Southwest Iberian Mediterranean sclerophyllous and mixed forests (11%) were among the most intensively studied ecological regions (Figure 2).

Our meta-analysis (using a fixed effect model) revealed that – when analyzed together – plant and animal species richness and abundance values slightly increased after land abandonment (Effect Size Point Estimate ES = 0.2109), and this overall effect was significant (Z = 5.5991, P<0.0001). However, heterogeneity was high (Q = 1048.89, P<0.0001) as outcomes were divided. In 91 (59%) of the 154 cases, species richness and abundance values were higher on abandoned land compared to managed farmland; in the remaining 63 cases (41%), values were lower. Fifty-one cases (33%) had positive effect sizes, indicating a significant increase of species richness or abundance after abandonment. Forty-four cases (29%) had negative effect sizes (indicating significant decreases of species richness or abundance), while in 59 cases (38%) effect sizes did not differ significantly from zero (i.e., SD included 0) (Table S3). Following Cohen's classification [47], 54% of the cases had large effect sizes (>0.8), 36% had medium effect sizes (0.2–0.8), and 8% had small (<0.2) effect sizes (Figure 3). Four cases were not replicated. Due to the heterogeneous effect size, mixed effect models were used to examine the different factors.

Differences in species richness were most pronounced in the fungi (Table 3). However, land abandonment also showed significant increases in animal and plant species richness. Among the taxa, we found significantly higher effect sizes for lichen and birds on abandoned land, while there was no global effect of land abandonment on arthropod and vascular plants (Table 3, Figure 4A). Thirty-four cases (38%) had a positive effect size for plant species richness, while 28 cases (31%) had a negative effect size and 27 cases (30%) were not significantly different from zero. Among studies of animal species richness, 14 cases (32%) had positive and 7 cases (16%) negative effect sizes, and in 23 cases (52%) effect sizes were not significantly different from zero. Animal abundance studies had positive effect sizes in 3 cases (14%), negative effect sizes in 9 cases (43%) and insignificant deviations from zero in 9 cases (43%).

As for spatial-temporal patterns, effect size of land abandonment (assessed for plant species richness only) was positive for small (<1 m^2) and large (>100 m^2) unit sizes, but insignificant for medium-sized units (1<10 m^2; 10<100 m^2) (Table 3, Figure 4B). No clear patterns emerged for the extent of the study areas (Figure 4C). The number of years since abandonment did have a significant impact on effect size; but only studies that covered an abandonment period of thirty to forty-nine years (not those with fifty or more years or less than thirty years of abandonment) showed significant increases in species richness and abundance (Table 3, Figure 4D).

Agroforestry, arable land, and pastures showed significantly different effect sizes between groups (Table 3, Figure 5A). On agroforestry and arable land, species richness and abundance increased after abandonment, while it decreased on pastures. Permanent crops did not exhibit significant effects.

Effect sizes of studies performed in mountains and lowlands significantly differed from each other (Table 3, Figure 5B), with abandonment in lowland areas showing stronger increases in plant and animal species richness and abundance. No differences were found comparing land inside and outside of the European network of protected areas (NATURA 2000, Table 3, Figure 5C). Temperature did not show significant effects, whereas areas with high precipitation showed significant declines of plant and animal species richness and abundance after abandonment (Table 3, Figure 5D).

Discussion

Land abandonment potentially has substantial environmental and socio-economic consequences [5]. This study presents the first formal meta-analysis that examines the particular impacts of land

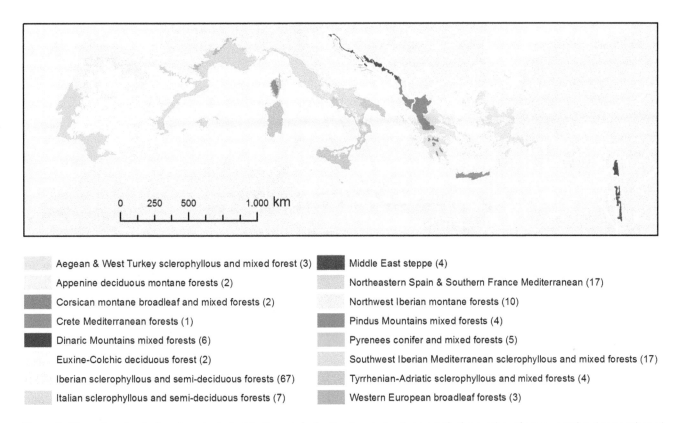

Aegean & West Turkey sclerophyllous and mixed forest (3)

Appenine deciduous montane forests (2)

Corsican montane broadleaf and mixed forests (2)

Crete Mediterranean forests (1)

Dinaric Mountains mixed forests (6)

Euxine-Colchic deciduous forest (2)

Iberian sclerophyllous and semi-deciduous forests (67)

Italian sclerophyllous and semi-deciduous forests (7)

Middle East steppe (4)

Northeastern Spain & Southern France Mediterranean (17)

Northwest Iberian montane forests (10)

Pindus Mountains mixed forests (4)

Pyrenees conifer and mixed forests (5)

Southwest Iberian Mediterranean sclerophyllous and mixed forests (17)

Tyrrhenian-Adriatic sclerophyllous and mixed forests (4)

Western European broadleaf forests (3)

Figure 2. Map of ecological regions included in the analysis. Numbers in brackets specify the number of cases considered per ecological region.

abandonment on biodiversity, using animal species richness, animal abundance, and plant species richness as indicators. The analysis focused on the Mediterranean Basin, an area of comparable climate where land abandonment is prevalent. The meta-analysis revealed that land abandonment has been shown to slightly but significantly result in increases in plant and animal species richness and abundance. However, heterogeneity in responses to abandonment was high. Among the 154 empirical cases used in the meta-analysis, many pointed to increases, and others to decreases, in biodiversity after farmland abandonment. For example, when a simply structured vineyard in Israel was abandoned, the mean species richness values of perennial plants (between vine rows) increased from 0.0 to 2.3 species per m^2 [48].

In contrast, mean plant species richness declined after abandonment from 16.4 to 12.4 species per m^2 in a multifunctional grazing system in Northern Spain [49], or from 38.4 to 21.8 species per 100 m^2 in a traditionally cultivated chestnut grove in Southern France [50]. In some empirical studies, effect sizes went in different directions when different species groups [51] or different farmland habitats [52] were investigated. Using a diversity of indicators, a qualitative, global review of land abandonment came to similar insight, with 77 studies pointing to biodiversity losses, but another 39 studies reporting increasing biodiversity [3]. Thus, the responses of species richness and abundance are not consistent enough to support general conclusions about biodiversity trends on abandoned lands in the Mediterranean. Rather, these

Table 2. Structure of the data set for comparing managed versus abandoned farmlands (number of cases).

Taxa	Agroforestry	Arable land	Pastures	Permanent crops
All	62	19	41	32
Fungi	4	0	0	0
Plants	20	4	33	28
Animals (richness)	28	8	5	3
...Arthropods (richness)	24	7	5	3
...Birds (richness)	4	1	0	0
Animals (abundance)	10	7	3	1
...Arthropods (abundance)	6	6	3	1
...Birds (abundance)	4	1	0	0

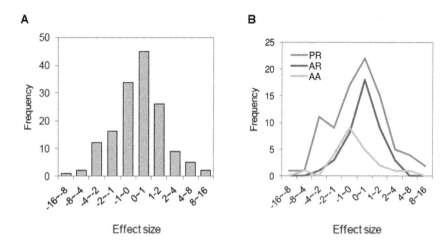

Figure 3. Frequency distribution of effect sizes for plant species richness, animal species richness and animal abundance (A) together and (B) separately. Mean difference effect size, g, and a mixed (random) effects model were used (PR – plant species richness; AR – animal species richness; AA – animal abundance).

responses seem strongly mediated by the specifics of each case study, whether they pertain to spatial-temporal scale, land-use, landforms, climate, or other parameters.

Variation in Land Abandonment Impacts

In regard to objective 1), the diverging views on increases or decreases in plant and animal populations that result from land

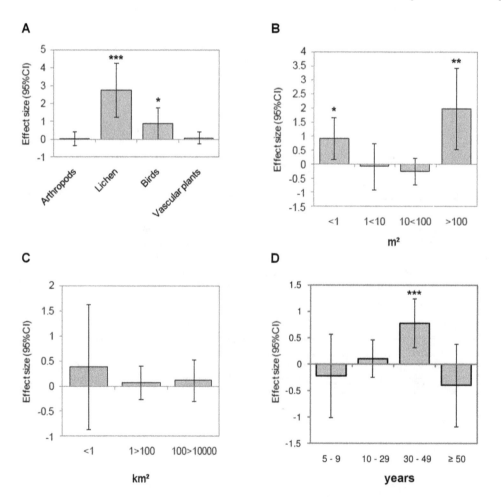

Figure 4. Effect size (95% CI) of land abandonment for (A) taxon; (B) unit size of study; (C) extent of study area; (D) time since abandonment. Q-test shows significant different effect sizes between groups (heterogeneity) for taxon (Q = 16.95, P = 0.002) and time since abandonment (Q = 12.68, P = 0.013), but not for extent (Q = 0.86, P = 0.8356).

Table 3. Summary of the meta-analysis.

Moderator (Q, P)	ES	SE	Z	P	95%CI Lower	95%CI Upper	N
Kingdom (95.36, <0.0001)							
Animals	0.1625	0.0628	2.5876	0.0097	0.0394	0.2856	65
Fungi	1.2717	0.1384	9.1895	<0.0001	1.0004	1.5429	4
Plants	0.1028	0.0501	2.0544	0.0399	0.0047	0.2010	85
Taxon (16.95, 0.002)							
Arthropods	0.0424	0.2026	0.2091	0.8344	−0.3548	0.4395	55
Birds	0.8829	0.4445	1.9862	0.0470	0.0117	1.7541	10
Lichen	2.7541	0.7738	3.5594	0.0004	1.2375	4.2706	4
Vascular plants	0.0909	0.1673	0.5432	0.5870	−0.2371	0.4189	85
Unit size(14.02, 0.0072)							
<1 m^2	0.9154	0.3793	2.4133	0.0158	0.1719	1.6589	18
1<10 m^2	−0.0827	0.4236	−0.1953	0.8452	−0.9130	0.7475	13
10<100 m^2	−0.2353	0.2421	−0.9722	0.3310	−0.7097	0.2391	47
>100 m^2	1.9763	0.7357	2.6862	0.0072	0.5343	3.4182	5
Extent (0.86, 0.8356)							
<1 km^2	0.3797	0.6310	0.6018	0.5473	−0.8570	1.6164	11
1>100 km^2	0.0713	0.1673	0.4263	0.6699	−0.2566	0.3992	91
100>10000 km^2	0.1196	0.2135	0.5601	0.5754	−0.2989	0.5381	48
Time since abandonment (12.68, 0.013)							
5–9 years	−0.2189	0.3996	−0.5477	0.5839	−1.0021	0.5644	16
10–29 years	0.1061	0.1807	0.5869	0.5572	−0.2481	0.4602	72
30–49 years	0.7822	0.2353	3.3244	0.0009	0.3210	1.2433	42
≥50 years	−0.3925	0.3963	−0.9903	0.3220	−1.1693	0.3843	13
Previous land use (18.72, 0.0009)							
Agroforestry	0.5765	0.1880	3.0666	0.0022	0.2080	0.9449	62
Arable land	0.7180	0.3559	2.0173	0.0437	0.0204	1.4155	19
Pastures	−0.5072	0.2254	−2.2500	0.0244	−0.9491	−0.0654	41
Permanent crops	0.1256	0.2938	0.4274	0.6691	−0.4502	0.7013	32
Landform (9.76, 0.0076)							
Mountain area	−0.0028	0.1471	−0.0191	0.9847	−0.2911	0.2854	107
Lowland area	0.7579	0.2426	3.1245	0.0018	0.2825	1.2333	47

Table 3. Cont.

Moderator (Q, P)	ES	SE	Z	P	95%CI Lower	95%CI Upper	N
Protected area status (0.31, 0.8553)							
Inside NATURA 2000 area	0.0430	0.1511	0.2843	0.7762	−0.2532	0.3392	107
Outside NATURA 2000 area	0.1242	0.2580	0.4814	0.6302	−0.3814	0.6298	37
Climate (12.18,0.0023)							
Intercept	2.1029	1.0552	1.9929	0.0463	0.0347	4.1711	150
Temperature	−0.0425	0.0602	−0.7054	0.4806	−0.1606	0.0756	150
Precipitation	−0.0021	0.0006	−3.2903	0.0010	−0.0034	−0.0009	150

Q = Q-test for heterogeneity (including P value); ES = effect size point estimate; SE = standard error; Z = two-tail Z-test; 95%CI Lower = lower confidence interval (95%); 95%CI Upper = upper confidence interval (95%); N = Number of records of each category. Categorical moderators are analyzed using mixed effect models; the last moderator "Climate" of two continuous variables is assessed by a meta-analytic model. The cursive fonts indicate the specific variables and their Q and P values.

abandonment can be partly explained by the different taxonomic groups involved. All three kingdoms (animals, fungi, plants) showed an overall positive effect size after abandonment, but the strongest one was found for lichen (remembering that all lichen cases were taken from one publication only). Bird species richness also showed clear increases in response to land abandonment. The finding that lichen and birds are more sensitive to land abandonment than other taxa may explain why the effect size for small (<1 m^2) and large study units (>100 m^2) was significant, as studies on lichen are conducted at finer scales while studies on birds are mainly carried out at broader scales. Responses of vascular plant richness were heterogeneous, with some plant communities favored by agricultural management (very likely those composed of ruderal, stress tolerant, and competitive farmland species) and some (very likely those composed of shrubland and woodland species) favored by abandonment. A meta-analysis of land abandonment effects on bird distribution changes also found such heterogeneous differences, with decreasing occurrence of farmland bird species and increasing occurrence of woodland and shrubland species after abandonment [19].

As for objective 2), our results showed that the temporal dimension of land abandonment studies is important [53]. Plant species richness often increases and exhibits strong dynamics in the first years after abandonment, but later species composition becomes more stable and species richness decreases. The intermediate disturbance hypothesis offers one potential explanation, as it predicts that plant competition has a greater influence on plant community development when it is not interrupted by disturbances such as cultivation, drought or grazing [50,54]. In a highly competitive environment, less successful competitors are often eventually suppressed. In our meta-analysis however, only studies considering an abandonment period of 30–49 years showed significant increases in species richness. Obviously, several decades are needed until colonizers in the regional species pool trigger community succession. Therefore, our results highlight that comparatively long time periods are required before general increases in species richness and abundance can be detected. However, a (non-significant) decline in species richness after an abandonment period of 50 or more years may indicate that exclusion processes eventually follow colonization processes in many of the case studies.

Substantially different outcomes were revealed for different agricultural systems [55] when objective 3) was investigated, confirming previous studies of the influence of farm-level attributes on biodiversity [56]. Species richness and abundance generally increased on cultivated habitats (arable land, agroforestry) after abandonment, and decreased in abandoned pastures. Effect sizes in Figure 5B suggest an order from increased to decreased species richness, from agroforestry (+), to arable land (+), to permanent crops (non-significant), and to pasture (-). This may be related to a gradient in plant height and in "naturalness" of the vegetation types. Cultivated habitats are generally more disturbed by agricultural activities and more distant in species composition from natural ecosystems than are pastures. Accordingly, stronger increases in species richness following land abandonment could be expected for cultivated lands as reduced soil disturbance allows longer-lived plants to become part of and "de-simplify" the available habitats [57]. However, mechanisms of species increase or decrease in these habitats are controlled by plot-level variation in landscape structure (in particular, the amount of semi-natural habitats), land-use legacies, and/or management intensity [58]. Effects on biodiversity values are also likely to vary within arable or pasture lands for the same reason. Biodiversity impacts after abandonment may differ between highly mechanized and

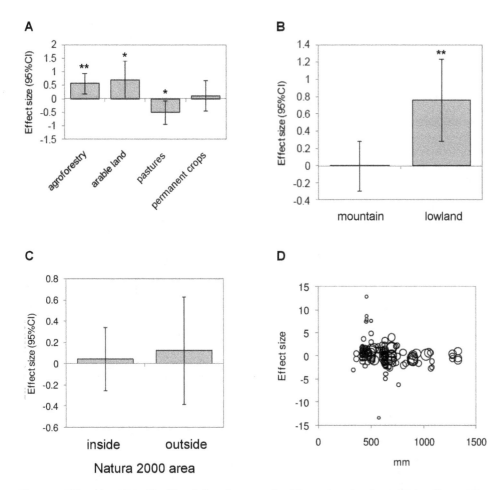

Figure 5. Effect size (95% CI) of land abandonment for (A) previous land use; (B) landform; (C) protected areas; (D) precipitation. Q-test shows significant different effect sizes (heterogeneity) between groups (A: Q = 18.72, P = 0.009 and B: Q = 9.76, P = 0.0076), but not for C: (Q = 0.31, P = 0.8553). D displays a bubble plot of the relationship between effect size and precipitation, with the size of the bubbles scaled according to the reciprocal of the standard deviation of the effect size.

simplified croplands for example, and traditionally grazed native pasture land where grazing may moderate competitive exclusion [59]. The "nature value" of farmlands prior to abandonment would be worth exploring as variables in the analyses, but spatially explicit information on the occurrence of high-nature value farmland is not available at the European level at present and therefore hard to consider in a meta-analysis. Surprisingly, we did not find differences between studies carried out inside and outside the network of protected areas that covers around 18% of the European Union.

Regarding objective 4), some effects of landforms, climate, and other contextual factors were revealed. Particularly influential was the ecological region of the Mediterranean Basin where the study was carried out. Land abandonment impacts were more negative to species richness in areas of higher precipitation (Table 3), i.e. in those environments of the Mediterranean where climatic conditions are less challenging. This pattern supports prevailing notions about non-equilibrium systems [60]; in accordance with non-equilibrium concepts, in areas where abiotic factors do not limit competition as a major driver, human disturbance may favor greater species richness. Therefore, higher precipitation may lead to high levels of competitive exclusion when disturbance from agriculture ceases.

Limitations of the Study

When interpreting the results of our meta-analysis, several caveats need to be taken into account. Although meta-analysis is acknowledged as a straightforward method that yields robust quantitative results, relevant information reported in the empirical studies used may be lost, and some relevant studies may be missed in the selection process. Our search found that many papers could not be included because necessary summary statistics were not provided. This information could be gained from some, but by no means all, authors. We tried to ensure comparability between primary studies by restricting our analysis to the Mediterranean Basin, and by adding standardized information (temperature, precipitation, landform, protected area status) from databases that covered the whole Basin (or at least the European part of it). However, the strong variability that we found indicates that the relationship between land use and biodiversity in the Mediterranean may be too complex for general conclusions.

Our analysis may be further limited because of publication bias, the idea that studies reporting significant differences are more frequently published than studies that do not find significant differences. We tried to minimize publication bias by not only including studies that were published in high impact journals but also results from theses, national-level periodicals, internet documents and other forms of "grey literature". The distribution of effect sizes of our 154 cases is rather symmetric and normal (Figure 3), so it does not indicate any obvious publication bias. Rather, most effect sizes are moderate, and only few are large. Before-after studies may be sensitive to random factors such good or bad years in terms of rainfall. However, only 2 out of the 154 cases were before-after studies, so we believe that the influence of such factors on the outcome of our meta-analysis is low.

Another issue to be discussed is whether the inclusion of several cases per published study leads to pseudoreplication. Having the number of cases exceed the number of studies is very common in meta-analysis studies [61]. It can be addressed by randomly selecting one case per study and examining whether the confidence interval of the effect size for these selected cases is different from the confidence interval for all cases. This is statistically equivalent to comparing the confidence intervals of the mean effect size for each study, calculated from a mixed-effect model using "study" as a factor, with the confidence interval of the effect size for all cases (from Figure 3A). A lack of overlap in 83% confidence intervals represents a significant difference at $P = 0.05$ (note that a lack of overlap in 95% confidence intervals indicates a statistically significant difference at much smaller thresholds of P value [62]). The 83% confidence interval in Fig. 3A is -2.115, 2.217, and the 83% confidence interval for the "ES section of 'study'" in Table S5 is -2.876, 1.497. Consequently, the effect of pseudoreplication is not significant and our way of analysis is acceptable.

Perhaps the most important limitation to address is the selection of biodiversity indicators. As have many other meta-analyses [57,63,64], we focused our study on species richness. However, species richness can be an unreliable indicator of biodiversity [53], and more sophisticated comparisons based on species composition would be more informative [65]. In addition to species richness, we considered species abundance for the assessment of the biodiversity impacts of land abandonment, as diminishing abundance may translate into reduced genetic diversity of populations [66]. Our approach did not allow us to consider studies of other dimensions of biodiversity, for example of differences in ecosystem diversity [67], or of population changes in individual species [19]. If these parameters were to be included, we suspect the overall impacts of land abandonment in Europe might more often lead to decreases in biodiversity [3,18]. In addition, our meta-analysis did not assess species composition. As a result losses in farmland biodiversity and especially agrobiodiversity that accompany many abandonment processes [27] might have been overlooked. Similarly, conceptions of what constitutes "land abandonment" can vary substantially [68].

Research Needs

Our sample of primary studies was not distributed evenly across the Mediterranean Basin. Most cases, 96%, were from Europe, with a single country – Spain – contributing 56% of all cases. In contrast, not a single one was assessed from the southern shore of the Mediterranean Basin. This might reflect a biased selection of study cases, but in large part can also be attributed to the fact that land abandonment is a particular phenomenon of the European Mediterranean, as land use pressure remains high in African and Asian regions [25,69]. Given that they represent regional-specific land-use systems, the *dehesa* and *montado* agroforestry systems of the Iberian Peninsula received a lot of attention in studies. In contrast, the biodiversity outcomes of land abandonment were comparatively little studied for arable land. Also, not all taxa received equal attention, and current studies do not allow identifying the specific kinds of plant and animal communities that are favored or hampered by land abandonment.

Future research studying land abandonment should strive to fill the gaps identified in this paper by focusing on neglected taxa and regions and by studying effects on species composition, turnover, and functional biodiversity rather than species numbers. Other promising directions might be closer examination of biodiversity outcomes under different intensities of land management (e.g., intensive crop cultivation versus high nature value farming versus organic agriculture) and land abandonment (e.g., complete versus "mild" abandonment of traditional olive cultivation), and under different land tenure regimes [70]. In particular, the landscape context of land abandonment and biodiversity needs much more attention.

Conclusions

It is challenging to explain the contrasting impacts of a complex and spatially diverse process such as land abandonment [18]. Synthesizing the results of 154 cases throughout the Mediterranean Basin, this meta-analysis indicates a slight increase in overall species richness and abundance after land abandonment. The effects of land abandonment on biodiversity were mediated by a broad set of drivers. As a result, the directions and intensities of response in species richness and abundance to land abandonment were heterogeneous across the Basin.

Our results point out that neither "rewilding" nor "high nature value farming" alone offer "one-size-fits-all" policy solutions for addressing biodiversity conservation following land abandonment in the Mediterranean Basin. Rather, agri-environmental and other approaches need to be tailored to the local ecosystem, landscape, and land-use context. For example, our study gives hints that abandonment of some plots within simply structured cultivated landscapes may increase landscape heterogeneity and habitat diversity and thus contribute to ecological restoration [71,72]. In contrast, land abandonment in pastoral landscapes may be detrimental, in particular to farmland biodiversity that is linked to active human intervention [14]. In fact, land abandonment affects extensively managed Mediterranean farmland much more than intensively managed farmland [73], so that the scenario of decreasing farmland biodiversity seems to be the more common case. An ideal configuration may be a landscape mosaic with patches of differing successional stages and management types, with each stage and type benefitting particular taxonomic groups [3]. An important task for the future will be to develop a typology of potential responses to land abandonment.

Given prevailing socio-economic conditions, land abandonment will continue in many parts of the Mediterranean Basin. An outcome of larger agricultural change, the process is unlikely to be efficiently addressed by broad agricultural policy and even less by biodiversity conservation programs [35]. Our results call, firstly, for spatially explicit targeting of policies toward specific hotspots of land abandonment [5,26]. Secondly, policies should address only those farmlands where otherwise uncontrolled abandonment would lead to socially undesired outcomes for biodiversity and ecosystem services. The results of this study point out that abandonment of pasture lands may need particular targeting by agri-environmental policies to maintain or restore biodiversity values. The strong role of factors at the farm and landscape scales that was revealed by the analysis indicates that purposeful management at these scales can have a powerful impact on biodiversity, for example in situations where ecological processes such as dispersal and regeneration are disrupted by surrounding industrial agriculture [7]. A context-specific approach requires assessments of broad sets of biodiversity and ecosystem services at the landscape scale as well as cross-sectoral rural development approaches.

Supporting Information

Figure S1 Flow diagram reporting the number of records identified, excluded, and added during the screening process.

Table S1 PRISMA (Preferred Reporting Items for Systematic Reviews and Meta-Analyses) checklist.

Table S2 Full references for the 51 studies included in the meta-analysis.

Table S3 Cases included in the meta-analysis: Dependent variables.

Table S4 Cases included in the meta-analysis: Independent variables.

Table S5 Summary of additional variables included in the meta-analysis.

Acknowledgments

Martin Mantel created maps and figures. We thank M. Price and B. Gebhardt for providing the "Map of European mountain massifs". We are grateful to the Collaboration for Environmental Evidence for providing guidance. Additional data were kindly given by the following authors of primary studies: H. Allen, J. Arroyo, P. Beja, G. Bonamomi, H. Castro, A. Catorci, T. Curt, M. Díaz, E. Farris, C. Gómez López, S. Gonçalves, I. Kosić, L. Lassaletta, F. López y Gelats, A. Morón-Ríos, F. Ojeda, B. Peco, S. Plexida, A.-H. Prieur-Richard, I. Santa Regina, J. Santana, A. Sfougaris, C. Sirami, R. Tárrega, and S. García Tejero. The research contributes to the Global Land Project (http://www.globallandproject.org) and the Programme on Ecosystem Change and Society (http://www.pecs-science.org).

Author Contributions

Conceived and designed the experiments: TP MG. Performed the experiments: TP MG. Analyzed the data: CH. Wrote the paper: TP LH. Contributed to the discussion: MG LH CH. Supervised experimental design: CH.

References

1. Smith P, Gregory PJ, van Vuuren D, Obersteiner M, Havlik P, et al. (2010) Competition for land. Philosophical Transactions of the Royal Society B-Biological Sciences 365: 2941–2957.
2. Sikor T, Auld G, Bebbington AJ, Benjaminsen TA, Gentry BS, et al. (2013) Global land governance: from territory to flow? Current Opinion in Environmental Sustainability 5: 522–527.
3. Rey Benayas JM, Martins A, Nicolau JM, Schulz JJ (2007) Abandonment of agricultural land: an overview of drivers and consequences. CAB Reviews: Perspectives in Agriculture, Veterinary Science, Nutrition and Natural Resources 2: 57.
4. Plieninger T, van der Horst D, Schleyer C, Bieling C (2014) Sustaining ecosystem services in cultural landscapes. Ecology and Society, in press.
5. Munroe DK, van Berkel DB, Verburg PH, Olson JL (2013) Alternative trajectories of land abandonment: causes, consequences and research challenges. Current Opinion in Environmental Sustainability 5: 471–476.
6. Ramankutty N, Foley JA (1999) Estimating historical changes in global land cover: Croplands from 1700 to 1992. Global Biogeochemical Cycles 13: 997–1027.
7. Cramer VA, Hobbs RJ, Standish RJ (2008) What's new about old fields? Land abandonment and ecosystem assembly. Trends in Ecology & Evolution 23: 104–112.
8. Hobbs RJ, Cramer VA (2007) Why old fields? Socioeconomic and ecological causes and consequences of land abandonment. In: Cramer VA, Hobbs RJ, Old Fields: Dynamics and Restoration of Abandoned Farmland. Washington D.C.: Island Press. pp. 1–14.
9. Bowen ME, McAlpine CA, House APN, Smith GC (2007) Regrowth forests on abandoned agricultural land: A review of their habitat values for recovering forest fauna. Biological Conservation 140: 273–296.
10. Navarro LM, Pereira HM (2012) Rewilding abandoned landscapes in Europe. Ecosystems 15: 900–912.
11. Blondel J, Aronson J, Bodiou J-Y, Boef G (2010) The Mediterranean Region: Biological Diversity in Space and Time. Oxford, New York: Oxford University Press. 376 p.
12. Peco B, Carmona CP, de Pablos I, Azcarate FM (2012) Effects of grazing abandonment on functional and taxonomic diversity of Mediterranean grasslands. Agriculture Ecosystems & Environment 152: 27–32.
13. Oppermann R, Beaufoy G, Jones G (2012) High Nature Value Farming in Europe - 35 European Countries, Experiences and Perspectives. Ubstadt-Weiher: Verlag Regionalkultur.
14. Plieninger T, Bieling C (2013) Resilience-based perspectives to guiding high nature value farmland through socio-economic change. Ecology and Society 18(4): 20, doi: 10.5751/ES-05877-180420.
15. Bignal EM, McCracken DI (2000) The nature conservation value of European traditional farming systems. Environmental Review 8: 149–171.
16. Phalan B, Onial M, Balmford A, Green RE (2011) Reconciling food production and biodiversity conservation: Land sharing and land sparing compared. Science 333: 1289–1291.
17. Tscharntke T, Clough Y, Wanger TC, Jackson L, Motzke I, et al. (2012) Global food security, biodiversity conservation and the future of agricultural intensification. Biological Conservation 151: 53–59.
18. Queiroz AI (2013) Managing for Biodiversity and Ecosystem Services in a Context of Farmland Abandonment. Doctoral Thesis in Sustainability Science. Stockholm: Stockholm University.
19. Sirami C, Brotons L, Burfield I, Fonderflick J, Martin JL (2008) Is land abandonment having an impact on biodiversity? A meta-analytical approach to bird distribution changes in the north-western Mediterranean. Biological Conservation 141: 450–459.
20. Myers N, Mittermeier RA, Mittermeier CG, da-Fonseca GAB, Kent J (2000) Biodiversity hotspots for conservation priorities. Nature 403: 853–858.
21. Underwood EC, Viers JH, Klausmeyer KR, Cox RL, Shaw MR (2009) Threats and biodiversity in the Mediterranean biome. Diversity and Distributions 15: 188–197.
22. Cuttelod A, García N, Abdul Malak D, Temple H, Katariya V (2008) The Mediterranean: a biodiversity hotspot under threat. In: Vié J-C, Hilton-Taylor C, Stuart SN, The 2008 Review of The IUCN Red List of Threatened Species. Gland: IUCN.
23. Plieninger T, Gaertner M, Hui C, Huntsinger L (2013) Does land abandonment decrease species richness and abundance of plants and animals in Mediterranean pastures, arable lands and permanent croplands? Environmental Evidence 2: 3.
24. Pauchard A, Cavieres LA, Bustamante RO (2004) Comparing alien plant invasions among regions with similar climates: where to from here? Diversity and Distributions 10: 371–375.
25. Weissteiner CJ, Boschetti M, Bottcher K, Carrara P, Bordogna G, et al. (2011) Spatial explicit assessment of rural land abandonment in the Mediterranean area. Global and Planetary Change 79: 20–36.
26. Sluiter R, de Jong SM (2007) Spatial patterns of Mediterranean land abandonment and related land cover transitions. Landscape Ecology 22: 559–576.
27. MacDonald D, Crabtree JR, Wiesinger G, Dax T, Stamou N, et al. (2000) Agricultural abandonment in mountain areas of Europe: Environmental consequences and policy response. Journal of Environmental Management 59: 47–69.
28. Mazzoleni S, di Pasquale G, Mulligan M, di Martino P, Rego F (2004) Recent dynamics of the Mediterranean vegetation and landscape. Chichester, West Sussex; Hoboken, NJ: John Wiley & Sons. 306 p.
29. Bugalho MN, Caldeira MC, Pereira JS, Aronson J, Pausas JG (2011) Mediterranean cork oak savannas require human use to sustain biodiversity and ecosystem services. Frontiers in Ecology and the Environment 9: 278–286.
30. Papanastasis VP (2007) Land abandonment and old field dynamics in Greece. In: Cramer VA, Hobbs RJ, Old Fields: Dynamics and Restoration of Abandoned Farmland. Washington DC: Island Press. pp. 225–246.
31. Grove AT, Rackham O (2001) The Nature of Mediterranean Europe: An Ecological History. New Haven, London: Yale University Press. 384 p.
32. Pereira E, Queiroz C, Pereira HM, Vicente L (2005) Ecosystem services and human well-being: a participatory study in a mountain community in Portugal. Ecology and Society 10(2): 14.
33. Santos-Martin F, Martin-Lopez B, Garcia-Llorente M, Aguado M, Benayas J, et al. (2013) Unraveling the relationships between ecosystems and human wellbeing in Spain. Plos One 8(9): e73249.

34. Keenleyside C, Tucker GM (2010) Farmland Abandonment in the EU: an Assessment of Trends and Prospects. Report prepared for WWF. London: Institute for European Environmental Policy.

35. Renwick A, Jansson T, Verburg PH, Revoredo-Giha C, Britz W, et al. (2013) Policy reform and agricultural land abandonment in the EU. Land Use Policy 30: 446–457.

36. Henrichs T, Zurek M, Eickhout B, Kok K, Raudsepp-Hearne C, et al. (2010) Scenario development and analysis for forward-looking ecosystem assessments. In: Ash N, Blanco H, Brown C, Garcia K, Henrichs T, et al., Ecosystems and Human Well-Being A Manual for Assessment Practicioners. Washington DC: Island Press. pp. 151–219.

37. Centre for Evidence-Based Conservation (2010) Guidelines for Systematic Review in Environmental Management. Version 4.0. Environmental Evidence: http://www.environmentalevidence.org/Authors.thm.

38. Magurran AE (2004) Measuring Biological Diversity. Malden: Blackwell Publishing. 256 p.

39. Noss RF (1990) Indicators for monitoring biodiversity - a hierarchical approach. Conservation Biology 4: 355–364.

40. Gotelli NJ, Colwell RK (2001) Quantifying biodiversity: procedures and pitfalls in the measurement and comparison of species richness. Ecology Letters 4: 379–391.

41. Bossard M, Feranec J, Otahel J (2000) CORINE land cover technical guide – Addendum 2000. Technical report No 40. Copenhagen: European Environment Agency.

42. Pickett S (1989) Space-for-time substitution as an alternative to long-term studies. In: Likens GE, Long-term Studies in Ecology: Approaches and Alternatives. New York: Springer. pp. 110–135.

43. Cohen J (1960) A coefficient of agreement for nominal scales. Educational and Psychological Measurement 20: 37–46.

44. Gaertner M, Den Breeyen A, Hui C, Richardson DM (2009) Impacts of alien plant invasions on species richness in Mediterranean-type ecosystems: a meta-analysis. Progress in Physical Geography 33: 319–338.

45. Viechtbauer W (2010) Conducting meta-analyses in R with the metafor package. Journal of Statistical Software 36: 1–48.

46. R Developement Core Team (2006) R: A language and environment for statistical computing. Vienna: R Foundation for Statistical Computing.

47. Cohen J (1988) Statistical Power Analysis for the Behavioral Sciences. Hillsdale: L. Erlbaum Associates. 567 p.

48. Neeman G, Izhaki I (1996) Colonization in an abandoned East-Mediterranean vineyard. Journal of Vegetation Science 7: 465–472.

49. Tarrega R, Calvo L, Taboada A, Garcia-Tejero S, Marcos E (2009) Abandonment and management in Spanish dehesa systems: Effects on soil features and plant species richness and composition. Forest Ecology and Management 257: 731–738.

50. Gondard H, Romane F, Grandjanny M, Li JQ, Aronson J (2001) Plant species diversity changes in abandoned chestnut (Castanea sativa) groves in southern France. Biodiversity and Conservation 10: 189–207.

51. Plexida S, Sfougaris A, Papadopoulos N (2012) Quantifying beetle and bird diversity in a Mediterranean mountain agro-ecosystem. Israel Journal of Ecology and Evolution 58: 1–25.

52. Peco B, Sanchez AM, Azcarate FM (2006) Abandonment in grazing systems: Consequences for vegetation and soil. Agriculture Ecosystems & Environment 113: 284–294.

53. Paillet Y, Berges L, Hjalten J, Odor P, Avon C, et al. (2010) Biodiversity differences between managed and unmanaged forests: Meta-analysis of species richness in Europe. Conservation Biology 24: 101–112.

54. Lavorel S, Lepart J, Debussche M, Lebreton JD, Beffy JL (1994) Small-scale disturbances and the maintenance of species-diversity in Mediterranean old fields. Oikos 70: 455–473.

55. Bonet A, Pausas JG (2007) Old field dynamics on the dry side of the Mediterranean Basin: Patterns and processes in semiarid Southeast Spain. In: Cramer VA, Hobbs RJ, Old Fields: Dynamics and Restoration of Abandoned Farmland. Washington DC: Island Press. pp. 247–264.

56. Billeter R, Liira J, Bailey D, Bugter R, Arens P, et al. (2008) Indicators for biodiversity in agricultural landscapes: a pan-European study. Journal of Applied Ecology 45: 141–150.

57. Batary P, Andras B, Kleijn D, Tscharntke T (2011) Landscape-moderated biodiversity effects of agri-environmental management: a meta-analysis. Proceedings of the Royal Society B-Biological Sciences 278: 1894–1902.

58. Grashof-Bokdam CJ, van Langevelde F (2005) Green veining: landscape determinants of biodiversity in European agricultural landscapes. Landscape Ecology 20: 417–439.

59. Perevolotsky A, Seligman NG (1998) Role of grazing in Mediterranean rangeland ecosystems. Bioscience 48: 1007–1017.

60. Westoby M, Walker B, Noymeir I (1989) Opportunistic management for rangelands not at equilibrium. Journal of Range Management 42: 266–274.

61. Vilà M, Espinar JL, Hejda M, Hulme PE, Jarošik V, et al. (2011) Ecological impacts of invasive alien plants: a meta-analysis of their effects on species, communities and ecosystems. Ecology Letters 14: 702–708.

62. Payton ME, Greenstone MH, Schenker N (2003) Overlapping confidence intervals or standard error intervals: what do they mean in terms of statistical significance? Journal of Insect Science 3: 34.

63. Humbert J-Y, Pellet J, Buri P, Arlettaz R (2012) Does delaying the first mowing date benefit biodiversity in meadowland? Environmental Evidence 1: 9.

64. Felton A, Knight E, Wood J, Zammit C, Lindenmayer D (2010) A meta-analysis of fauna and flora species richness and abundance in plantations and pasture lands. Biological Conservation 143: 545–554.

65. Sax DF, Kinlan BP, Smith KF (2005) A conceptual framework for comparing species assemblages in native and exotic habitats. Oikos 108: 457–464.

66. Davis MA (2009) Invasion Biology. Oxford, New York: Oxford University Press. 244 p.

67. Mottet A, Ladet S, Coque N, Gibon A (2006) Agricultural land-use change and its drivers in mountain landscapes: A case study in the Pyrenees. Agriculture Ecosystems & Environment 114: 296–310.

68. Plieninger T, Gaertner M (2011) Harnessing degraded lands for biodiversity conservation. Journal for Nature Conservation 19: 18–23.

69. Plieninger T, Schaich H, Kizos T (2011) Land-use legacies in the forest structure of silvopastoral oak woodlands in the Eastern Mediterranean. Regional Environmental Change 11: 603–615.

70. Schaich H, Plieninger T (2013) Land ownership drives stand structure and carbon storage of deciduous temperate forests. Forest Ecology and Management 305: 146–157.

71. Keenleyside C, Tucker G, McConville A (2010) Farmland Abandonment in the EU: an Assessment of Trends and Prospects. London: WWF, IEEP.

72. Pointereau P, Coulon F, Girard P, Lambotte M, Stuczynski T, et al. (2008) Analysis of Farmland Abandonment and the Extent and Location of Agricultural Areas that are Actually Abandoned or are in Risk to be Abandoned. Ispra: European Commission-JRC-Institute for Environment and Sustainability.

73. Cocca G, Sturaro E, Gallo L, Ramanzin M (2012) Is the abandonment of traditional livestock farming systems the main driver of mountain landscape change in Alpine areas? Land Use Policy 29: 878–886.

12

Intercropping with Shrub Species That Display a 'Steady-State' Flowering Phenology as a Strategy for Biodiversity Conservation in Tropical Agroecosystems

Valerie E. Peters*¤

Odum School of Ecology, University of Georgia, Athens, Georgia, United States of America

Abstract

Animal species in the Neotropics have evolved under a lower spatiotemporal patchiness of food resources compared to the other tropical regions. Although plant species with a steady-state flowering/fruiting phenology are rare, they provide predictable food resources and therefore may play a pivotal role in animal community structure and diversity. I experimentally planted a supplemental patch of a shrub species with a steady-state flowering/fruiting phenology, *Hamelia patens* Jacq., into coffee agroforests to evaluate the contribution of this unique phenology to the structure and diversity of the flower-visiting community. After accounting for the higher abundance of captured animals in the coffee agroforests with the supplemental floral resources, species richness was 21% higher overall in the flower-visiting community in these agroforests compared to control agroforests. Coffee agroforests with the steady-state supplemental floral patch also had 31% more butterfly species, 29% more hummingbird species, 65% more wasps and 85% more bees than control coffee agroforests. The experimental treatment, together with elevation, explained 57% of the variation in community structure of the flower-visiting community. The identification of plant species that can support a high number of animal species, including important ecosystem service providers, is becoming increasingly important for restoration and conservation applications. Throughout the Neotropics plant species with a steady-state flowering/fruiting phenology can be found in all aseasonal forests and thus could be widely tested and suitable species used throughout the tropics to manage for biodiversity and potentially ecosystem services involving beneficial arthropods.

Editor: Jordi Moya-Larano, Estacion Experimental de Zonas Áridas (CSIC), Spain

Funding: Funding for this study was provided by the EarthWatch Institute (www.earthwatch.org) and the Andrew W. Mellon Foundation (www.mellon.org). The funders had no role in study design, data collection and analysis, decision to publish, or preparation of the manuscript.

Competing Interests: The author has declared that no competing interests exist.

* E-mail: petersve@miamioh.edu

¤ Current address: Department of Biology and Institute for the Environment and Sustainability, Miami University, Oxford, Ohio, United States of America

Introduction

The relatively low degree of spatiotemporal variation in flower and fruit availability in Neotropical forests has favored the evolution of more diverse communities of frugivores and pollinators compared to the other tropical regions [1–4]. One unique component of the Neotropical flora that contributes to reducing the spatiotemporal patchiness of resources is an Andean-centered radiation of epiphytes, understory shrubs, and palmetto-like monocots [5]. Not only does this group contribute to providing a more abundant and species-rich food resource base in the Neotropics, but also some plant species in this group provide their pollinators or dispersal agents with a year-round food supply. This is accomplished either by a single species through a continual [6] or 'steady-state' [7] flowering/fruiting phenology (hereafter, both terms are used interchangeably, as different authors have presented these terms to describe the same phenology) at the individual or population level [8] or at the guild level with individual species in the guild having a staggered phenology [1]. The overall effect of either strategy is to maintain their animal dispersers and pollinators in residence in the community [9].

The importance of plant species with extended, continuous or steady-state resource production has been most broadly studied for their role in maintaining frugivore populations during times of resource scarcity [10–14] and to a lesser extent for their role (a) in the evolution of a more diverse specialized pollinator group in the Neotropics [4] and (b) in dampening fluctuations in pollinator abundances [15]. The concept of 'bridging plants', which does not include the duration of flowering, has been employed to describe plant species that could potentially be used to restore pollinator communities successfully because they provide nectar and pollen resources during otherwise resource-limited times [16,17]. Only recently, though, have studies begun to identify and experimentally test bridging species [17].

Plant species with continuous resource production at the individual level are rare and most frequently early successional species [18,19]. In addition, some Neotropical plant species either alter resource production or only produce resources in early successional habitats, such as treefall gaps, where resource density is higher than in undisturbed forest [1,11]. Not only do these habitats play a key role in structuring Neotropical communities [20,21] but they also provide critical resources to animal species, especially during times of resource scarcity [11,22]. However, early successional habitat may not be sufficient in landscapes where development and agriculture are the dominant land use, and where natural disturbance regimes are suppressed [23]. Restoring early successional habitat or characteristics of early successional habitats in landscapes where they are lacking due to human impact may contribute to the conservation of biodiversity [24].

The role that the early successional habitat characteristic of steady-state resource production has in supporting Neotropical organisms has not been evaluated. While ecologists working in undisturbed Neotropical forests still struggle to accurately estimate how animal populations naturally fluctuate in response to natural variation in food resources [25], agroecologists must move ahead to experimentally test potential management practices that can decrease the spatiotemporal patchiness of food resources in fragmented landscapes. The community-wide demand for any given resource should be frequency-dependent and inversely related to the number of alternative resources that are simultaneously available [12]. Therefore, the predictable and extended food resources provided by any plant species with a steady-state phenology, regardless of resource quality, should have a community-wide impact, especially in agricultural lands where gaps in food availability are more frequent and of longer duration. Despite this, plant species with a steady-state phenology at the individual level have not been studied for their contribution to biodiversity conservation and structuring animal communities in agricultural lands. To determine whether steady-state resource production is important for species in fragmented landscapes, I conducted a manipulative field study in coffee agroforests. The coffee agroforest provides an ideal framework for experimental manipulation [26] and can be managed for biodiversity conservation [27–30]. I hypothesized that coffee agroforests with intercropped steady-state flowering shrubs would host a greater diversity and abundance of flower-visiting species compared to control coffee agroforests. Specifically, I selected one shrub species, *Hamelia patens* Jacq., to exemplify the steady-state flowering phenology, experimentally planted a supplemental patch of this species in several coffee agroforests, and quantified the response of the flower-visiting community, i.e. hummingbirds, butterflies, bees and wasps.

Materials and Methods

Ethics statement

All research in this study conforms with the legal requirements for field work in Costa Rica and the United States of America. A permit to conduct research in Costa Rica was approved by the Costa Rican Ministerio de Ambiente y Energia (MINAE; Permit Number 014-2010-ACAT). Hummingbird mist-net capture and handling techniques were reviewed and approved by the University of Georgia's Animal Care and Use Committee (A2008 03-061-Y3-A0). All land accessed for this study was privately owned and all six landowners gave permission for this study to be conducted on their land. No listed endangered or protected species were involved in this study.

Study sites

Study sites were located in the Monteverde region of the Puntarenas province of Costa Rica (10°N, 84°W; Fig. S1). Six coffee agroforests that shared similar management regimes and were in close proximity to the University of Georgia's research station in the Upper San Luis Valley (elevation 925–1100 m) were selected. The six agroforests were intercropped with *Musa*, *Citrus*, and *Psidium* spp., had high shade tree diversity (19–23 tree species ha^{-1}) and a high shade canopy cover (62–80% canopy cover). The

herbaceous ground cover was removed via machete, monthly, during all months of the rainy season (May to November), and farmers did not use pesticides. Each agroforest was separated by at least 100 m from another agroforest. Extensive mark-recapture data from the agroforests revealed that most bird species, including hummingbirds, did not move among the agroforests [31,32].

Study species

Hamelia patens Jacq. (Rubiaceae) occurs in secondary growth from Mexico to Bolivia. Demonstrating tolerance of a wide range of environmental conditions, *H. patens* has been recorded from 0 to 2000 m elevation in Costa Rica and its phenology has been documented from lowland wet, lowland dry and cloud forests in Costa Rica [18,33]. In lowland wet forests of Costa Rica where only about 7% of shrub and treelet species exhibit continuous flowering, *H. patens* individuals in secondary growth had flowers throughout the year [18,34]. In these forests each inflorescence produces a total of 0 to 5 open flowers per day at an average rate of 1.5 per day, producing from 30 to more than 100 flowers over its "lifetime" [35]. In Costa Rican cloud forests, 33% of shrub and treelet species were identified as extended flowerers, although *H. patens* was the only species recorded with flowers during all months of the year. In contrast, *H. patens* plants in Costa Rican lowland dry forests flower only during early wet season months [18]. In the landscape where our study was conducted, *H. patens* was the only plant species with flowers and fruit during all months of the year.

Hypothesized selective forces for plant species with the continuous flowering and fruiting strategy are (1) its primary pollen vector, hummingbirds, and (2) its association with early successional habitats, which are ephemeral, because the likelihood of successful colonization will increase with more frequent seed production [19]. In the literature, *H. patens* has been most notably associated with hummingbird pollination [35,36], and its flowers which are odorless, orange, narrow and tubular fit the bird-pollination syndrome. However, pollination syndromes can lead researchers to focus only on floral visitors that conform to the 'correct' pollinator. In reality, though, flowers conforming to a particular syndrome can receive visits from opportunistic insects belonging to different orders that contribute to the fitness of the plant [37]. In fact, *H. patens* individuals across the fragmented landscape where this study was conducted attracted a generalist assemblage of pollinators and floral visitors (Fig. 1), with some moving visibly more pollen than the hummingbird visitors (Fig. 2). However, the aim of this study was not to evaluate pollinator effectiveness, but instead to evaluate the impact of plant species with a continuous flowering strategy on the animal community that uses floral resources in agroecosystems.

Experimental design

In May 2007 fifteen *H. patens* shrubs were experimentally planted within a 400-m² plot in three out of the six coffee agroforests selected for the study. All shrubs were planted via either (1) transplanting shrubs < 1m from nearby roadsides where they were growing wild or (2) flagging seedlings naturally occurring within the coffee farm so that the farmer would not remove them when weeding in between coffee rows. Agroforests with supplementation plots were not selected randomly but instead determined by the farmer's willingness to permit me to supplement their farms with *H. patens* patches. Moran's *I* analyses were used to test for spatial dependence in the response variables among sample locations. No spatial autocorrelation in the response variables after treatments was found (Table 1). Prior observation of *H. patens* plants in the region revealed that individuals located in shaded conditions produced few flowers, and therefore *H. patens* plots were

Figure 1. *Hamelia patens* **attracts a generalist assemblage of insect species in the study area.** (a) Nectar-robbing assassin bug and ant species, (b) *Aphrissa* sp. foraging for nectar, (c) *Calephelis* sp. robbing nectar, (d) Diptera species robbing nectar, (e) *Heliconius* sp. foraging for nectar, (f-h) Wasp species were observed to systematically visit *H. patens* floral ovaries after flowers had fallen off, (i) Coleoptera species inside corolla.

located both close to the middle of the coffee agroforests and in full sun to maximize the number of flowers. Individual *H. patens* plants began producing flowers by August 2007. Floral resource availability (i.e. the number of all open, non-coffee flowers, including *H. patens* flowers) was estimated for all trees, shrubs and herbaceous plants within each 1–2 ha agroforest (Table S1) for

Figure 2. Bees pollinating *Hamelia patens.* (a) *Trigona fulviventris* with collected *H. patens* pollen (b) Halictidae species emerging from *H. patens* corolla, covered in pollen (c) *Ceratina* species, emerging from corolla and covered in pollen (d) *Bombus pullatus* with collected *H. patens* pollen.

Table 1. Moran's I statistics for testing spatial autocorrelation in response variables.

Variable	Observed value	Expected value	Standard deviation	p-value
Wasp richness	−0.016	−0.034	0.032	0.57
Wasp abundance	−0.001	−0.034	0.033	0.31
Butterfly richness	−0.015	−0.043	0.041	0.48
Butterfly abundance	−0.041	−0.043	0.041	0.95
Hummingbird richness	−0.021	−0.008	0.008	0.10
Hummingbird abundance	−0.016	−0.008	0.008	0.30
Bee richness	−0.004	−0.029	0.028	0.37
Bee abundance	−0.004	−0.029	0.027	0.37

each month, except for September, from February to December 2008.

Bird and insect sampling

Hummingbirds were the only bird pollinators in the study area that used floral resources of *H. patens*. Mist-nets were used to quantify hummingbird abundance and species richness in the coffee agroforests. Hummingbirds were captured during three sampling periods from 23 March to 22 May 2009, 14 July to 6 August 2009 and 9 June to 27 July 2010 using 30-mm, 34-mm and 60-mm mesh mist-nets. All coffee agroforests were sampled in each of the three sampling periods. Three mist-nets were placed along windbreaks in each agroforest for 5 to 11 days per sampling period. Nets were opened daily from 700 to 1400 hours except during periods of heavy rain. Hummingbird species richness and abundances were totaled on a daily basis for each agroforest.

Insects were sampled with Malaise traps, with one trap placed near the center of each coffee agroforest per sampling period. Malaise traps were left open for 15 days and sampling was conducted in July 2008, Nov 2008, May 2009, July 2009, Nov 2009, and July 2010. Each agroforest was sampled during each of the six sampling periods. At the end of each sampling period, all insects from the Malaise traps were collected and identified to order in the laboratory. All captured butterflies were identified to species and categorized as either nectar-feeding or fruit-feeding, based on DeVries [38,39]. Among Hymenoptera the wasps were distinguished from all other families, and were classified into morphospecies. Identification to morphospecies-level has been shown to serve as a good proxy in the estimation of species richness [40]. Although identifications to lower taxonomic categories would have been desirable, the large number of insects precluded greater taxonomic precision. Captured bees were identified to species or genus. Because bees are the most effective pollinators of coffee [27], they were analyzed in a separate paper that focused on assessing the potential role of sowing plants with a steady-state flowering phenology into agricultural lands to increase coffee yield [41]. The potential benefit of adding plant species with a steady-state flowering phenology into agricultural lands for (a) biodiversity conservation and (b) the ecosystem service of pollination were evaluated separately, as both are important goals of management in agricultural lands, but management actions will not always synergistically improve both biodiversity and ecosystem services [31,42]. However, because this paper deals with conservation of the flower-visiting community as a whole, some new analyses of bees not included in the previous publication are presented, such as (a) habitat specificity, (b) species evenness and (c) species richness after removing the effect of abundance (see

Statistical analyses section below for details of these analyses), and bees are also included in all analyses where the variable is 'flower-visiting community'.

Statistical analyses

Generalized linear and linear mixed models were used to compare the number of open, non-coffee flowers, open *H. patens* flowers, and hummingbird abundance and species richness between treatments, hereafter H+ for agroforests with supplemental plantings and C for control agroforests. For hummingbirds, models were fit by the Laplace approximation with a Poisson distribution in R version 2.11.1, package 'lme4' [43]. All models included treatment as a fixed effect and site and sampling period within site as random effects. To test for statistical significance of the treatment effect, a likelihood ratio test was performed to compare models with and without the treatment as an explanatory variable. To account for overdispersion in the model for hummingbird abundance it was necessary to add an observation-level random effect.

Generalized linear (GLM) and linear models (LM) were used to test how insect groups responded to the experimental plantings of *H. patens*. Models included site as the explicit error term and sampling period as the within error term. GLM and LM was used for malaise trap data because (a) the data were balanced and (b) the sampling phase (within) error term was larger than the site (source) error term, which causes mixed models to set the site error term to zero thus providing an unreliable estimate of the treatment effect. GLMs were fit with a Poisson distribution, and when LMs were used, data were log transformed to meet the conditions of normality.

Likelihood ratio tests comparing models with and without the linear variables of the X-Y coordinates of the plots were used to assess whether there was any linear dependency in the placement of the experimentally applied treatments. Likelihood ratio test results indicate that adding linear variables to the models did not improve model fit (Table S2). Results were not affected by the inclusion of the X-Y coordinates in the models either (Table S3).

Sample-based rarefaction curves were generated with EstimateS version 8.2.0 [44] to compare butterfly, wasp and hummingbird species richness between the treatments. Curves were calculated from 100 randomizations of sample order, without sample replacement. For butterflies and wasps, abundance-based data from Malaise trap capture were used to estimate species richness and a sample represents one 15-day period. For hummingbirds, incidence-based data from mist-net capture were used to estimate species richness and a sample represents one day. The sample-based rarefaction curves for butterflies and wasps were rescaled to

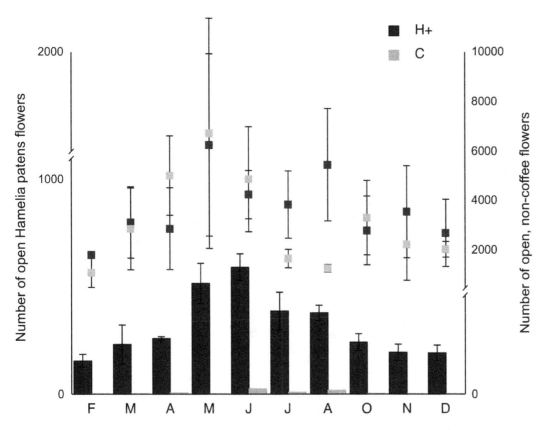

Figure 3. Seasonality of floral resources in coffee agroforests. *Hamelia patens* blooms (first y-axis) and open, non-coffee flowers (second y-axis). H+ represents agroforests with supplemental *H. patens* plantings and C represents control agroforests.

the number of individuals to compare species richness between treatments [45], whereas hummingbird rarefaction curves, scaled by sample, compare species density between treatments. Rarefaction curves are presented with the Mao Tau estimate and 95% confidence intervals so that statistical comparisons can be made between treatments.

To remove the effect of abundance on the estimates of species richness in C and H+ agroforests, sample-based randomizations were used. This procedure shuffled all samples across control and treatment agroforests, retaining the number of observed samples during each randomization, and tests whether or not the observed species richness in C and H+ agroforests could have been obtained by random allocation of samples among treatments. Pseudo F-statistics were calculated from the randomized species richness values and ranked F-statistics were compared to the observed to determine P-values [46]. The same procedure was also carried out to compare the relative abundance of species, or evenness, between C and H+ agroforests. Species evenness was calculated by dividing the Shannon diversity index for each site by the natural logarithm of the site's species richness.

A distance-based RDA (Redundancy Analysis), using the Bray-Curtis distance measure with the capscale function in the R version 2.11.1, package 'vegan', was performed to examine variation in community structure of the entire flower-visiting assemblage, including hummingbirds, butterflies, wasps and bees. The analysis was focused on evaluating whether the relationship between the experimental plantings and community structure was greater than what would be expected by chance, and elevation was included in the model because although the experiment was designed to minimize the effect of confounding environmental

variables other than the treatment, a slight elevation gradient (925–1100 m elevation) existed among the sites.

Finally, habitat specificity was estimated to determine the number of species out of the total species pool that was more frequently associated with either the H+ or the C agroforests. Habitat specificity was calculated using an area unweighted index calculation that divides the number of species in each treatment by the harmonic mean of species abundances, or by the harmonic mean of the total number of samples for which each species is present [47]. Samples were then randomized to test whether the observed specificity among H+ and C sites could have also been expected by a random allocation of samples among treatments. The randomization procedure shuffled all samples across treatments with the constraint that the number of samples randomly allocated to each treatment was the same as the observed number of samples. Observed values of habitat specificity were then compared to the null distribution to determine whether observed values were significantly different than those expected by chance. Unlike the pseudo F-statistic, this test is two-tailed because observed specificity can either be higher or lower than the expected [48].

Results

Flowering phenology

The coefficient of variation (CV) in monthly floral resource availability showed that patchiness in floral resource availability was higher in coffee agroforests without the supplemental patch of *H. patens* (0.88, 0.96 and 1.05) compared to the agroforests with the experimental plantings (0.56, 0.57 and 0.65). All coffee agroforests had >1000 non-coffee flowers available monthly for at least 80% of months for which floral resource availability was estimated. Treatment and control agroforests did not differ in the mean number of open, non-coffee flowers (Likelihood ratio $= 0.34$, $P = 0.56$) but did differ in the number of open *H. patens* flowers (Likelihood ratio $= 13.18$, $P = 0.0002$); (Fig. 3).

Flower-visiting community

A total of 7174 potential flower visitors representing 278 species, including hummingbirds, butterflies, wasps and bees, were captured across all H+ and C agroforests during this study. Almost twice as many individuals were captured in H+ agroforests compared to C agroforests, and after accounting for the difference in abundance, species richness of the flower-visiting community was significantly higher in H+ compared to C agroforests (Table 2). The presence of the *H. patens* supplemental floral resource patch was also found to be a significant factor influencing the composition of the flower-visiting community, with elevation and treatment together explaining 57% of the variation in structure of the flower-visiting community ($P = 0.015$, Fig. 4).

A total of 112 hummingbirds from seven species were captured across both types of coffee agroforests. Both hummingbird abundance and species richness were greater in H+ coffee agroforests compared to C coffee agroforests (Table 3). After accounting for the effect of abundance on richness, the slightly higher hummingbird species richness in H+ agroforests was not statistically significant ($P = 0.06$, Table 2). However, species richness curves scaled by sample period showed higher species richness in H+ coffee agroforests (7 species) compared to C coffee (5 species) agroforests (Fig. 5). The number of open, non-coffee flowers of all species in an agroforest was not a significant variable in the models for either hummingbird richness or abundance (richness: likelihood ratio test $= 3.0$, $P = 0.08$; abundance: likelihood ratio test $= 2.28$, $P = 0.13$).

A total of 942 butterflies from 49 species were captured across both types of coffee agroforests. Overall butterfly richness and richness within the nectar-feeding butterfly guild were both higher in H+ agroforests, but there was no difference between agroforest types for the fruit-feeding butterfly guild (Table 3). Overall butterfly abundance and within both feeding guilds was higher, though not significantly so, in C agroforests, however when we removed the most common nectar-feeding species, the generalist *Anthanassa ardys*, from our data the direction of higher abundance shifted to H+ coffee agroforests. After accounting for the effect of abundance on richness, H+ agroforests were not found to have higher butterfly richness compared to C agroforests (Table 2). However, rarefaction curves with 95% CI showed a significantly higher number of butterfly species overall in H+ coffee agroforests compared to C agroforests (Fig. 5). Butterfly richness and abundance were not related to the number of open, non-coffee flowers of all plant species in an agroforest (overall butterfly richness: $z = -1.0$, $P = 0.32$; nectarivorous butterfly richness: $z = -1.3$, $P = 0.21$; frugivorous butterfly richness: $z = 0.1$, $P = 0.90$; butterfly abundance: $t = -0.6$, $P = 0.57$).

A total of 4862 wasps from 160 morphospecies were captured in Malaise traps in both types of coffee agroforests. Species richness of wasps was higher in H+ coffee agroforests, however this result was marginally statistically significant ($P = 0.055$, Table 3). Sixty-five percent more wasps were captured in H+ coffee agroforests compared to C coffee agroforests (Table 3). After accounting for the effect of abundance on richness, species richness of wasps was statistically significantly higher in H+ agroforests compared to C agroforests (Table 2). Wasp morphospecies richness and abundance were not related to the number of open, non-coffee flowers

Table 2. Total abundance, mean total species richness, species evenness and habitat specificity measures for flower-visiting community in coffee agroforests in Costa Rica.

Response variable	H+[a]	C[b]	F	P
Community[c]				
Abundance	4354	2820		
Species richness	150	124	9.66	0.02
Species evenness	0.78	0.78	0.01	0.93
Habitat specificity	157.7	120.3		*
Hummingbirds				
Abundance	89	23		
Species richness	5	3	4	0.06
Species evenness	0.93	0.59	1.30	0.22
Habitat specificity	4.7	2.3		*
Butterflies				
Abundance	412	530		
Species richness	24	18.3	6.28	0.10
Species evenness	0.73	0.60	12.37	0.10
Habitat specificity	29.9	19.1		*
Wasps				
Abundance	3030	1832		
Species richness	87.3	74	3.23	<0.01
Species evenness	0.74	0.76	0.24	0.54
Habitat specificity	86.6	72.4		
Bees				
Abundance	823	435		
Species richness	34	28.6	6.92	0.17
Species evenness	0.78	0.79	0.24	0.43
Habitat specificity	36.5	26.5		

[a]Coffee agroforests with supplemental *H. patens* patch.
[b]Coffee agroforests without *H. patens*.
[c]Flower- visiting community includes hummingbirds, butterflies, wasps, and bees.
*Denotes that the observed habitat specificity was significantly different than the expected for both H+ and C sites.

of all plant species in an agroforest (richness: $t = 1.3$, $P = 0.22$; abundance: $t = 0.3$, $P = 0.77$).

Bee summary data and results of treatment effects on bees and coffee fruit set were presented in [41]. Pseudo F-tests conducted in this paper, however, remove the confounding effect of abundance on species richness, and show no statistically significant difference in species richness of bees between H+ and C coffee agroforests (Table 2).

Species evenness and habitat specificity

Species evenness was similar for all taxonomic groups between H+ and C coffee agroforests (Table 2), indicating that the observed increases in abundance in H+ agroforests were evenly distributed across species and not just the result of a few dominant species.

Observed habitat specificity was higher in H+ coffee agroforests for the flower-visiting community and for all groups within the flower-visiting community (Table 2), indicating that agroforests with the supplemental patch of steady-state floral resources represented a larger fraction of the total sampled richness from both types of coffee agroforests. This contribution of higher species

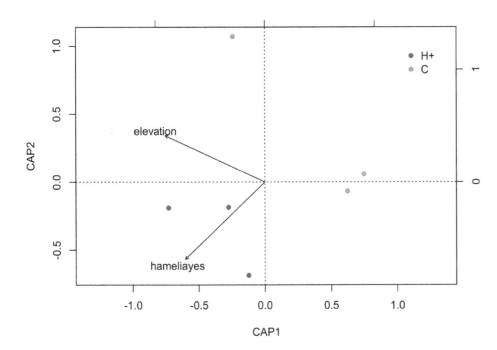

Figure 4. Ordination of the flower-visiting community in coffee agroforests. Distance-based Redundancy Analysis (RDA) of the entire flower-visiting community in coffee agroforests in Costa Rica. H+ represents agroforests with supplemental *H. patens* plantings and C represents control agroforests. The ordination model included elevation and the presence of *H. patens* (represented by hameliayes arrow in graph) supplemental patches ($P=0.015$).

richness to the regional species pool was significantly different than what would be expected by chance for the entire flower-visiting community when analyzed as a whole, and for both humming-birds and butterflies, but not for bees or wasps (Fig. 6).

Table 3. Flower-visiting community in coffee agroforests in Costa Rica, comparing agroforests with and without experimentally planted shrubs with a steady-state flowering phenology.

Response variable	Mean ± SE		Model	
	H+[a]	C[b]		P
Hummingbirds				
Species richness	1.02±0.13	0.32±0.08	Likelihood ratio	0.015
Abundance	1.39±0.26	0.35±0.09	Likelihood ratio	0.006
Butterflies				
Overall species richness	10.67±0.93	8.33±1.02	GLM	0.04
Nectarivore species richness	7.75±0.83	5.42±0.87	GLM	0.01
Frugivore species richness	2.92±0.34	2.92±0.34	GLM	0.97
Overall abundance	34.1±4.4	44.2±8.5	LM	0.57
Nectarivore abundance	15.8±2.4	16.8±5.1	LM	0.41
Frugivore abundance	18.2±3.5	27.3±4.6	LM	0.15
Wasps				
Morphospecies richness	30.9±3.6	27.3±2.8	LM	0.055
Abundance	202±46.0	122.1±26.4	LM	0.045

[a]Coffee agroforests with supplemental *H. patens* patch.
[b]Coffee agroforests without *H. patens*.

Discussion

This study reveals that even when agroecosystems have high plant diversity (approx. 20 tree species ha^{-1}), intercropping with plant species with a steady-state flowering phenology can have an impact on the diversity and structure of the flower-visiting community. All coffee agroforests evaluated during this study had a monthly average of >1600 open, non-coffee flowers, and each had >1000 non-coffee flowers available monthly for at least 80% of months for which floral resource availability was estimated. Despite this, the coffee agroforests with the supplemental patch of steady-state floral resources supported 21% more species overall, and this difference was even greater when examining species affinities for H+ versus C coffee agroforests (32% more species showed an affinity for H+ agroforests). Furthermore, the analysis of community composition highlighted the important role that the experimental treatment had in the structure of the flower-visiting community. For example, only 38% of all butterflies captured in C agroforests were species that feed primarily on nectar whereas almost half (46%) of all butterflies captured in H+ coffee agroforests were nectarivorous. In contrast to the majority of plant species, species with a steady-state flowering strategy produce new flowers every day of the year-consequently offering predictable food resources, in a predictable location, and with a considerable amount of nectar becoming available on a daily basis [e.g. one *H. patens* flower can produce up to 50 µl in 24 h [49]; and individual *H. patens* plants in the supplemental patches in this study yielded 20–100 open flowers per day].

Although most previous work with *H. patens* has focused on its relationship with hummingbirds, diffuse interactions in pollination may be far more prevalent than previously recognized [50]. Two other studies have highlighted a more generalist assemblage of pollinators at *H. patens* [51,52] while others only mention that arthropod visitation does occur [36]. During this study many

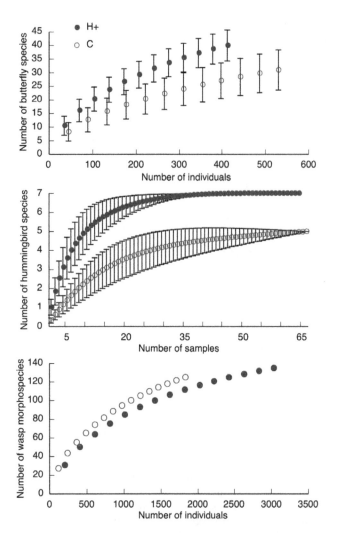

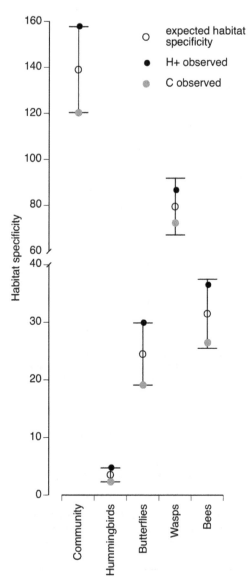

Figure 5. Sample-based rarefaction curves of control and treatment coffee agroforests. Curves compare treatment (H+; supplemental steady-state floral resources) and control (C) coffee agroforests using Mao Tau expected richness in EstimateS. Rarefaction was performed on presence/absence data for hummingbirds (a) and abundance data for butterflies (b) and wasps (c). Curves were rescaled by the number of individuals for butterflies and wasps to compare species richness between agroforest types, and show the mean ± 95% CI. Non-overlapping CI show statistically significant group differences.

Figure 6. Habitat specificity of flower-visiting species in control and treatment coffee agroforests. Expected habitat specificity and 95% CI were obtained from 1000 sample-based randomizations. Observed habitat specificity is an area unweighted index obtained by dividing the number of species in each treatment by the harmonic mean of species abundances. Observed habitat specificity is shown as either significantly higher or lower than expected by chance if the observed value falls on the outer limits of the 95% CI of the null distribution.

arthropod species were observed to forage on nectar and pollen resources from *H. patens* flowers, both legitimately and as robbers, as well as to systematically visit other floral related parts, post-flowering (see Fig. 1f-h). Although I was unable to observe which, if any, resources were obtained during these post-flowering visits, I observed ants, flies, and wasps participating in this behavior. It is likely that these resources also display a steady-state phenology if they are related to floral production. One possibility is that just after flowers fall off, nectar is still available at the style base where the nectaries are located [52] and arthropods with short-tongues can take advantage of these resources. One other potential explanation is that the wasp species are searching for *Proctolaelaps kirmsei*, a flower mite that is monophagous on *H. patens* [35].

Within individuals, extended duration of flowering may be advantageous for spreading the risk of uncertain pollination, or reflect sparse or unpredictable resources in the understory [33]. In addition, reducing the number of flowers per day increases

cross-pollination rates by promoting pollinator movement among plants [3]. Thus, individuals of plants with a continual or steady-state flowering strategy should be widely dispersed to promote the traplining behavior of their pollinators to increase outcrossing [53]. When individuals are clumped, however, pollinators can become territorial instead of traplining [53]. In the study sites I observed territorial behavior by hummingbirds on *H. patens*, and what appeared to be traplining behavior by euglossine bees, although I did not directly study euglossine bees and Malaise traps do not effectively capture euglossine bees. These observations suggest that *H. patens* or similar self-incompatible species [35] could have reduced fitness when planted in agricultural lands as a management action for biodiversity conservation. Although I did not compare fruit set rates among individuals in different land use types, fruit counts in *H. patens* supplemental patches averaged 198±20.1 ripe fruits per day (counted on one day per each month of 2008)- and many bird species were observed foraging on the fruits of *H. patens* in supplemental patches [29]. These observed fruit set rates were comparable to those reported from a more natural area in lowland Costa Rica- an average of 11–15 ripe fruits per infructescence produced over a two-wk period [54]. In this study I was not concerned though with the reproductive fitness of *H. patens*; instead, I wanted to understand the flower-visiting community response to resource predictability and whether this could be an effective management action for biodiversity conservation in agricultural lands.

A somewhat controversial term, keystone species can potentially be identified as those species whose removal is expected to result in the loss of at least half the assemblage considered [55], while another method includes identifying those plant species with low consumer specificity [12]. In this study, if the flower-visiting assemblage represents the entire consumer assemblage, then the potential keystone role of *H. patens* can be evaluated. I did not conduct a removal experiment, but species richness of the entire flower-visiting community was only 21% higher in the experimental plots, far less than half of the assemblage. However, casual observation revealed that specificity of flower-visitors was extremely low: just in the study sites I observed at least 80 species of arthropods and 9 hummingbird species foraging on floral resources and 15 bird species foraging on fruit resources. The identification of keystone or strong interactor species has long held appeal for restoration and conservation applications [55]. Menz and colleagues [17] suggest that the highest priority of plant species for restoration includes species that support a large number of pollinator species. This is especially true where the amount of space available for conservation actions is constrained, such as in small reserves or agroecosystems. If strong interactor plant species were preferentially planted in agroecosystems to enhance their conservation value, then perhaps fewer plants and plant species would be needed to obtain a conservation benefit. This management action would be more likely to be implemented by farmers reluctant to dedicate land to non-crop species. Identification of strong interactor species has proven difficult, however, in the Neotropics a good starting point is with the few plant species that have a steady-state flowering/fruiting phenology. The following are examples of plant families that are likely to have some shrub species with continual or steady-state phenology in the Neotropics: for both flowers and fruit resources- Rubiaceae (In Bawa et al. [34] four of 15 of the continually flowering species belonged to the Rubiaceae), Onagraceae (*Fuchsia* spp.), Melastomataceae, Annonaceae; for flower resources only- Verbenaceae (*Stachytarpheta* spp.), Lamiaceae. Future studies should aim to evaluate other species that produce steady-state food resources in

different aseasonal environments to test whether the flowering/fruiting phenology has broad applicability for biodiversity conservation.

Finally, a recent meta-analysis aimed at understanding why agri-environmental measures vary in their effectiveness for pollinators found that measures were more effective when they were implemented in structurally simple versus cleared or complex landscapes, and when they increased contrast in floral resource availability compared to surrounding lands [56]. This study was conducted in the Monteverde region of Costa Rica, an important conservation area, and thus the landscape is structurally complex with high floral resource availability in the matrix surrounding the coffee agroforests. Therefore, the results of this study suggest that in more simplified landscapes and in landscapes where agriculture is more intensive, intercropping with plant species with a steady-state flowering phenology could be an even more effective management action for conservation of the flower-visiting community.

Supporting Information

Figure S1 Map of the study area. White circles depict coffee agroforests that received the treatment of a supplemental patch of steady-state floral resources and black circles depict control farms.

Table S1 List of all plant species occurring in the coffee agroforests, excluding herbaceous ground cover species.

Table S2 AIC scores and likelihood ratio tests comparing models for significance of X,Y coordinates.

Table S3 Model results for treatment effect, after adding X,Y coordinates, and site included as a random variable.

Acknowledgments

The farmers, Oldemar Salazar, Alvaro Vega, Gilberth Lobo, Rafael Leiton, Odilio Ramirez and Olivier Garro, graciously allowed me to conduct this experiment and collect data within lands that provided their livelihoods. Many EarthWatch volunteers helped collect data. T.O. Crist provided help with data analysis. T.O. Crist, K.U. Campbell and two anonymous reviewers provided useful comments on a previous version of this manuscript.

Author Contributions

Conceived and designed the experiments: VEP. Performed the experiments: VEP. Analyzed the data: VEP. Contributed reagents/materials/analysis tools: VEP. Wrote the paper: VEP.

References

1. Stiles G (1975) Ecology, flowering phenology and hummingbird pollination of some Costa Rican *Heliconia* species. Ecology 56: 285–301.
2. Fleming TH, Breitwisch R, Whitesides GH (1987) Patterns of tropical vertebrate frugivore diversity. Annu Rev Ecol Syst 18: 91–109.
3. Sakai S (2001) Phenological diversity in tropical forests. Popul Ecol 43: 77–86.
4. Fleming TH, Muchhala N (2008) Nectar-feeding bird and bat niches in two worlds: pantropical comparisons of vertebrate pollination systems. J Biogeogr 35: 764–780.
5. Gentry AH (1982) Neotropical floristic diversity: phytogeographical connections between Central and South America, Pleistocene climatic fluctuations, or an accident of the Andean orogeny? Ann Mo Bot Gar 69: 557–593.
6. Newstrom LE, Frankie GW, Baker HG (1994) A new classification for plant phenology based on flowering patterns in lowland tropical rain forest trees at La Selva, Costa Rica. Biotropica 26: 141–159.

7. Gentry AH (1974) Flowering phenology and diversity in tropical Bignoniaceae. Biotropica 6: 64–68.
8. Sakai S (2000) Reproductive phenology of gingers in a lowland dipterocarp forest in Borneo. J Trop Ecol 16: 337–354.
9. Baker HG (1963) Evolutionary mechanisms in pollination biology. Science 139: 877–883.
10. Terborgh J (1986) Keystone plant resources. In: Soule ME, ed. Conservation biology: the science of scarcity and diversity. Sinauer, Sunderland, Massachusetts. pp 330–344.
11. Levey DJ (1990) Habitat-dependent fruiting behaviour of an understorey tree, *Miconia centrodesma*, and tropical treefall gaps as keystone habitats for frugivores in Costa Rica. J Trop Ecol 6: 409–420.
12. Peres CA (2000) Identifying keystone plant resources in tropical forests: the case of gums from *Parkia* pods. J Trop Ecol 16: 287–317.

13. Kinnaird MF, O'Brien TG (2005) Fast foods of the forest: the influence of figs on primates and hornbills across Wallace's line. In: Dew JL, Boubli JP, eds. Tropical fruits and frugivores: the search for strong interactors. Springer, The Netherlands. pp 37–57.

14. Stevenson P (2005) Potential keystone plant species for the frugivore community at Tinigua Park, Colombia. In: Dew JL, Boubli JP, eds. Tropical fruits and frugivores: the search for strong interactors. Springer, The Netherlands. pp 155–184.

15. Ackerman JD (1983) Diversity and seasonality of male euglossine bees (Hymenoptera: Apidae) in central Panama. Ecology 64: 274–283.

16. Dixon KW (2009) Pollination and restoration. Science 325: 571–573.

17. Menz MHM, Phillips RD, Winfree R, Kremen C, Aizen MA, Johnson SD, Dixon KW (2011) Reconnecting plants and pollinators: challenges in the restoration of pollination mutualisms. Trends Plant Sci 16: 4–12.

18. Opler PA, Frankie GW, Baker HG (1980) Comparative phenological studies of treelet and shrub species in tropical wet and dry forests in the lowlands of Costa Rica. J Ecol 68: 167–88.

19. Kang H, Bawa KS (2003) Effects of successional status, habit, sexual systems, and pollinators on flowering patterns in tropical rain forest trees. Am J Bot 90: 865–876.

20. Schnitzer SA, Carson WP (2001) Treefall gaps and the maintenance of species diversity in a tropical forest. Ecology 82: 913–919.

21. Dalling JW, Muller-Landau HC, Wright SJ, Hubbell SP (2002) Role of dispersal in the recruitment limitation of neotropical pioneer species. J Ecol 90: 714–727.

22. Wunderle JM, Willig MR, Henriques LMP (2005) Avian distribution in treefall gaps and understorey of *terra firme* forest in the lowland Amazon. Ibis 147: 109–129.

23. Russell KN, Ikerd H, Droege S (2005) The potential conservation value of unmowed powerline strips for native bees. Biol Conserv 124: 133–148.

24. Hopwood JL (2008) The contribution of roadside grassland restoration to native bee conservation. Biol Conserv 141: 2632–2640.

25. Milton K, Giacalone J, Wright SJ, Stockmayer G (2005) Do frugivore population fluctuations reflect fruit production? Evidence from Panama. In: Dew JL, Boubli JP, eds. Tropical fruits and frugivores: the search for strong interactors. Springer, The Netherlands. pp 5–35.

26. Greenberg R, Perfecto I, Philpott SM (2008) Agroforests as model systems for tropical ecology. Ecology 89: 913–914.

27. Klein AM, Steffan-Dewenter I, Tscharntke T (2003) Fruit set of highland coffee increases with the diversity of pollinating bees. Proc R Soc Lond B 270: 955–961.

28. Perfecto I, Vandermeer J, Mas A, Soto Pinto L (2006) Biodiversity, yield and shade coffee certification. Ecol Econ 54: 435–446.

29. Peters VE, Mordecai R, Carroll CR, Cooper RJ, Greenberg R (2010) Bird community response to fruit energy. J Anim Ecol 79: 824–835.

30. Philpott SM, Bichier (2011) Effects of shade tree removal on birds in coffee agroecosystems in Chiapas, Mexico. Agri, Ecosyst, Environ

31. Peters VE, Greenberg R (2013) Fruit supplementation affects birds but not arthropod predation by birds in Costa Rican agroforestry systems. Biotropica 45: 102–110.

32. Hernandez SM, Mattson BJ, Peters VE, Cooper RJ, Carroll CR (2013) Coffee agroforests remain beneficial for Neotropical bird community conservation across seasons. PLOSone 8: e65101.

33. Koptur S, Haber WA, Frankie GW, Baker HG (1988) Phenological studies of shrub and treelet species in tropical cloud forests of Costa Rica. J Trop Ecol 4: 323–346.

34. Bawa KS, Kang H, Grayum MH (2003) Relationships among time, frequency, and duration of flowering in tropical rain forest trees. Am J Bot 90: 877–887.

35. Collwell RK (1995) Effects of nectar consumption by the hummingbird flower mite, *Proctolaelaps kirmsei* on nectar availability in *Hamelia patens*. Biotropica 27: 206–217.

36. Lasso E, Naranjo ME (2003) Effect of pollinators and nectar robbers on nectar production and pollen deposition in *Hamelia patens* (Rubiaceae). Biotropica 35: 57–66.

37. Johnson SD, Steiner KE (2000) Generalization versus specialization in plant pollination systems. Trends Ecol Evol 15: 140–143.

38. DeVries PJ (1987) The butterflies of Costa Rica and their natural history. Princeton University Press. Princeton, NJ.

39. DeVries PJ (1997) The butterflies of Costa Rica and their natural history, Vol. II: Riodinidae. Princeton University Press. Princeton, NJ.

40. Oliver I, Beattie AJ (1996) Invertebrate morphospecies as surrogates for species: a case study. Conserv Biol 10: 99–109.

41. Peters VE, Carroll CR, Cooper RJ, Greenberg R, Solis M (2013) The contribution of plant species with a steady-state flowering phenology to native bee conservation and bee pollination services. Insect Conserv Diver 6: 45–56.

42. Power AG (2010) Ecosystem services and agriculture: tradeoffs and synergies. Phil Trans R Soc B 365: 2959–2971.

43. R Development Core Team (2008) R: A Language and Environment for Statistical Computing. R Foundation for Statistical Computing, Vienna, Austria.

44. Colwell RK (2009) EstimateS: Statistical estimation of species richness and shared species from samples. Version 8.2. User's Guide and application published at: http://purl.oclc.org/estimates.

45. Gotelli NJ, Colwell RK (2001) Quantifying biodiversity: procedures and pitfalls in the measurement and comparison of species richness. Ecol Lett 4: 379–391.

46. Anderson MJ, Crist TO, Chase JM, Vellend M, Inouye BD, et al. (2011) Navigating the multiple meanings of β diversity: a roadmap for the practicing ecologist. Ecol Lett 14: 19–28.

47. Halvorsen R, Edvardsen A (2009) The concept of habitat specificity revisited. Landscape Ecol 24: 851–861.

48. Diekötter T, Crist TO (2013) Quantifying habitat-specific contributions to insect diversity in agricultural mosaic landscapes. Insect Conserv Diver DOI: 10.1111/icad.12015

49. Feinsinger P (1978) Ecological interactions between plants and hummingbirds in a successional tropical community. Ecol Monogr 48: 269–287.

50. Borrell BJ (2005) Long tongues and loose niches: evolution of euglossine bees and their nectar flowers. Biotropica 37: 664–669.

51. Thomas CD, Lackie PM, Brisco MJ, Hepper DN (1986) Interactions between hummingbirds and butterflies at a *Hamelia patens* bush. Biotropica 18: 161–165.

52. Chauhan S, Galetto L (2009) Reproductive biology of the *Hamelia patens* Jacq. (Rubiaceae) in Northern India. J Plant Reprod Biol 1: 63–71.

53. Newstrom LE, Frankie GW, Baker HG, Colwell RK (1994) Diversity of long-term flowering patterns. In: McDade LA, Bawa KS, Hespenheide HA, Hartshorn GS, eds. La Selva: Ecology and natural history of a neotropical rain forest. The University of Chicago Press, Chicago. pp 142–160.

54. Levey DJ (1987) Facultative ripening in *Hamelia patens* (Rubiaceae): effects of fruit removal and rotting. Oecologia 74: 203–208.

55. Mills LS, Soule ME, Doak DF (1993) The keystone-species concept in ecology and conservation. BioScience 43: 219–224.

56. Scheper J, Holzschuh A, Kuussaari M, Potts SG, Rundlof M, et al. (2013) Environmental factors driving the effectiveness of European agri-environmental measures in mitigating pollinator loss- a meta-analysis. Ecol Lett 16: 912–920.

13

Spatial Characteristics of Tree Diameter Distributions in a Temperate Old-Growth Forest

Chunyu Zhang[1][*][9], **Yanbo Wei**[1][9], **Xiuhai Zhao**[1], **Klaus von Gadow**[2]

1 Key Laboratory for Forest Resources and Ecosystem Processes of Beijing, Beijing Forestry University, Beijing, China, **2** Faculty of Forestry and Forest Ecology, Georg-August-University Göttingen, Göttingen, Germany

Abstract

This contribution identifies spatial characteristics of tree diameter in a temperate forest in north-eastern China, based on a fully censused observational study area covering 500×600 m. Mark correlation analysis with three null hypothesis models was used to determine departure from expectations at different neighborhood distances. Tree positions are clumped at all investigated scales in all 37 studied species, while the diameters of most species are spatially negatively correlated, especially at short distances. Interestingly, all three cases showing short-distance attraction of *dbh* marks are associated with light-demanding shrub species. The short-distance attraction of *dbh* marks indicates spatially aggregated cohorts of stems of similar size. The percentage of species showing significant *dbh* suppression peaked at a 4 m distance under the heterogeneous Poisson model. At scales exceeding the peak distance, the percentage of species showing significant *dbh* suppression decreases sharply with increasing distances. The evidence from this large observational study shows that some of the variation of the spatial characteristics of tree diameters is related variations of topography and soil chemistry. However, an obvious interpretation of this result is still lacking. Thus, removing competitors surrounding the target trees is an effective way to avoid neighboring competition effects reducing the growth of valuable target trees in forest management practice.

Editor: Gil Bohrer, The Ohio State University, United States of America

Funding: This research is supported by the National Special Research Program for Forestry Welfare of China (200904022), the 12th five-year National Science and Technology plan of China (2012BAC01B03), and the National Natural Science Foundation of China (31200315). The funders had no role in study design, data collection and analysis, decision to publish, or preparation of the manuscript.

Competing Interests: The authors have declared that no competing interests exist.

* E-mail: zcy_0520@163.com

[9] These authors contributed equally to this work.

Introduction

A forest is composed of a set of trees, which are characterized by their locations and sizes. Tree diameter distributions can provide the necessary information about tree sizes, ignoring tree locations. Tree diameters are, however, associated with tree positions and growth is sensitive to spatial interaction between trees [1–2] as well as local habitat characteristics [3]. Continuous functions of the spatial coordinates including tree size have been implemented in a method known as Mark Correlation analysis, in which tree locations and diameters are regarded as a realization of a marked point stochastic process [4–7].

Spatial dependence of topography, drainage and soil characteristics can create different spatial structures of a forest community [8]. Most processes affecting forest trees occur at short neighborhood distances, such as seed dispersal, some pollination and competition for light and nutrients [9–11]. Forest soils provide nutrition and moisture for tree growth and the spatial heterogeneity of soil nutrients will thus affect the spatial structure of tree diameter distributions at particular spatial scales. Specific null models, such as a random labeling model and a heterogeneous Poisson model, are required to factor out the effects of habitat associations at varying spatial scales.

The theory of marked point processes provides a formal framework for analysis of spatial characteristics of tree diameter

distributions, in which the points indicate tree locations, and the marks might denote particular tree characteristics, such as diameter at breast height (*dbh*), tree height and growth during a given time span [6,12–13]. Assessment of the proportion of species exhibiting aggregation, regularity or randomness of *dbh* at different spatial scales provides important insights into the spatial structures of temperate communities. Studies investigating neighborhood effects indicated that direct plant–plant interactions worked strongly at local plant neighborhoods but faded away at larger scales in tropical forests [14–16] and temperate forests [7,17]. We therefore assume departures from expectations at local neighborhood distances while assuming randomness at larger distances.

These concepts were applied to data from a fully censused 30ha old-growth forest plot, in which the tree locations were mapped and the soil chemistry and topography were available spatially explicit. The objectives of this study are (*i*) to ascertain the spatial characteristics of species-specific *dbh*'s at various distances, and (*ii*) to assess the particular effects of habitat heterogeneity.

Materials and Methods

Ethics statement

All necessary permits were obtained for the field studies. The study was approved by the Ethics Committee of *Jiaohe* Administration Bureau of the Forest Experimental Zone in Jilin province, in Northeastern China.

Field sample

The observations for this study were obtained in a fully censused, unmanaged old-growth forest plot (OGF), covering 30 ha (500×600 m). The OGF plot was established in a temperate mixed broadleaf-conifer forest in the summer of 2010, and is located at 43°57.928′ ~ 43°58.214′N, 127°45.287′ ~127°45.790′E. The OGF study area is situated in a protected locality at *Jiaohe* Administration Bureau of the Forest Experimental Zone, far away from villages, where human disturbance has been virtually unknown. The elevation of the OGF plot ranges from 576 to 784 m above sea level and the topography is characterized by a valley between two slopes. Altogether 37 tree species with more than 20 individuals are included in this analysis (Table S1).

Point pattern analysis

Habitat heterogeneity and plant interactions may cause increasing variation of the local point densities. Accordingly, *Ripley*'s $L(r)$ functions were used to analyze the spatial distribution of point positions [18–19]. To correct edge effects, each count was weighted by the inverse of the proportion of the circle that falls within the study plot [18,20]. $L(r)$ is defined as follows:

$$L(r) = \sqrt{K(r)/\pi} - r \qquad (1)$$

where, $K(t)$ is *Ripley*'s univariate K-function [21], and r is the distance category.

Marked point pattern and null hypothesis models

The spatial correlations of tree diameters were determined using the mark correlation function (MCF). The mark correlation function takes account of a quantitative characteristic (such as tree breast height diameter in this study) which is associated with tree locations, and then calculates the spatial correlation of these marks in the observed point pattern. The spatial correlation of marks in the marked point process is characterized by a test function $k[mm](r)$ for two marks M_i and M_j. This test function describes the correlation between the marks M_i and M_j at different points which are a distance r apart. According to Stoyan and Stoyan [22], for a point process X with numeric marks (*dbh*), the mark correlation function $k[mm](r)$ is defined as:

$$k[mm](r) = E[0u](M(0) * M(u))/E(M * M') \qquad (2)$$

where, $E[0u]$ denotes the conditional expectation given that there are points of the process at the locations 0 and u separated by a distance r, and where $M(0), M(u)$ denote the marks attached to these two points. M, M' are random marks drawn independently from the marginal distribution of marks, and E is the expectation.

The implemented edge correction method is *Ripley's* isotropic correction [22]. Three null hypothesis models were used to compute the confidence intervals. When the observed $k[mm](r)$ was greater than, equal to, or less than the 99% confidence interval calculated from the predicted $k[mm](r)$ from 99 realizations of the adopted model, the *dbh* values of the focal species were assumed to be positive, independent or negative correlation, respectively.

One of the adopted models is a homogeneous Poisson model (denoted by *HomP*). In the homogeneous Poisson process, the assumption is that the spatial location of a given point (tree) is independent of any other points (trees). Thus, *HomP* was used to examine the effect of a pure random process. This process implies

identical average tree density per unit area for one given focal species. The locations of the focal species were randomized using 99 realizations of the homogeneous Poisson process. Then, the observed values of the marks (*dbhs*) were randomly assigned to the completely random locations.

The second model that was used in this study is a random labeling model (*RLM*) which assumes that environmental conditions play a decisive role in the spatial distribution of the trees. The observed values of the marks (*dbhs*) were assigned randomly to the positions of the focal species, while the positions were kept unchanged. Then, 99 random labeling realizations were generated to obtain confidence envelopes.

In addition, a heterogeneous Poisson model (*HetP*) was used to examine the effect of habitat heterogeneity. The *HetP* process assumes that the density of each tree species is associated with the specific habitat conditions in the study area. Thus, the *dbh* values are randomized among trees, while the tree positions are simulated and distributed in accordance with the intensity of the focal species. The pixel image of the tree density of the focal species was estimated using a Gaussian kernel. 99 realizations of the inhomogeneous Poisson process with intensity equal to the pixel values of the image were generated to obtain confidence envelopes.

Significant departure from the above three null models was determined using the lowest and highest value of the 99 simulations of the null models to generate 99% simulation envelopes.

Soil chemical properties

Soil sampling included lattice sampling and random sampling to assess the spatial variation of soil chemical properties at different spatial scales (for details see Figure S1). A total of 540 sampling position was available in the 30ha plot. The soil samples were taken from the upper (0–10 cm), middle (10–20 cm) and lower layer (>20 cm) to analyze total nitrogen (using the Kelvin Digestion method), total potassium (fusion with NaOH method), organic matter (exterior heating potassium dichromate-density method) and soil pH in August of 2009.

A semivariogram, which indicates spatial correlations in observations measured at sample locations, was used to fit a model of the spatial correlations of the soil chemical properties. Box-Cox transformation was applied for above soil chemical variables to meet the normality assumptions required in the semivariogram model [23] (Table S2). Several theoretical models were used to simulate the spatial variance of soil chemical properties (Table S3). On the basis of the optimizing models, kriging interpolation was used to estimate the total nitrogen, total potassium, organic matter and pH value for each tree position. Pearson's correlation analysis was used to measure the relationships between tree diameter and soil chemical properties.

Diameter differentiation and topography

The 30-ha study area was subdivided into 750 cells, each covering 20×20 m (400 m^2). Then the coefficient of variation of tree diameters ($CVd = dbh$ standard deviation/mean *dbh*) was calculated for all trees with a *dbh* greater or equal to 1 cm in each cell. The spatial distribution of CVds is shown in Figure S2.

To quantify topography, the heights at the intersections of the cell grid lines, called "nodes", were measured. The elevation of a particular cell was calculated as the mean of the elevations of its four corner nodes. The cell slope for each of the five cell sizes was estimated as the mean angular deviation from the horizontal plane of each of the four triangular planes which were formed by connecting three of its adjacent corners [24]. The convexity of a cell was calculated as the elevation of the focal cell minus the mean elevation of the eight surrounding cells [25]. For the edge cells, convexity was taken as the elevation of the center point minus the mean of the four corners. Positive and negative convexity values respectively indicate convex (ridge) and concave (valley) land surfaces. The aspect of a cell can be obtained from the average angle of the four triangular planes that deviate from the north direction. Pearson's correlation coefficient was used to test the correlations between CVd and four topographical factors.

All calculations were performed using the "spatstat" and "geoR" packages in the comprehensive R environment [26]. The main code sections used in our analyses are included in the attached Text S1.

Results

Spatial characteristics of diameter distributions

To analyze the spatial characteristics of the diameter distributions of each of the 37 tree species, the heterogeneous Poisson process ($HetP$) was used, involving three steps. First, for each of the tree species, the pixel image of the tree density was estimated using a Gaussian kernel. Figure 1 is an example of the kernel smoothed intensity map of the point distribution for the species $Syringa$ $reticulata$ var. $amurensis$. The tree locations were then simulated and distributed in accordance with the intensity of the focal species. Finally, the dbh values were randomly assigned to the simulated tree locations.

The spatial characteristics of the tree locations were then evaluated using the complete spatial randomness model while their dbh marks were analysed using the $HomP$, RLM and $HetP$ models. The results for the species $Syringa$ $reticulata$ var. $amurensis$ show, for example, that tree positions were significantly aggregated at distances between 0 and 50 m. The dbh's, however, are spatially negatively correlated at much smaller distances: between 0 and 7 m using the $HomP$ model; between 0 and 11 m using the RLM model and between 0 and 10 m using the $HetP$ model (Figure 2).

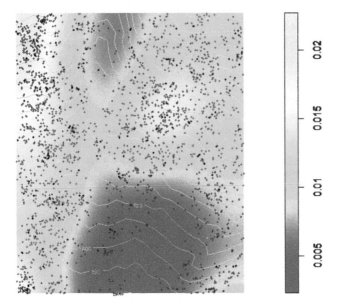

Figure 1. Example of a kernel smoothed intensity map showing the point pattern. The map colours show the intensities (number of trees per m²) of $Syringa$ $reticulata$ var. $amurensis$ and the elevation contours at 10-m intervals within the 30-ha study area. The unit of the axes is meters.

Interestingly, the results of the three models are very similar for the example species. However, this is not always the case. For example, the results differ greatly in $Betula$ $platyphylla$.

Scale-dependent distribution of species and diameters

The spatial point patterns of each species were analysed using $Ripley's$ L-function to test departure from complete spatial randomness. The positions of all 37 species showed significant aggregation, even at greater distances up to 50 m (Table 1).

For each species, the spatial distribution of tree diameters was analysed separately (see Table 1). For most species, the results of the $HomP$ model, which only simulates random points without changing dbh's, are significantly different from those of the RLM and $HetP$ models. In the RLM approach, the observed values of the marks are assigned randomly to tree positions (which are kept unchanged), assuming that the spatial distribution is influenced by habitat. The $HetP$ model is also designed to take into account the effect of habitat heterogeneity by randomizing the dbh values, while the tree positions are simulated in accordance with the intensity of the focal species. These basic differences between $HomP$ one the one hand and RLM and $HetP$ on the other hand, explains the different results in Table 1.

While tree positions are clumped at all studied scales, the diameters associated with them are significantly negatively correlated, especially at short distances, for most species under the RLM (24 of 37 species) and $HetP$ model (25 of 37 species). At relatively small distances, the mark correlation function for three species, $Cerasus$ $maximowiczii$, $Euonymus$ $macropterus$ and $Acanthopanax$ $senticosus$, shows significant positive correlation. Positive correlations of dbh marks may reflect the effects of historical gap formation of these shrub species. Figure 3 further highlights the distance-dependent effects of the spatial distribution of tree dbh's by presenting the proportions of positive, independent and negative effects at distances of up to 50 m for all species combined. The observational area is very large and it would be possible to evaluate distances up to 200 m. However, distances exceeding 50 m are not considered to be relevant.

Negative correlation indicates inhibition and suppression. For all tree species, the percentage of species showing significant dbh suppression peaks at a distance of 4 m, and decreases sharply with increasing distances between 4 to 13 m distances under the $HetP$ model. In contrast, the percentage of species showing dbh independence increases with increasing distance (Figure 3).

Effects of soil chemistry and topography

Total nitrogen, total potassium, organic matter and the soil pH value of three soil layers showed significant spatial heterogeneity in the study area. Table S3 shows the semivariogram models of soil chemical properties in the study area while Figure 4 presents the semivariogram maps of the soil chemical properties which present the basis for kriging and the spatial estimates of total nitrogen, total potassium, organic matter and pH value for each tree position. Figure S3 shows the contour plots for soil chemical properties, based on the kriging models in Figure S4. There are some significant relationships between tree diameters and specific soil properties (Table S4). However, the underlying causes for these correlations are not known.

Pearson's correlation analysis was used to quantify the relationships between the coefficient of variation of tree diameters (CVd) and four topographical variables in the 20 m×20 m cells. Figure 4 shows that CVd is significantly associated with elevation and slope but not with aspect and convexity.

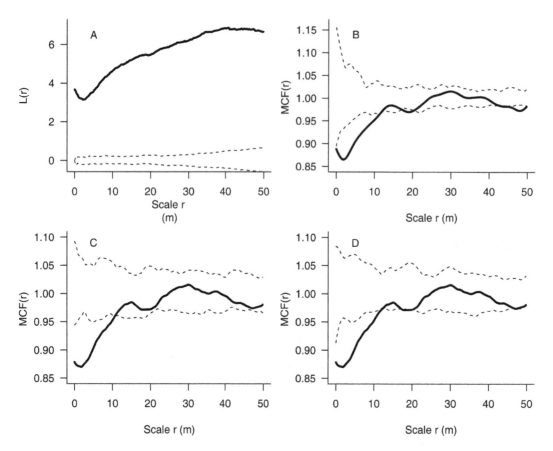

Figure 2. Exemplary results for *Syringa reticulata* var. *amurensis* to illustrate the analysis. Diagram a presents the spatial distribution pattern using the *L* function. Diagrams b, c and d show the spatial characteristics of the diameter distribution evaluated by the mark correlation function. The significance of a and b was tested by the homogeneous Poisson model, c by the random labeling model, and d by the heterogeneous Poisson model. Dashed lines indicate the confidence envelopes, while solid lines indicate the *L* or *MCF* values calculated from the observations. When the solid line was below the lower envelope, inside both envelopes, or above the upper envelope, the pattern was assumed to be significantly regular, spatially random, or significantly aggregated in the *L*-function analysis. Correspondingly, a solid line above the upper envelope, inside both envelopes, or below the lower envelope, indicates significant positive, independent, or negative correlation of *dbh* marks in the *MCF* analysis.

Discussion

Species aggregation and diameter suppression

Previous studies in subtropical and tropical forests have reported that individual tree species tend to be spatially aggregated and the degree of aggregation is scale dependent [27–31]. This has also been observed in temperate forests [32–33]. The spatial aggregation of tree species is usually thought to be caused by seed dispersal limitation [34] and environmental heterogeneity [32,35–36]. In this study, all 37 species were found to be spatially clumped at all investigated scales from 0 to 50 metres. This result is not entirely unexpected if one assumes that the regeneration of seedlings is clustered in the vicinity of parent trees. Only long-term observation can reveal whether such clustering will be sustained over several tree generations. Habitat heterogeneity may also cause spatial aggregation, at least for particular species that show specific habitat preferences.

Neighborhood competition is an important characteristic of stand structure which can be defined as an interaction between neighboring individuals for territory or specific resources. The available resources are usually concentrated on an impact zone surrounding the trees. The bigger the tree, the greater the impact zone. Thus, the competition between two trees depends on their sizes and the distance between them. The "competition effect" between any two trees will increase with increasing size of the

competitor and decrease with increasing tree-to-tree distance. Schlather et al. [37] tested the negative interactions in a spruce stand in the Fichtelgebirge, in northern Bavaria. He found a competitive radius of up to 6 m when using stem diameters as marks. A neighborhood competition effect up to a range of about 6 m was also found when using upper crown surface areas as marks in two deciduous and two coniferous forests in central Germany [38]. Our study also identified mostly independence of *dbh*'s at greater distances, but highly significant negative correlation at short distances.

All significant short-distance attraction of *dbh*'s was surprisingly associated with the light-demanding shrub species (for example, *Euonymus macropterus*, *Acanthopanax senticosus* and *Cerasus maximowiczii*). An obvious explanation for the spatially aggregated cohorts of stems of similar size would be that after establishment of a canopy gap, the open area is rapidly colonised by these light-demanding shrubs. These gap colonisation cohorts are likely to have similar diameters. However, the question remains why the tree size-clumping mainly occurs at short distances. Large-scale fires do not occur in the area and large scale wind damage is also unknown. Our study was carried out in an old-growth forest, in which most trees in the canopy layer are more than 150 years old. The old-growth forest has a mix of tree ages due to a distinct regeneration pattern. The uneven-aged structure often indicates that the forest

Table 1. Spatial characteristics of tree locations and tree *dbhs* at 0–50 m distances.

Species name	Light shade	Canopy undergrowth	L function at 0–50 m	Marked point pattern at different distances		
				Homogeneous Poisson model (*HomP*)	Random labeling model (*RLM*)	Heterogeneous Poisson model (*HetP*)
Betula platyphylla	L	C	0–50 (+)	0–50 (r)	0–50 (r)	0–6 (−)
Acer mandshuricum	L	C	0–50 (+)	0–50 (−)	0–32 (−)	0–42,47–50 (−)
Syringa reticulata var. amurensis	L	U	0–50 (+)	0–7 (−)	0–11 (−)	0–10 (−)
Euonymus macropterus	L	U	0–50 (+)	20–28 (+)	0–16,20–29 (+)	0–15,20–30,39–46 (+)
Padus racemosa	L	U	0–50 (+)	8–26,41–50 (−)	0–26 (−)	0–27 (−)
Abies nephrolepis	S	C	0–50 (+)	0–50 (r)	0–50 (r)	0–50 (r)
Ulmus davidiana var. japonica	L	C	0–50 (+)	0–50 (r)	0–3 (−)	0–8 (−)
Acanthopanax senticosus	S	U	0–50 (+)	0–50 (r)	0–5 (+),35–50 (−)	0–15 (+),35–50 (−)
Acer barbinerve	S	U	0–50 (+)	22–29,32–34,42–50 (−)	44–50 (−)	43–50 (−)
Ulmus macrocarpa	L	C	0–50 (+)	0–50 (r)	0–15 (−)	0–15 (−)
Philadelphus schrenkii	M	U	0–50 (+)	0–50 (r)	0–50 (r)	0–50 (r)
Betula costata	L	C	0–50 (+)	0–5 (−)	0–7 (−)	0–7 (−)
Betula dahurica	L	C	0–50 (+)	0–50 (r)	0–50 (r)	0–2 (−)
Cerasus maximowiczii	S	U	0–50 (+)	0–22 (+)	0–19 (+)	0–20 (+)
Pinus koraiensis	L	C	0–50 (+)	41–50 (−)	0–5,17–22,41–50 (−)	0–5,15–25,35–50 (−)
Juglans mandshurica	L	C	0–50 (+)	0–2 (−)	0–6 (−)	0–6 (−)
Acer ukurunduense	L	U	0–50 (+)	0–50 (r)	0–50 (r)	0–50 (r)
Sorbus pohuashanensis	L	U	0–50 (+)	0–50 (r)	0–10 (−)	0–12 (−)
Fraxinus rhynchophylla	L	C	0–50 (+)	0–50 (r)	36–42 (−)	0–50 (r)
Phellodendron amurense	L	C	0–50 (+)	48–50 (+)	0–50 (r)	0–50 (r)
Lonicera praeflorens	M	U	0–50 (+)	0–50 (r)	0–50 (r)	0–50 (r)
Lonicera maackii	L	U	0–50 (+)	39–50 (+)	0–14 (−)	0–17 (−)
Tilia mandshurica	L	C	0–50 (+)	0–50 (r)	18–45 (−)	18–45 (−)
Ulmus laciniata	L	C	0–50 (+)	0–50 (−)	0–50 (−)	0–50 (−)
Euonymus pauciflorus	L	U	0–50 (+)	5–14,19–26 (+)	0–26 (−)	0–26 (−)
Aralia elata	M	U	0–50 (+)	0–50 (r)	0–50 (r)	0–50 (r)
Corylus mandshurica	S	U	0–50 (+)	15–50 (−)	17–26,48–50 (−)	17–50 (−)
Quercus mongolica	L	C	0–50 (+)	0–50 (r)	0–50 (r)	0–50 (r)
Carpinus cordata	L	C	0–50 (+)	0–50 (−)	0–41 (−)	0–45 (−)
Acer tegmentosum	S	U	0–50 (+)	0–50 (r)	0–4 (−)	0–4 (−)
Acer mono	L	C	0–50 (+)	0–24 (−)	0–19 (−)	0–24 (−)
Abies holophylla	S	C	0–50 (+)	3–8 (−)	0–12 (−)	0–12 (−)
Rhamnus davurica	S	U	0–50 (+)	0–50 (r)	0–50 (r)	0–50 (r)
Fraxinus mandshurica	L	C	0–50 (+)	0–50 (r)	0–4 (−)	0–4 (−)
Sorbus alnifolia	M	C	0–50 (+)	0–50 (r)	0–50 (r)	0–50 (r)

Table 1. Cont.

Species name	Light shade	Canopy undergrowth	L function at 0–50 m	Marked point pattern at different distances		
				Homogeneous Poisson model (*HomP*)	Random labeling model (*RLM*)	Heterogeneous Poisson model (*HetP*)
Populus koreana	L	C	0–50 (+)	0–50 (r)	36–46 (−)	4–7,36–46 (−)
Tilia amurensis	L	C	0–50 (+)	0–50 (r)	3–11 (−)	3–11 (−)

Note: L means light; S means shade and M means middle. C means canopy trees and U means understory.
Spatial point patterns were tested for randomness using the *L*-function. The spatial characteristics of tree *dbhs* were analyzed by the homogeneous Poisson (*HomP*), random labeling (*RLM*) and heterogeneous Poisson (*HetP*). Spatial distances at which tree locations show significant aggregation, regularity and randomness are indicated by the symbols "+" in parenthesis in *L*-function. Spatial distances at which tree locations and tree *dbh* marks show significant positive, independent and negative correlation are indicated by the symbols "+", "r" and "−" in parenthesis, respectively.

represents a relatively stable ecosystem in the long term. Canopy gaps are essential in creating and maintaining mixed-age stands.

Research concerning the forest gap structure in this particular study area showed that almost all gaps have areas less than 1000 m² [39]. A circular gap which occupies 1000 m², has a radius of 17.8 m. Therefore, a likely explanation for the short distance clumping of *dbh*'s is that gaps are small, i.e. caused by the death and decay of large trees.

Effects of soil chemistry and topography

Environmental variables such as ground cover vegetation, light conditions, microclimate, soil characteristics and ecological history play an important role in the spatial distribution of tree diameter at the study site. Topography and soil chemistry may significantly affect the spatial distribution of particular tree species. Valencia et al. [35], for example, found that tree aggregation patterns in tropical forests could be related to specific topographic features. Zhang et al. [40] could identify similar species-habitat associations in a 660×320 m temperate forest in north-eastern China. This study was based on the assumption that habitats are spatially autocorrelated and that species are spatially aggregated due to

seed dispersal limitations. The relationships between species richness and topographic variables were found to be scale-dependent, while the great majority of the species showed distinct habitat-dependence in that study.

Regarding the effects of soil chemistry, John et al. [41] compared distribution maps of 10 essential plant nutrients in the soils to species maps of all trees to test plant-soil associations in three neotropical forest plots. They found that the spatial distributions of 36–51% of the tree species showed strong associations with soil nutrient distributions. Zhang et al. [42] determined the effects of different soil chemical variables on the variation of tree sizes in three study areas of uniform topography, each covering an area of 5.2ha and representing a specific forest developmental stage in the Changbai Mountain region. The results showed that over 14 percent of the spatial variation of tree diameters could be explained by soil chemistry in two secondary forests, and only 4.2 percent in the virgin forest.

These studies have shown effects of soil chemistry and topography on the tree distribution, but several questions regarding the effects on the spatial variation of tree diameters remain unanswered. The results of this study show that there are

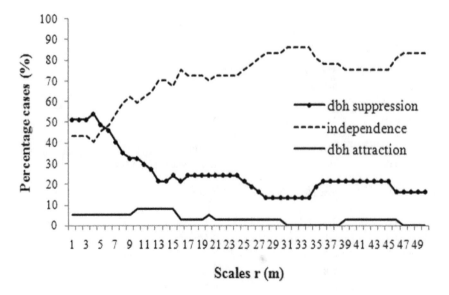

Figure 3. Proportion of species exhibiting significant departures from randomness. Diagram presents the proportion of species showing significant suppression of *dbh* marks (lines with solid circles), attraction of *dbh* marks (solid lines) and independence of *dbh* marks (dashed lines) under the heterogeneous Poisson model. For each distance, the three values add up to 100.

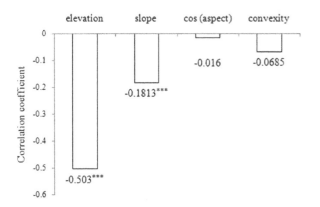

Figure 4. Pearson's correlations between the coefficient of variation of tree diameters and topographical variables. *** indicated a significance at the 0.001 level.

some significant relationships between tree diameters and specific soil properties (Table S4). For example, total N in the upper soil layer is positively correlated with the *dbh*'s of canopy species like *Acer mandshuricum*, *Ulmus macrocarpa*, *Philadelphus schrenkii*, *Juglans mandshurica*, *Acer ukurunduense*, *Acanthopanax senticosus*, *Fraxinus rhynchophylla* and *Carpinus cordata*. The correlations are highly significant and positive in the case of the last-named three species and negative in the others. There are also highly significant correlations between total K in the upper soil layer and the diameters of *Acer mandshuricum*, *Syringa reticulata* var. *amurensis*, *Acer barbinerve*, *Ulmus macrocarpa*, *Betula costata*, *Cerasus maximowiczii*, *Pinus koraiensis*, *Juglans mandshurica*, *Acer ukurunduense*, *Carpinus cordata*, *Acer tegmentosum*, *Acer mono*, *Sorbus alnifolia* and *Populus koreana*. Negative correlations between the pH value in the lower soil layer and tree diameters, which are also highly significant, are also found in *Syringa reticulata* var. *amurensis*, *Euonymus macropterus*, *Padus racemosa*, *Acer barbinerve*, *Pinus koraiensis*, *Carpinus cordata* and *Ulmus macrocarpa*.

The small-scale variations of soil chemical variables may have some influence on the spatial distribution of tree diameters. Suppression of *dbh*'s at close range may be caused by specific topographical and soil properties. However, these results are difficult to interpret. A young tree growing on a nutrient rich spot in a shaded and suppressed position will be small when compared with an old tree growing in a less favorable location. Tree diameters are related to tree age and past competition effects. Past interactions between neighboring trees would have influenced tree diameters. Furthermore, different species have significantly different growth rates. For example, light demanding pioneer species usually grows faster than shade-tolerant species if conditions are favorable. These physiological differences have an effect on tree diameters. For this reason, the correlations between *dbh*'s and soil chemical variables may be meaningless. Thus, the spatial characteristics of tree diameters may be interpreted by the variations of topography and soil chemical variables to some degree, but the underlying causes are not always obvious. Soil chemistry might give some species an advantage over others, but small-scale neighboring competition and historical gap formation are likely to have a significant influence.

Conclusion

The major objective of this study was to improve our knowledge about the spatial correlations between tree diameters. Based on a large dataset of mapped trees, our research has shown that the diameters of most species are negatively correlated in space. The

percentage of species showing significant *dbh* suppression peaked at short distances especially. In the beginning of the 20th century, the original virgin forest has been subjected to large-scale industrial logging, and then replaced by a secondary forests and plantations [43]. During the past 60 years, the protection of these unique ecosystems has been ensured and is still a matter of great concern to ecologist and local foresters.

However, it may be possible that selective harvest in continuous cover forest management systems [44] will be introduced in the future. In that case, our research may provide some guidance to local management. This study has shown that neighboring competition effects which may reduce the growth of valuable target trees, can be avoided if competing individuals are removed within a radius of less 10 m around the target individuals. However, this result could be merely a local effect, involving the particular species community in the observational study area and further research in other ecosystems is required to obtain a broader understanding of competition effects in different communities.

Supporting Information

Figure S1 Soil sampling map in the old-growth forest plot. Grids are 40 m×40 m and 40 m×20 m. The intersections of 40 m×40 m grid lines are regarded as base points (Black points). Based on theses base points, we randomly selected one of eight directions to sample at 2 m and 8 m, or 2 m and 15 m or 8 m and 15 m from base points. Red points are 2-m extra points, blue points are 8-m extra points and green points are 15-m extra points. Total 540 sample points were determined in the research plot.

Figure S2 Spatial distributions of size differentiation and topographic variables. The shading from light to dark shows an increase from low (0.6537) to high (2.0699) size differentiation, from low (577.8) to high (780.7) elevation (m), from low (2.4) to high (45.7) slope (degrees), from low (72.8) to high (299.9) aspect (degrees) and from low (−6.8) to high (8.6) convexity.

Figure S3 The image and contour plots for soil chemical properties.

Figure S4 Semi-variogram maps of soil chemical properties.

Table S1 The list of growth forms and the number of individuals for examined species in the 30-ha OGF plot.

Table S2 Data transform based on parameter estimation for the Box-Cox transformation.

Table S3 Semivariogram models of soil chemical properties in old-growth forest plot.

Table S4 Person correlation coefficients between tree diameter and soil chemical properties.

Text S1 The main codes used in our analyses.

Author Contributions

Conceived and designed the experiments: XZ. Performed the experiments: CZ KG. Analyzed the data: CZ KG. Contributed reagents/materials/analysis tools: CZ. Wrote the paper: CZ YW KG XZ.

References

1. Shi H, Zhang L (2003) Local Analysis of Tree Competition and Growth. For Sci 49: 938–955.
2. Kunstler G, Albert CH, Courbaud B, Lavergne S, Thuiller W, et al. (2011) Effects of competition on tree radial-growth vary in importance but not in intensity along climatic gradients. J Ecol 99: 300–312.
3. Kariuki M, Rolfe M, Smith RGB, Vanclay JK, Kooyman RM (2006) Diameter growth performance varies with species functional-group and habitat characteristics in subtropical rainforests. For Ecol Manag 225: 1–14.
4. Stoyan D, Kendall WS, Mecke J (1987) Stochastic Geometry and its Applications. Hoboken, NJ: Wiley.
5. König D, Carvajal-Gonzalez S, Downs AM, Vassy J, Rigaut JP (1991) Modelling and analysis of 3-D arrangements of particles by point processes with examples of application to biological data obtained by con-focal scanning light microscopy. J Microsc 161: 405–433.
6. Parrott L, Lange H (2004) Use of interactive forest growth simulation to characterise spatial stand structure. For Ecol Manage 194: 29–47.
7. Schlather M, Ribeiro PJ Jr, Diggle PJ (2004) Detecting dependence between marks and locations of marked point processes. J R Stat Soc B 66: 79–93.
8. Lejeune O, Tlidi M (1999) A model for the explanation of tiger bush vegetation stripes. J Veg Sci 10: 201–208.
9. Baack EJ (2005) To succeed globally, disperse locally: effects of local pollen and seed dispersal on tetraploid establishment. Heredity 94: 538–546.
10. Ewel JJ, Mazzarino MJ (2008) Competition from below for light and nutrients shifts productivity among tropical species. PNAS 105:18836–18841.
11. Niggemann M, Wiegand T, Robledo-Arnuncio JJ, Bialozyt R (2012) Marked point pattern analysis on genetic paternity data for uncertainty assessment of pollen dispersal kernels. J Ecol 100: 264–276.
12. Lancaster J, Downes BJ (2004). Spatial point pattern analysis of available and exploited resources. Ecography 27: 94–102.
13. Lancaster J (2006) Using neutral landscapes to identify patterns of aggregation across resource patches. Ecograpy 29: 385–395.
14. Stoll P, Newbery DM (2005) Evidence of species-specific neighborhood effects in the *Dipterocarpaceae* of a Bornean rain forest. Ecology 86:3048–3062.
15. Peters HA (2003) Neighbour-regulated mortality: the influence of positive and negative density dependence on tree populations in species-rich tropical forests. Ecol Lett 6:757–765.
16. Uriarte M, Condit R, Canham CD, Hubbell SP (2004) A spatially explicit model of sapling growth in a tropical forest: does the identity of neighbours matter? J Ecol 92:348–360.
17. Getzin S, Wiegand K, Schumacher J, Gougeon FA (2008) Scale-dependent competition at the stand level assessed from crown areas. For Ecol Manage 255: 2478–2485.
18. Ripley BD (1976) The second-order analysis of stationary point processes. J Appl Prob 13: 255–266.
19. Diggle PJ (1983) Statistical Analysis of Spatial Point Patterns. New York: Academic Press.
20. Haase P (1995) Spatial pattern analysis in ecology based on Ripley's K-function: Introduction and methods of edge correction. J Veg Sci 6: 575–582.
21. Besag J (1977) Contribution to the discussion of Dr Ripley's paper. J Roy Stat Soc (Ser B) 39: 193–195.
22. Stoyan D, Stoyan H (1994) Fractals, random shapes and point fields: methods of geometrical statistics. Hoboken, NJ: John Wiley and Sons.
23. Box GEP, Cox DR (1964) An analysis of transformations. JRSS B 26:211–246.
24. Harms KE, Condit R, Hubbell SP, Foster RB (2001) Habitat associations of trees and shrubs in a 50-ha neotropical forest plot. J Ecol 89: 947–959.
25. Yamakura T, Kanzaki M, Ohkubo T, Ogino K, Chai EOK, et al. (1995) Topography of a large-scale research plot established within a tropical rain forest at Lambir, Sarawak. Tropics 5: 41–56.
26. R Core Team (2012). R: A language and environment for statistical computing. R Foundation for Statistical Computing, Vienna, Austria. ISBN 3-900051-07-0. Available: http://www.R-project.org/. Accessed 2013 Feb 25.
27. Hubbell SP (1979) Tree dispersion, abundance and diversity in a dry tropical forest. Science 203:1299–1309.
28. He F, Legendre P, LaFrankie JV (1997) Distribution patterns of tree species in a Malaysia tropical rain forest. Journal of Vegetation Science (Washington, DC) 8: 105–114.
29. Condit R, Ashton PS, Baker P, Bunyavejchewin S, Gunatilleke S, et al. (2000) Spatial patterns in the distribution of tropical tree species. Science 288:1414–1418.
30. Plotkin JB, Potts MD, Leslie N, Manokaran N, LaFrankie J, et al. (2000) Species-area curves, spatial aggregation, and habitat specialization in tropical forests. J Theor Biol 207:81–99.
31. Zhang Z, Hu G, Zhu J, Ni J (2012) Aggregated spatial distributions of species in a subtropical karst forest, southwestern China. J Plant Ecol. doi:10.1093/jpe/rts027.
32. Zhang C, Gao L, Zhao X (2009) Spatial structures in a secondary forest in Changbai Mountains, Northeast China. *Allgemeine Forst Und Jagdzeitung* 180:45–55.
33. Wang X, Ye J, Li B, Zhang J, Lin F, et al. (2010) Spatial distribusions of species in an old-growth temperate forest, north-eastern China. Can J For Res 40: 1011–1019.
34. Seidler TG, Plotkin JB (2006) Seed dispersal and spatial pattern in tropical trees. PLoS Biol 4(11): e344. doi:10.1371/journal.pbio.0040344.
35. Valencia R, Foster RB, Villa G, Condit R, Svenning J, et al. (2004) Tree species distributions and local habitat variation in the Amazon: large forest plot in eastern Ecuador. J Ecol 92: 214–229.
36. Réjou-Méchain M, Flores O, Bourland N, Doucet J-L, Fétéké RF, et al. (2011) Spatial aggregation of tropical trees at multiple spatial scales. J Ecol 99: 1373–1381.
37. Schlather M, Ribeiro PJ Jr, Diggle PJ (2004) Detecting dependence between marks and locations of marked point processes. J R Stat Soc B 66, 79–93.
38. Getzin S, Wiegand K, Schumacher J, Gougeon FA (2008) Scale-dependent competition at the stand level assessed from crown areas. Forest Ecol Manag 255: 2478–2485.
39. Zhao X, Zhang C, Zheng J (2005) Correlation between gap structure and tree diversity of mixed-broadleaved Korean pine forests in northeast China. Chin J Appl Ecol 16: 2236–2240.
40. Zhang C, Zhao Y, Zhao X, Gadow K (2012) Species-habitat associations in a northern temperate forest in China. Silva Fenn.In press.
41. John R, Dalling JW, Harms KE, Yavitt JB, Stallard RF, et al. (2007) Soil nutrients influence spatial distributions of tropical tree species. PNAS 104: 864–869.
42. Zhang C, Zhao X, Gadow K (2010) Partitioning temperate plant community structure at different scales. Acta Oecol 36: 306–313.
43. Chen D, Zhou X, Zhu N (1994) Natural Secondary Forest-Structure, Function, Dynamic, and Management. Harbin, China: Northeast Forestry University Press.
44. Pukkala T, Gadow K (2012) Continuous cover forestry. Managing forest ecosystems, Volume 23. Berlin: Springer Science+Business Media B.V.

14

Coffee Agroforests Remain Beneficial for Neotropical Bird Community Conservation across Seasons

Sonia M. Hernandez[1][*][¤a][◑], Brady J. Mattsson[2][*][¤b][◑], Valerie E. Peters[1][¤c], Robert J. Cooper[2], C. Ron Carroll[1]

1 Odum School of Ecology, University of Georgia, Athens, Georgia, United States of America, 2 Daniel B. Warnell School of Forestry and Natural Resources, University of Georgia, Athens, Georgia, United States of America

Abstract

Coffee agroforestry systems and secondary forests have been shown to support similar bird communities but comparing these habitat types are challenged by potential biases due to differences in detectability between habitats. Furthermore, seasonal dynamics may influence bird communities differently in different habitat types and therefore seasonal effects should be considered in comparisons. To address these issues, we incorporated seasonal effects and factors potentially affecting bird detectability into models to compare avian community composition and dynamics between coffee agroforests and secondary forest fragments. In particular, we modeled community composition and community dynamics of bird functional groups based on habitat type (coffee agroforest vs. secondary forest) and season while accounting for variation in capture probability (i.e. detectability). The models we used estimated capture probability to be similar between habitat types for each dietary guild, but omnivores had a lower capture probability than frugivores and insectivores. Although apparent species richness was higher in coffee agroforest than secondary forest, model results indicated that omnivores and insectivores were more common in secondary forest when accounting for heterogeneity in capture probability. Our results largely support the notion that shade-coffee can serve as a surrogate habitat for secondary forest with respect to avian communities. Small coffee agroforests embedded within the typical tropical countryside matrix of secondary forest patches and small-scale agriculture, therefore, may host avian communities that resemble those of surrounding secondary forest, and may serve as viable corridors linking patches of forest within these landscapes. This information is an important step toward effective landscape-scale conservation in Neotropical agricultural landscapes.

Editor: Martin Krkosek, University of Toronto, Canada

Funding: The authors thank the staff at the UGA San Luis Research Station, and the UGA Costa Rica office for logistical and financial support. This work was funded by the National Science Foundation Graduate Research Fellowship (Hernandez). The funders had no role in the study design, data collection and analysis, decision to publish or preparation of the manuscript.

Competing Interests: The authors have declared that no competing interests exist.

* E-mail: shernz@uga.edu (SMH); brady.mattsson@gmail.com (BJM)

◑ These authors contributed equally to this work.

¤a Current address: Daniel B. Warnell School of Forestry and Natural Resources and the Southeastern Cooperative Wildlife Disease Study, University of Georgia, Athens, Georgia, United States of America
¤b Current address: Department of Integrative Biology and Biodiversity Research, University of Natural Resources and Life Sciences, Vienna, Austria
¤c Current address: Department of Zoology and the Institute for Environment and Sustainability, Miami University, Oxford, Ohio, United States of America

Introduction

Habitat loss and fragmentation continue to be the leading threats to biodiversity in the Neotropics [1]. Previous conservation efforts have focused on protecting large tracts of undisturbed forest, a solution that is often not possible or feasible. In response, the term "gardenification" was coined to emphasize a philosophical shift in conservation away from total reliance upon unpopulated preserves and towards encouraging people to appropriately manage their "gardens", or small parcels of privately owned land, to improve biodiversity [2]. In contrast with more traditional conservation approaches that emphasize protection of uninhabited landscapes, conserving ecological integrity within human-dominated landscapes can be achieved through focusing on biodiversity benefits to humanity, such as an 'ecosystem services' approach or the use of innovative markets for specialized products, e.g. shade-grown coffee [3]. As an extension of gardenification, sustainable agriculture has been forwarded as a

means to ensure that private land, while providing economic support for people, sustains biodiversity and buffers protected areas [4]. For example, there is evidence that coffee plantations managed under a floristically and structurally diverse tree canopy provide important habitat for diverse avian communities [5,6,7]. Thus, shade-grown coffee may be viewed as a surrogate for forest habitats in the tropics by maintaining avian communities that reflect those found in large tracts of forest (henceforth, habitat-surrogate hypothesis).

However, some authors have cautioned that the conservation value of shade-coffee may be overstated and that although avian biodiversity may be conserved in these systems, bird assemblages are not identical to those in secondary forests [8,9]. Furthermore, most coffee studies have not sufficiently quantified observer bias or detectability differences between habitats, making abundance estimates problematic [8,10]. Many of the pioneering studies assessing differences between coffee agroforestry systems and secondary forests compared the two habitat types using indices of species diversity (e.g. Shannon diversity index) but did not present habitat-specific community compositions or consider factors affecting detectability or capture probability [11,12]. Furthermore, as tropical avian community dynamics are highly dependent on insect and plant phenology, studies comparing shade-coffee and secondary forests should span multiple seasons. Thus, a more informative approach to determine the suitability of shade-coffee for protecting forest bird species would be to describe and compare the dynamics of avian communities between shade-coffee and forest with respect to the changing environmental conditions across seasons [13,14] while incorporating the factors influencing detectability of species such as breeding phenology, foraging preferences and habitat vegetation structure [15,16].

The objective of our study was to estimate and compare avian community composition and dynamics between coffee agroforests and secondary forests while accounting for seasonal dynamics and differences in detectability between habitats and among species functional groups. We chose to model avian community composition with respect to functional groups (e.g. dietary guilds) to reflect the ecosystem services provided by birds in agroforestry systems including pest control by insectivores and seed dispersal by frugivores, and such an ecosystem-services approach has been promoted for biodiversity conservation in human-dominated landscapes [3].

We predicted that turnover in the bird species community would differ by habitat type, because shade-coffee plantations can have a higher temporal patchiness in food resource availability compared to forests, and birds are highly mobile species that track resources [17,18]. For example, some food resources in coffee plantations are dependent upon the annual crop cycle and relate to tilling, pruning of trees or farmer selection in planting of crop/shade tree species that affect insect and fruit phenology. Neotropical secondary forests, in comparison, have a higher plant species richness and therefore higher spatio-temporal predictability of food resource availability [19]. Based on these habitat type comparisons, we predict that managed shade-coffee plantations, when compared to secondary forest, provide a more dynamic source of food resources for birds across seasons, leading to lower patch persistence by birds between seasons.

Materials and Methods

Study Area

Our study took place in the San Luis valley, located in the northwest region of Costa Rica approximately 7 km from the town of Santa Elena in the Monteverde region, in the province of Puntarenas. In the Monteverde region life zones change dramatically with only small changes in altitude, such that six life zones occur within a 600 m elevational range [20]. In addition to altitude, topographic position and exposure to trade winds and trade-wind driven clouds and rain shape a variety of microclimates in the Monteverde region [21]. Mean annual precipitation in the Monteverde region is estimated at 2500 mm; however, precipitation measurements underestimate moisture from wind-driven clouds. Two distinct weather seasons are defined in the region: (1) *wet season*- May to November; mornings are clear, yet cloud accumulation throughout the day results in rain in the afternoon and evening, and (2) *dry season*- December to April; begins with very strong trade winds, clouds and wind-driven rain giving way to moderate winds, clear skies and wind-driven mist, particularly at night and early morning [21]. The San Luis valley is located on the Pacific, leeward side of the volcanic Cordillera de Tilarán mountain range, and therefore the valley experiences a more well-defined, i.e. longer and drier, dry season than the broader

Monteverde region [21]. Despite the long dry season, due to its high elevation, the forests of the San Luis valley are primarily evergreen (<10% of the canopy is leafless during the dry season), and have a moderate epiphyte diversity and abundance. The understory in these forests is fairly open with few shrubs and tree saplings [20].

Seven sites in the upper San Luis valley (10°16′N, 84°47′W; 950–1100 m elevation) were selected to represent the two habitat types (shade-coffee and secondary forest) (Fig. 1). Four study sites were privately owned, shade-grown coffee plantations of approximately 1–2 ha, and the other three sites were located within large patches (>5 ha) of secondary forest. All forest patches used in the study were approximately 50–75 yrs old. All coffee sites share a similar history, with coffee mainly being sun grown in the region during the 60's and 70's, but with the promotion of conservation-friendly agriculture in the region, most coffee in the San Luis valley now incorporates many shade trees. All coffee sites shared the following management practices: a moderate canopy cover [percent canopy cover range: 31–60%; mean tree height: 7.9 m (range: 6.9–9.6 m); and mean distance between trees: 7.3 m (range: 6.7–8.3 m)], a high species richness of shade trees (mean: 21 species ha^{-1}), no pesticide use, and mechanical removal of weeds via machete. All secondary forest sites also shared similar vegetation structural characteristics, with percent canopy cover ranging from 87–91%, a mean tree height of 24.2 m (range: 16.6–31.0 m), a mean distance between trees of 2.6 m (range: 1.3–4 m), and a mean of 59 tree species ha^{-1}. All study sites were separated by at least 250 m (max distance: 1.5 km), and all shade coffee sites were located within an agricultural matrix typical of those found throughout the neotropics, that included small forest patches, monoculture windbreaks (*Croton niveus* Jacq. or *Montanoa guatemalensis* B. L. Rob), and cattle pasture.

Avian capture events and seasons

Birds were sampled in study sites across four pre-defined seasons from 2005–2008. Pre-defined seasons represented combinations of two biologically relevant factors: climatic cycle (i.e. alternating wet and dry seasons) and reproductive activity (peak breeding and non-peak breeding seasons; Table 1). The annual cycle of seasons was defined in the following chronological sequence: (1) dry, non-peak breeding (December–January); (2) dry, peak breeding

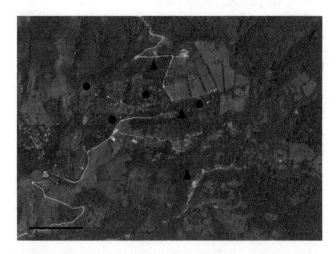

Figure 1. Study area as viewed from an airplane in the San Luis Valley near the Monteverde Reserve in Costa Rica. Shade coffee sites are indicated by circles, and secondary forest sites are indicated by triangles. Black bar indicates approximately 1 km.

Table 1. Annual avian capture bouts during four seasons in seven sites near San Luis, Costa Rica, 2005–2008.

	Dry						Wet					
	Non-peak breeding			Peak breeding			Peak breeding			Non-peak breeding		
Sites	2005	2006–7	2007–8	2005	2006	2007	2005	2006	2007	2005	2006	2007
Shade coffee												
Gilberth	×			×		×	×	×	×	×		
Joel	×	×		×			×	×		×		
Alvaro	×	×		×		×	×	×		×		
Vargas				×			×					
Secondary forest												
Zapote	×		×	×			×	×	×	×		
Nenes	×		×	×		×	×	×		×		
Pena	×		×	×		×	×	×	×	×		

The annual cycle of seasons follow a chronological sequence: (1) dry, non-peak breeding (December–January); (2) dry, peak breeding (February–April); (3) wet, peak breeding (May–August); and (4) wet, non-peak breeding (September–November). An × in a cell indicates that a particular site was sampled during a particular season-year combination.

(February–April); (3) wet, peak breeding (May–August) and (4) wet, non-peak breeding (September–November). These seasons distinguish the biologically relevant time periods for birds in this region [22]. For modeling purposes migratory birds were assumed to be absent during season 3 (i.e. wet, peak breeding).

Within each season, birds were sampled across multiple days in a site using mist-nets, and following a standard methodology [23]. Eight to fourteen nylon mesh mist-nets (38-mm mesh, 2 m high × either 6 or 9 m in length) were placed within a 1-ha area of each site, and at least 20 m from any bordering habitat type to ensure that captured birds were representative of the habitat type in which they were captured. The number of nets placed per site was multiplied by each net's length to ensure that total net length per site was equal. During each capture day, nets were opened starting between 0530 and 0700 h, and nets remained open until 1300 to 1400 h, depending on weather conditions (i.e. mist-nets were closed during periods of heavy rain, or when wind speed exceeded 20 kph). During each season, mist-nets were placed in two sites at a time for 4- to 7-d periods, and capture days were alternated between the two sites on consecutive days to prevent net shyness (i.e. decrease in daily capture probability) [24,25]. Efforts were made to sample all seven sites during each sampling period, although this was not always feasible. Captured birds were extracted from mist-nets at 30-min intervals, and birds were processed immediately to minimize holding time before they were released near the point of capture. All birds >10 g were held for biological sample collection as part of a concurrent study [26] and were processed within 20 min of extraction from the nets. Birds were identified to species according to the available field guides [27,28]. Following published species accounts, all bird species captured were classified according to the following life history attributes: migratory behavior (migratory, non-migratory), foraging strata preference (ground/understory or middle/upper canopy), and diet (frugivore, omnivore, insectivore; Table 2) [27,28]. All avian capture and handling techniques were reviewed and approved by the University of Georgia's Animal Care and Use Committee (A2008 03-061-Y3-A0).

Modeling approach

Seasonal composition and interseasonal dynamics of avian communities in shade-coffee and in secondary forest were examined using a Bayesian implementation of a multi-species, multi-scale occupancy model [29] and a multi-species dynamic occupancy model [30,31], (referred to as "multi-scale" and "dynamic" models from here forward). These occupancy modeling frameworks provided estimates of parameters allowing us to compare avian community composition and dynamics between habitat types. The multi-scale model framework provided estimates of (1) occupancy, defined as the probability that a patch is occupied by any species in a particular functional group at least once during the annual cycle; and (2) seasonal use, defined as the probability that any species of a particular functional group occurs in a patch during a season given that it occupies that patch at least once during the annual cycle. The dynamic model by contrast provided estimates of (1) initial use, defined as the probability of a species occupying a given patch during an initial season unconditional on occupancy throughout the annual cycle; (2) interseasonal patch extinction, defined as the probability that a species using a given patch does not use that patch the following season; and (3) interseasonal patch colonization, defined as the probability that a species uses a patch that it did not use in the previous season. In the dynamic model, interseasonal transitions (i.e. patch extinction and colonization, henceforth 'turnover') were estimated between each successive season in the annual cycle: 1–2, 2–3, 3–4, 4–1. This required an altered parameterization from the original dynamic model described by [30]. Specifically, each of the four transitions was included in the model, rather than just the first three, which would be standard for a chronological sequence of time periods (e.g. years). As such, this dynamic model becomes a cyclical dynamic occupancy model.

Both multi-scale and dynamic models account for biases induced by variation in seasonal capture probability to evaluate differences in avian community composition and dynamics. Seasonal capture probability, pooled across individual sampling days within a season, was estimated instead of daily capture probability because daily capture rates were too low for estimating daily capture probabilities and occupancy metrics. The submodel for seasonal capture probability for both the dynamic and multi-scale occupancy models included the following factors: (1) species by season interaction, (2) habitat type, (3) diet guild, (4) migratory status, (5) foraging strata preference, and (6) habitat-type by diet-guild interaction. The submodel for occupancy within the multi-

Table 2. Total number birds captured in shade coffee and secondary forest sites near San Luis, Costa Rica, 2005–2008.

Common name	Scientific name	Migratory status	Diet	Foraging	Forest	Coffee
Alder Flycatcher	Empidonax alnorum	M	O	Y	0	3
Baltimore Oriole	Icterus galbula	M	O	Y	0	4
Barred Antshrike	Thamnophilus doliatus	R	I	Y	0	2
Northern Barred-Woodcreeper	Dendrocolaptes sanctithomae	R	I	Y	6	0
Black-and-white Warbler	Mniotilta varia	M	I	Y	1	3
Black-headed Nightingale-Thrush	Catharus mexicanus	R	O	Y	1	0
Blue-crowned Motmot	Momotus momota	R	O	N	16	26
Blue-gray Tanager	Thraupis episcopus	R	O	N	0	5
Boat-billed Flycatcher	Megarynchus pitangua	R	O	N	0	1
Brown Jay	Psilorhinus morio	R	O	N	0	1
Buff-throated Saltator	Saltator maximus	R	O	Y	1	11
Canada Warbler	Cardellina canadensis	M	I	N	1	2
Chiriqui Quail-Dove	Geotrygon chiriquensis	R	O	Y	0	1
Clay-colored Thrush	Turdus grayi	R	O	Y	23	84
Common Bush-Tanager	Chlorospingus ophthalmicus	R	O	Y	10	0
Dusky-capped Flycatcher	Myiarchus tuberculifer	R	O	Y	4	22
Emerald Toucanet	Aulacorhyncus prasinus	R	O	N	5	9
Eye-ringed Flatbill	Rhynchocyclus brevirostris	R	O	N	1	0
Golden-crowned Warbler	Basileuterus culicivorus	R	O	Y	25	1
Grayish Saltator	Saltator coerulescens	R	O	Y	0	2
Great Kiskadee	Pitangus sulphuratus	R	O	N	0	1
House Wren	Troglodytes aedon	R	I	Y	4	19
Keel-billed Toucan	Ramphastos sulfuratus	R	O	N	1	0
Kentucky Warbler	Geothlypis formosa	M	I	Y	10	1
Least Flycatcher	Empidonax minimus	M	O	Y	0	1
Lesser Elaenia	Elaenia chiriquensis	R	O	N	0	1
Lesser Greenlet	Hylophilus decurtatus	R	O	N	9	4
Long-tailed Manakin	Chiroxiphia linearis	R	F	N	151	49
Louisiana Waterthrush	Parkesia motacilla	M	I	N	0	1
Paltry Tyrannulet	Zimmerius vilissimus	R	O	N	0	4
Mountain Elaenia	Elaenia frantzii	R	O	N	0	1
Mountain Thrush	Turdus plebejus	R	O	Y	0	5
Ochre-bellied Flycatcher	Mionectes oleagineus	R	O	Y	33	11
Olivaceous Woodcreeper	Sittasomus griseicapillus	R	I	Y	10	0
Olive-striped Flycatcher	Mionectes olivaceus	R	O	Y	3	2
Orange-bellied Trogon	Trogon aurantiiventris	R	O	N	1	0
Orange-billed Nightingale-Thrush	Catharus aurantiirostris	R	O	Y	90	73
Ovenbird	Seiurus aurocapilla	M	I	Y	10	5
Philadelphia Vireo	Vireo philadelphicus	M	O	N	0	2
Plain Wren	Thryothorus modestus	R	I	Y	10	29
Plain Xenops	Xenops minutus	R	I	N	1	0
Red-crowned Ant-Tanager	Habia rubica	R	O	Y	19	0
Red-eyed Vireo	Vireo olivaceus	M	O	Y	0	2
Rufous-breasted Wren	Thryothorus rutilus	R	I	Y	1	0
Ruddy Woodcreeper	Dendrocincla homochroa	R	I	N	48	6
Rufous-browed Peppershrike	Cyclarhis gujanensis	R	I	N	0	3
Rufous-capped Warbler	Basileuterus rufifrons	R	O	Y	55	41
Rufous-and-white Wren	Thryothorus rufalbus	R	I	Y	62	49
Passerini's Tanager	Ramphocelus passerinii	R	O	Y	1	0

Table 2. Cont.

Common name	Scientific name	Migratory status	Diet	Foraging	Forest	Coffee
Silver-throated Tanager	*Tangara icterocephala*	R	O	N	0	1
Streak-headed Woodcreeper	*Lepidocolaptes souleyetii*	R	I	N	2	5
Swainson's Thrush	*Catharus ustulatus*	M	O	Y	13	24
Tennessee Warbler	*Oreothlypis peregrina*	M	O	N	1	3
Wedge-billed Woodcreeper	*Glyphorynchus spirurus*	R	I	Y	0	1
Eastern Wood-Pewee	*Contopus virens*	M	I	Y	0	1
White-eared Ground-Sparrow	*Melozone leucotis*	R	O	Y	80	79
White-throated Thrush	*Turdus assimilis*	R	O	Y	27	11
White-tipped Dove	*Leptotila verreauxi*	R	O	Y	5	8
Wilson's Warbler	*Cardellina pusilla*	M	I	N	4	6
Wood Thrush	*Hylocichla mustelina*	M	O	Y	11	10
Worm-eating Warbler	*Helmitheros vermivorum*	M	I	Y	2	0
Yellow-bellied Flycatcher	*Empidonax flaviventris*	M	I	N	0	4
Yellow Tyrannulet	*Capsiempis flaveola*	R	O	Y	2	1
Yellow-bellied Elaenia	*Elaenia flavogaster*	R	O	Y	0	2
Yellow-billed Cacique	*Amblycercus holosericeus*	R	O	Y	1	4
Yellow-crowned Euphonia	*Euphonia luteicapilla*	R	O	N	0	1
Yellow-faced Grassquit	*Tiaris olivaceus*	R	O	Y	1	26
Yellow-green vireo	*Vireo flavoviridis*	M	O	N	0	13
Yellowish Flycatcher	*Empidonax flavescens*	R	O	Y	1	1
Yellow-margined Flycatcher	*Tolmomyias assimilis*	R	O	Y	0	1
White-naped Brush-Finch	*Atlapetes albinucha*	R	O	Y	8	12
Yellow-throated Euphonia	*Euphonia hirundinacea*	R	O	Y	12	66
Yellow-throated Vireo	*Vireo flavifrons*	M	I	N	0	2

Following Stiles and Skutch (1989) migratory status is listed as Neotropical-nearctic migrant (M) or year-round neotropical resident (R); dietary guilds are listed as frugivorous (F), omnivorous (O), or insectivorous (I); and foraging strata preferences are listed as ground/understory (Y) or middle/upper canopy (N).

scale occupancy model included the following factors: (1) species, (2) habitat type, (3) diet guild, and (4) habitat-type by diet-guild interaction. The submodel for species turnover in the dynamic occupancy model and the submodel for seasonal use in the multi-scale occupancy model included: (1) species, (2) season, (3) habitat type, (4) diet guild, and (5) habitat-type by diet-guild interaction. In each of these submodels, predictors were specified as fixed effects except for the species effect and species-by-season effect interaction. These random effects were specified such that parameters for species or species-season combinations with sparse captures were informed by species or species-season combinations with more frequent captures [31].

We used a Bayesian approach to fit the models using programs R and WinBUGS [32,33], whereby we specified vague logit-scale priors using a normal distribution with a mean of 0 and a precision of 0.4 for the fixed effects and means for the hyperdistributions on the random effects. For the standard deviation of the random effect, we specified a vague uniform prior with a minimum of 0.001 and a maximum of 10. We confirmed convergence of three Markov-chain Monte Carlo (MCMC) chains each of length 15,000 by visually inspecting the plotted chains and ensuring that Gelman-Rubin diagnostic values were below 1.5 for all estimates [34]. Predicted estimates from these models enabled us to compare avian community composition and dynamics between shade-coffee and secondary forest. In particular, inferences were based on 95% Bayesian credibility intervals (BCIs; 2.5th and 97.5th percentiles

from MCMC output) surrounding probability estimates. If a BCI excluded the mean of a contrasting estimate, then this was considered a statistically significant difference. When comparing a higher probability to a lower probability, the following standard formula was used: (higher-lower)/higher.

Results

Avian captures

Including recaptures, a total of 1,561 captures representing 73 bird species and 13 guilds were made. Of those, 773 captures representing 61 species were in shade coffee and 788 captures representing 46 species were in secondary forest (Table 2). Of the total birds captured, 217 captures were birds that had been previously captured and banded. This included 115 birds that were re-captured once, 31 birds that were re-captured twice, 6 birds that were re-captured three times, 4 birds that were re-captured 4 times and 1 bird that was re-captured 6 times. The total number of individuals banded during our study was 1,344 and 1,192 of these were only captured once. Fifty-three percent of all bird species captured were captured in only one of the two habitat types: 36% were captured only in shade coffee (27 bird species) and 16% were captured only in secondary forest (12 bird species). Sixty percent of the highly forest dependent species captured were forest insectivores (e.g. Dendrocolaptidae: *Sittasomus griseicapillus*, *Glyphorhynchus spirurus*, and *Dendrocolaptes certhia*; Tham-

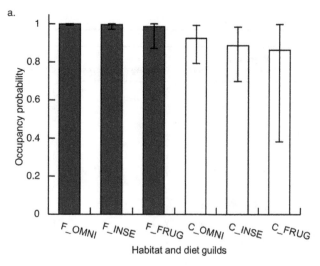

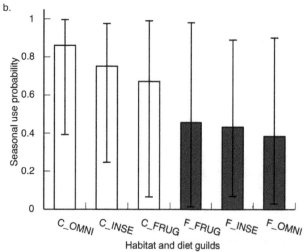

Figure 2. Variation in patch occupancy (A) and seasonal patch use (B) among habitats and avian dietary guilds based on a multi-scale occupancy model fit to mist-net data from four shade coffee plantations and three secondary forest fragments near San Luis, Costa Rica, 2005–2008. Habitat types included forest (F) and shade coffee (C); avian dietary guilds included frugivores (FRUG), insectivores (INSE), and omnivores (OMNI). Whiskers represent 95% Bayesian credibility intervals.

nophilidae: *Thamnophilus doliatus*; Tyrannidae: *Rhynchocyclus brevirostris*, and *Tolmomyias assimilis*). We captured one frugivorous species (*Chiroxiphia linearis*) during the study, which was captured in both shade-coffee and secondary forest. Compared to secondary forest, in shade-coffee we captured a higher number of omnivorous (43 in shade-coffee vs. 30 in secondary forest) and insectivorous species (17 in shade-coffee vs. 15 in secondary forest).

Occupancy and seasonal use

Occupancy estimates were quite high across habitats and avian functional groups (>0.85; Fig. 2a). Omnivores and insectivores were more common in secondary forest than they were in shade-coffee, and these differences were statistically significant. Mean occupancy of frugivores was greater in secondary forest than in shade coffee, but this difference was not significant. Seasonal use was moderate across most of the diet-habitat type combinations (range: 0.38–0.75), except for seasonal use by omnivores of shade-

coffee, which was quite high (0.86; BCI: 0.39–1.00; Fig. 2b). Although means for seasonal use were greater in shade coffee than they were in secondary forest, none of the differences were statistically significant because of wide BCIs.

Community turnover

Omnivores tended to have higher mean interseasonal colonization rates in shade-coffee (0.38; BCI: 0.04–0.88) compared to forest (0.09; BCI: 0.00–0.42), but this difference was not statistically significant. Other functional groups had moderate colonization rates across habitats (range: 0.21–0.59), and comparisons of these rates between habitats were not statistically significant. Estimates for interseasonal persistence were moderate (range: 0.57–0.73), and differences among habitat types and functional groups were not significant.

Capture probability

Capture probabilities were quite variable between shade-coffee and secondary forest, and among functional groups based on the multi-scale occupancy model (range: 0.13–0.67; Fig. 3a). As expected, capture probability estimates based on the dynamic occupancy model were similar to the multi-scale occupancy model (Fig. 3b). For frugivores and insectivores, estimates for capture probabilities were moderate (range: 0.43–0.60) and were not significantly different between the two habitat types. Frugivores and insectivores tended to have higher detectability in secondary forest compared to shade-coffee, but the differences were not statistically significant. For omnivores, estimates for capture probabilities showed a lower mean detectability of omnivores compared to insectivores and frugivores, for both habitat types.

Discussion

Our study and analyses of bird community composition and dynamics did not refute the hypothesis that coffee agroforestry systems serve as surrogate habitats for forest bird species. The modeling approach we used allowed us to consider the important factors of seasonal use and detectability differences while comparing the ecosystem services provided by birds in shade-coffee to those by birds in secondary forest. Specifically, the analyses revealed that bird community composition and dynamics in shade-grown coffee plantations that are small and embedded in the typical tropical countryside landscape do, in fact, reflect those of secondary forest, and these findings are consistent with the habitat-surrogate hypothesis. With few exceptions, patch occupancy, seasonal use, species turnover, and even capture probability were mostly similar between habitat types.

Species composition and turnover

Based on raw captures, our study indicated that species richness was higher in shade-coffee compared to secondary forest across the avian community and within the omnivore and insectivore guilds. When we accounted for differences in capture probability, however, species occupancy was higher for omnivores and insectivores in secondary forest compared to shade-coffee. The probability of occupancy for individual species by guild and habitat type is a function of the number of species within a guild occupying a given habitat type and their distribution across patches (i.e., study sites) of that habitat type. As the number of species in patches of a given habitat type increases, the estimated probability of occupancy within that habitat type increases. Probability of occupancy in this context can therefore be considered a proxy for species richness both within and across patches of a particular habitat type.

Although the difference in occupancy between habitat types was statistically significant, occupancy rates in both habitat types were all very high (>0.85) and differences were less than 15%, which we interpret as not a biologically meaningful difference. Thus, we conclude that the high occupancy rates across all dietary-guild and habitat type combinations lends support for the habitat-surrogate hypothesis that bird communities in coffee agroforests represent those in secondary forest. Our examination of species turnover and seasonal use demonstrated that avian communities were quite dynamic between seasons, especially in shade-coffee when compared to secondary forest, but we found no statistically significant differences in these dynamics between habitats. Omnivores, however, displayed evidence that led us to hypothesize a greater likelihood they would colonize a shade-coffee plantation than they would colonize a secondary forest fragment. Despite the one potential exception with respect to omnivore colonization, species turnover patterns were consistent with the habitat-surrogate hypothesis.

Capture probability

Based on our analysis of bird community composition and dynamics, bird species within functional guilds have similar levels of detectability in shade-coffee and in secondary forest when using mist-nets as a sampling method. Although one recent study has incorporated detection probability to compare migratory bird species densities between shade-coffee and forested habitats [35], our study is the first to account for detection probability when assessing how well these habitat types support the entire bird community [10]. Additionally, omnivores displayed some differences in detectability compared to the other functional groups irrespective of habitat types (i.e. omnivores had a lower detectability in both shade-coffee and secondary forest compared to frugivores and insectivores). Determining which traits of omnivores predispose them to be less likely to be detected in these two habitat types may further demonstrate the need for a standard methodology and analyses for studies of bird community changes in human-dominated landscapes. If we had ignored capture probability, we would have found that species richness, especially within the insectivore guild, was greater in shade-coffee than in secondary forest. When accounting for capture probability, however, we found the opposite to be the case albeit the difference was probably too small to be biologically significant. Nonetheless, our finding that omnivores had lower detectability compared to frugivores and insectivores underscores the importance of considering detection probability when evaluating the contribution of a particular habitat type toward conservation of multiple functional groups.

Conservation and management implications

In order to make sound conservation and management recommendations for shade-coffee plantations, we must first be confident about their role in supporting forest bird communities. This requires conducting more detailed studies that not only measure apparent species richness, but also aim to evaluate relative abundance, community dynamics, and community composition. Although we did not measure landscape-scale variables, we did conduct our study in a landscape that represents the typical landscape where shade-coffee farms are located, as most coffee agroforestry production systems are smallholder farms. Based on our findings, therefore, shade-coffee still appears to provide a suitable surrogate for secondary forest across the annual cycle, with two exceptions. First, we found that bird occupancy

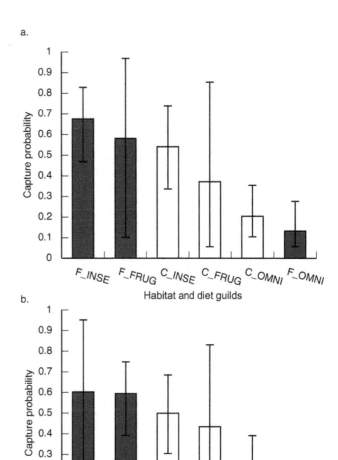

Figure 3. Variation in avian capture probabilities among habitats and avian dietary guilds based on a multi-scale occupancy model (A) and a dynamic occupancy model (B) fit to mist-net data from four shade coffee plantations (white bars) and three secondary forest fragments (gray bars) near San Luis, Costa Rica, 2005–2008. Habitat types included forest (F) and shade coffee (C); avian dietary guilds included frugivores (FRUG), insectivores (INSE), and omnivores (OMNI). Whiskers represent 95% Bayesian credibility intervals.

was slightly higher in secondary forest compared to shade-coffee, and this difference was statistically significant for omnivores and insectivores. However, some caution should be taken in extrapolating our results to coffee agroforests that are either very extensive or are within landscapes that do not include forest patches.

Nature tourism has increased dramatically in the last ten years in the Monteverde region of Costa Rica [36] and several economically-important bird species are found in the Monteverde region (e.g. Resplendent Quetzal, Three-wattled Bellbird, and Black-faced Solitaire). Although this region originally gained worldwide acclaim for its three private reserves, most of the region, outside of the reserves, is more typical of tropical agricultural landscapes [21,37]. Within these areas outside of the reserves in Monteverde, the economic condition of many families is directly related to income from bird-related nature tourism. Therefore, bird research aimed at understanding how well various land use types in the agricultural matrix support bird communities would not only help to guide conservation efforts in the region, but would also be economically beneficial to local human communities. Much research has already been conducted on birds in the Monteverde region, however several factors limit the broad applicability of this research to conservation/management of bird communities throughout the entire region. First, most previous research in the Monteverde region has focused on single bird species [38,39], limiting application of management recommendations to entire avian communities. Second, study sites have primarily been within protected areas, not in agricultural land use types. Lastly, avian community composition changes rapidly with elevation in the region [22,40]. For example, many species of the avian community of the San Luis valley, where our study was conducted, do not range into the higher elevations of the Monteverde region [37,38,41,42]. Thus, our study provides important conservation information for the region, demonstrating that agroforestry systems can protect forest bird communities throughout the annual cycle.

Although we did not include a more intensified agricultural system in our study, agroforestry systems can protect more native biodiversity than more simplified agricultural systems [10]. For example, cacao and banana agroforestry systems contain bird assemblages that may be as abundant, species-rich and diverse as secondary forests [9]. Thus, conserving and managing shade-coffee and forest remnants for bird species should be considered as a complementary rather than a contradictory strategy for bird conservation. Management efforts should therefore focus on conserving all forested habitats, including agroforestry systems, to maintain bird species composition and dynamics.

Acknowledgments

We thank the staff at the UGA San Luis Research Station, and the UGA Costa Rica office for logistical and financial support. We thank the shade-coffee parcel owners for their participation; P. Logan Weygandt, Adan Fuentes, Avery Tomlinson, Raj Joshi, Jason Norman, Kate Spear, Paul Scarr, Ryan Malloy, Andrea Smith and several other volunteers provided field assistance. We appreciate reviews of earlier versions of this manuscript by Viviana Ruiz-Gutierrez and Rua Mordecai.

Author Contributions

Conceived and designed the experiments: SMH BJM. Performed the experiments: SMH VEP. Analyzed the data: SMH BJM VEP. Contributed reagents/materials/analysis tools: SMH BJM. Wrote the paper: SMH BJM VEP RJC CRC.

References

1. Wassenaar T, Gerber P, Verburg PH, Rosales M, Ibrahim M, et al. (2007) Projecting land use changes in the Neotropics: The geography of pasture expansion into forest. Global Environmental Change 17: 86–104.
2. Janzen D (1998) Gardenification of wildland nature and the human footprint. Science 279: 1312–1313.
3. Sanchez-Azofeifa GA, Pfaff A, Robalino JA, Boomhower JP (2007) Costa Rica's payment for environmental services program: Intention, implementation, and impact. Conservation Biology 21: 1165–1173.
4. Perfecto I, Vandermeer J (2008) Biodiversity conservation in tropical agroecosystems - A new conservation paradigm. Pages 173–200 In: Year in Ecology and Conservation Biology 2008. Blackwell Publishing, Oxford.
5. Perfecto I, Rice RA, Greenberg R, Van der Voort ME (1996) Shade coffee: A disappearing refuge for biodiversity. Bioscience 46: 598–608.
6. Perfecto I, Armbrecht I, Philpott SM, Dietsch T, Soto-Pinto L (2007) Shaded coffee and the stability of rainforest margins in Latin America. Pages 227–264 in Tscharntke T, Zeller M, Guhudja E, Bidin A, editor. The stability of tropical rainforest margins: linking ecological, economic, and social constraints of land use and conservation. Springer, Environmental Science Series, Heidelberg.
7. Moguel P, Toledo VM (1999) Biodiversity conservation in traditional coffee systems of Mexico. Conservation Biology 13: 11–21.
8. Komar O (2006) Ecology and conservation of birds in coffee plantations: a critical review. Bird Conservation International 16: 1–23.
9. Harvey CA, Gonzalez-Villalobos JA (2007) Agroforestry systems conserve species-rich but modified assemblages of tropical birds and bats. Biodiversity and Conservation 16: 2257–2292.
10. Sekercioglu CH (2012) Bird functional diversity and ecosystem services in tropical forest, agroforests and agricultural areas. Journal of Ornithology DOI 10.1007/s10336-012-0869-4
11. Greenberg R, Bichier P, Sterling J (1997) Bird populations in rustic and planted shade coffee plantations of Eastern Chiapas, Mexico. Biotropica 29: 501–514.
12. Greenberg R, Bichier P, Cruz-Angon A (2000) The conservation value for birds of cacao plantations with diverse planted shade in Tabasco, Mexico. Animal Conservation 3: 1367–9430.
13. Peltonen A, Hanski I (1991) Patterns of island occupancy explained by colonization and extinction rates in shrews. Ecology 72: 1698–1708.
14. Hines JE, Boulinier T, Nichols JD, Sauer JR, Pollock KH (1999) COMDYN: software to study the dynamics of animal communities using a capture-recapture approach. Bird Study 46: 209–217.
15. Royle JA, Nichols JD (2003) Estimating abundance from repeated presence-absence data or point counts. Ecology 84: 777–790.
16. Moore JE, Swihart RK (2005) Modeling patch occupancy by forest rodents: incorporating detectability and spatial autocorrelation with hierarchically structured data. Journal of Wildlife Management 69: 933–949.
17. Carlo TA, Collazo JA, Groom MJ (2003) Avian fruit preferences across a Puerto Rican forested landscape: pattern consistency and implications for seed removal. Oecologia 134: 119–131.
18. Peters VE, Mordecai R, Carroll CR, Cooper RJ, Greenberg R (2010) Bird community response to fruit energy. Journal of Animal Ecology 79: 824–835.
19. Fleming TH, Breitwisch R, Whitesides GH (1987) Patterns of tropical vertebrate frugivore diversity. Annual Review of Ecology and Systematics 18: 91–109.
20. Haber WA (2000) Plants and vegetation. In: Nadkarni N, Wheelwright N, editors. Monteverde: Ecology and conservation of a tropical cloud forest. Oxford University Press, New York. pp. 39–94.
21. Clark KJ, Lawton RO, Butler PR (2000) The physical environment. In: Nadkarni N, Wheelwright N, editors. Monteverde: Ecology and conservation of a tropical cloud forest. Oxford University Press, New York. pp. 15–38.
22. Young BE, McDonald DB (2000) Birds. In: Nadkarni N, Wheelwright N, editors. Monteverde: Ecology and conservation of a tropical cloud forest. Oxford University Press, New York. pp. 179–122.
23. Bibby CJ, Burgess ND, Hill DA, Mustoe S (2000) Bird Census Techniques. 2nd edition. Academic Press, London.
24. Karr JR (1991) Biological integrity-a long-neglected aspect of water-resource management. Ecological Applications 1: 66–84.
25. Ralph CJ, Dunn EH, eds (2004) Monitoring bird populations using mist nets. Cooper Ornithological Society, Lawrence.
26. Hernandez-Divers SM (2008) Investigating the differences in health, pathogen prevalence and diversity of wild birds inhabiting shade-grown coffee plantations and forest fragments in San Luis, Costa Rica. University of Georgia, Athens.
27. Stiles GF, Skutch AF (1989) A guide to the birds of Costa Rica. Cornell University Press, Ithaca.
28. Garrigues R, Dean R (2007) The Birds of Costa Rica: A Field Guide. Cornell University Press, Ithaca.
29. Mordecai RS, Mattsson BJ, Tzilkowski CJ, Cooper RJ (2011) Addressing challenges when studying mobile or episodic species: Hierarchical Bayes estimation of occupancy and use. Journal of Applied Ecology 48: 56–66.
30. Ruiz-Gutiérrez V, Zipkin EF, Dhont AA (2010) Occupancy dynamics in a tropical bird community: unexpectedly high forest use by birds classified as non-forest species. Journal of Applied Ecology 47: 621–630.
31. Zipkin EF, Dewan A, Royle JA (2009) Impacts of forest fragmentation on species richness: a hierarchical approach to community modelling. Journal of Applied Ecology 46: 815–822.
32. Spiegelhalter DA, Thomas A, Best N, Lunn D (2003) WinBUGS user manual, version 1.4. MRC Biostatistics Unit, Cambridge.
33. Sturtz S, Ligges U, Gelman A (2005) R2WinBUGS: a package for running WinBUGS from R. Journal of Statistical Software 12: 1–16.
34. Gelman A, Rubin DB (1992) Inference from iterative simulation using multiple sequences. Statistical Science 7: 457–472.
35. Bakermans MH, Vitz AC, Rodewald AD, Rengifo CG (2009) Migratory songbird use of shade coffee in the Venezuelan Andes with implications for conservation of Cerulean warbler. Biological Conservation 142: 2476–2483.
36. Koens JF, Dieperink C, Miranda M (2009) Ecotourism as a development strategy: experiencesfrom Costa Rica. Environ Dev Sustain 11: 1225–1237.
37. Guindon C (1996) The importance of forest fragments to the maintenance of regional biodiversity in Costa Rica. In: Schelhas R, editor. Forest patches in tropical landscapes. Island Press, Washington DC.
38. Lawton MF, Lawton RO (1985) The breeding biology of the Brown jay in Monteverde, Costa Rica. Condor 87: 192–204.

39. Powell GVN, Bjork R (1995) Implications of intratropical migration on reserve design- A case study using *Pharomachrus mocinno*. Conservation Biology 9: 354–362.

40. Blake JG, Loiselle BA (2000) Diversity along an elevational gradient in the Cordillera Central, Costa Rica. The Auk 117: 663–686.

41. Lawton MF, Lawton RO (1980) Nest-site selection in the Brown jay. Auk 97: 631–633.

42. Lawton MF, Guindon CF (1981) Flock composition, breeding success, and learning in the Brown jay. Condor 83: 27–33

15

Can Joint Carbon and Biodiversity Management in Tropical Agroforestry Landscapes Be Optimized?

Michael Kessler[1], Dietrich Hertel[2]*, Hermann F. Jungkunst[3], Jürgen Kluge[1,4], Stefan Abrahamczyk[1,5], Merijn Bos[6,7], Damayanti Buchori[8], Gerhard Gerold[9], S. Robbert Gradstein[10], Stefan Köhler[11], Christoph Leuschner[2], Gerald Moser[2,12], Ramadhanil Pitopang[13], Shahabuddin Saleh[14], Christian H. Schulze[15], Simone G. Sporn[16], Ingolf Steffan-Dewenter[17], Sri S. Tjitrosoedirdjo[18], Teja Tscharntke[6]

1 Systematic Botany, University of Zurich, Zurich, Switzerland, 2 Plant Ecology, University of Göttingen, Göttingen, Germany, 3 Geoecology/Physical Geography, Institute for Environmental Sciences, University of Koblenz-Landau, Landau, Germany, 4 Faculty of Geography, University of Marburg, Marburg, Germany, 5 Systematic Botany and Mycology, Department of Biology, University of Munich, Munich, Germany, 6 Agroecology, University of Göttingen, Göttingen, Germany, 7 Louis Bolk Institute, LA Driebergen, The Netherlands, 8 Department of Plant Protection, Faculty of Agriculture, IPB, Bogor Agricultural University, Kampus Darmaga, Bogor, Indonesia, 9 Landscape Ecology, Institute of Geography, University of Göttingen, Göttingen, Germany, 10 Muséum National d'Histoire Naturelle, Departement Systématique et Evolution (UMS 602), C.P. 39, Paris, France, 11 Landscape Ecology and Land Evaluation, Faculty for Agricultural and Environmental Sciences, University of Rostock, Rostock, Germany, 12 Department of Plant Ecology, University of Giessen, Giessen, Germany, 13 Department of Biology, Faculty of Mathematics and Natural Sciences, Tadulako University, Palu, Indonesia, 14 Department of Agrotechnology, Faculty of Agriculture, University of Tadulako, Palu, Indonesia, 15 Department of Animal Biodiversity, Faculty of Life Sciences, University of Vienna, Vienna, Austria, 16 FEMS Central Office, CL Delft, The Netherlands, 17 Department of Animal Ecology and Tropical Biology, Biocenter, University of Würzburg, Würzburg, Germany, 18 Department of Biology, Faculty of Mathematics and Natural Sciences, IPB, Bogor Agricultural University, Kampus Darmaga, Bogor, Indonesia

Abstract

Managing ecosystems for carbon storage may also benefit biodiversity conservation, but such a potential 'win-win' scenario has not yet been assessed for tropical agroforestry landscapes. We measured above- and below-ground carbon stocks as well as the species richness of four groups of plants and eight of animals on 14 representative plots in Sulawesi, Indonesia, ranging from natural rainforest to cacao agroforests that have replaced former natural forest. The conversion of natural forests with carbon stocks of 227–362 Mg C ha^{-1} to agroforests with 82–211 Mg C ha^{-1} showed no relationships to overall biodiversity but led to a significant loss of forest-related species richness. We conclude that the conservation of the forest-related biodiversity, and to a lesser degree of carbon stocks, mainly depends on the preservation of natural forest habitats. In the three most carbon-rich agroforestry systems, carbon stocks were about 60% of those of natural forest, suggesting that 1.6 ha of optimally managed agroforest can contribute to the conservation of carbon stocks as much as 1 ha of natural forest. However, agroforestry systems had comparatively low biodiversity, and we found no evidence for a tight link between carbon storage and biodiversity. Yet, potential win-win agroforestry management solutions include combining high shade-tree quality which favours biodiversity with cacao-yield adapted shade levels.

Editor: Ben Bond-Lamberty, DOE Pacific Northwest National Laboratory, United States of America

Funding: This study was funded by the German Research Foundation (DFG), grant SFB-552 (Collaborative Research Centre 'Stability of Rainforest Margins in Indonesia'). The funders had no role in study design, data collection and analysis, decision to publish, or preparation of the manuscript.

Competing Interests: The authors have declared that no competing interests exist.

* E-mail: Dietrich.Hertel@biologie.uni-goettingen.de

Introduction

Carbon storage in above- and belowground forest vegetation and in the soil plays a crucial role in the terrestrial greenhouse gas balance [1,2,3]. After fossil fuel use, tropical deforestation and forest degradation represent the second largest source of carbon emissions, contributing about 12–20% of the annually released CO_2 [4,5]. Accordingly, conservation of tropical forests and reforestation of formerly forested habitats are viewed as important components of global strategies to reduce CO_2 emissions [6]. At the same time, tropical forests harbour some of the highest levels of biodiversity on Earth as well as the largest number of species threatened with global extinction [7]. This dual role of tropical forests as carbon and biodiversity repositories presents a potential win-win situation, in which management of habitats for carbon storage may in parallel result in biodiversity conservation [8,9,10]. Over the last decade, there have been several moves towards establishing payment schemes, in which tropical forest conservation or reforestation initiatives are remunerated [11]. Politically, such proposals are hotly debated, as both the controlling mechanisms and the potential benefits are unclear [12,13,14].

There is little doubt that the preservation of large tracts of natural tropical forests will safeguard both large carbon stocks and the habitats of much threatened fauna and flora, especially for those species dependent on undisturbed habitats and with low population densities, therefore requiring large habitat tracts to persist [15,16]. However, human-impacted ecosystems, including logged natural forest and secondary forests as well as agricultural areas, cover ever-increasing areas and play a crucial role both in carbon and biodiversity management [17,18]. Indeed, several of the proposed carbon payment schemes exclusively focus on the management of impacted ecosystems as carbon sinks [19]. The Kyoto Protocol, for example, explicitly excluded the reduction of emissions by avoiding deforestation because of both political and technical obstacles.

In agricultural systems, the relationship of carbon stocks and biodiversity is far from clear [16,20,21]. Globally, 46% of the agricultural area has at least 10% tree cover, and can thus be classified as agroforests [22]. Amongst them, agroforests holding a substantial tree cover (at least 30%) still account for as much as 374 million hectares [22]. In 2007, agroforests for coffee and cacao production in tropical landscapes, which are the second- and third-largest international trade commodities after petroleum, covered no less than 17.7 million hectares worldwide [23]. Because agroforests are tree-dominated ecosystems, they potentially play an important role for carbon management [24]. At the same time, agroforests can harbour significant levels of biodiversity [25,26,27]. Yet, the potential links between carbon stocks and biodiversity levels in tropical agroforests as a basis for environmentally optimized agroforestry management remain unexplored.

In the present study, we assessed the potential to optimize carbon management in agroforests while at the same time safeguarding high levels of biodiversity. Because natural forest-based biodiversity is commonly considered to be the most threatened in tropical forest ecosystems [5,28], we placed a special focus on the richness of species that we recorded in the natural forest habitats.

Materials and Methods

Our study complies with the current laws of Indonesia and Germany and with international rules. Permissions for fieldwork in Indonesia and collecting and exporting samples have been provided by national and local authorities.

Study Area and Site Selection

The study took place around the village of Toro (1°30'24'' S, 120°2'11'' E) located at the western border of Lore Lindu National Park, about 100 km south of Palu, the capital city of Central Sulawesi, Indonesia. The natural vegetation around the village is submontane rainforest. The agricultural landscape in the region consists of pastures, paddy fields and cacao-dominated agroforests. Cacao production in the region increased strongly in the 1990s. The cacao agroforests are managed by small-scale farmers. Shade tree management in the region is dynamic and farmers tend to remove shade trees in mature agroforestry systems to increase cacao production [26].

We defined four habitat types with different shade tree diversity [29]: (1) natural forest sites, situated at least 300 m away from forest sites where selective logging occurred; (2) cacao agroforests with diverse, natural shade trees, retained after thinning of the previous forest cover, underplanted with cacao trees and few fruit trees (high shade agroforests); (3) cacao agroforests with shade tree stands dominated by various species of planted fruit and timber trees (medium shade); (4) cacao agroforests with a low diversity of planted shade trees, predominantly non-indigenous, nitrogen-fixing leguminous trees and a few native fruit tree species (low shade).

We randomly selected 3–4 replicates from a larger subset of each habitat type. Natural forest sites were chosen that were representative for rain forests in this region and elevation belt in terms of forest structure and tree species composition. Agroforest sites were selected based on the age of the cacao trees, which was at all sites 4–17 years. At the time of this study, farmers regularly pruned trees and weeded the plantations and only rarely treated them with fertilizers and pesticides.

Distance between study sites ranged between 0.3–5 km. All sites were at 850–1100 m above sea level. The agroforests did not have sharp borders, but gradually changed into other forms of land-use and at the landscape scale formed a continuous band along the forest margin. We marked core areas of $50 \times 50 \text{ m}^2$ in the middle of each site, whose land-use and shade tree composition was as constant as possible. Sites belonging to the different habitat types were not spatially clustered, but geographically interspersed.

As for all observational studies regarding natural forest conversion, our comparative study approach could have been affected by confounding effects such as that the chosen natural forest plots were not typical for the area and therefore not representative for those forest stands that have been converted prior to the recent farmer agroforest management or that farmers preferentially converted low-biodiversity forest for some reasons. However, recent conversion of former natural forest into agroforestry systems is still common in the study area, and there are no apparent biases in where clearing takes place (except that more accessible and less steep sites are preferred, but this was taken into account by our sampling design). This together with the fact that the natural forest structure and composition is quite homogenous in the entire study area (rather species-rich with comparable carbon stocks) makes it unlikely that our results have been affected by such confounding effects.

Carbon Stock Estimation in Above- and Belowground Tree Biomass

The estimation of above-ground tree biomass of the natural forest trees was conducted according to the methods described in [30]. The procedure followed common standard procedures and is based on stand inventories of above-ground tree dimensions, data on wood-specific density of the present genera, and the application of allometric equation models from the literature [e.g. 31,32]. The below-ground biomass of all forest trees in each plot was estimated using the root/shoot ratio from [33] for tropical-subtropical moist forest and plantations. For trees in plots with an above-ground biomass (AGB) $<125 \text{ Mg ha}^{-1}$, the applied root:shoot ratio was 0.205, while for trees from plots with AGB $>125 \text{ Mg ha}^{-1}$ it was 0.235. Above- and belowground biomass of cacao and planted *Gliricidia* trees were estimated from stem diameter records using allometric relationships for above-ground biomass as well as root/shoot ratios established in the nearby plots [34]. The root/shoot ratios in this study were 0.394 for *Theobroma* and 0.488 for *Gliricidia*. For the calculation of the above- and below-ground biomass of *Coffea* trees, we used the allometric equation from [35].

The above- and below-ground biomass sums of all inclined plots were transformed to the horizontal projection and are given as Mg ha^{-1}. Carbon stock sizes were calculated from the biomass data applying data on carbon contents of 42% for above-ground and 46% for below-ground biomass that were measured in nearby forest plots [30].

Soil Carbon Stock Estimation

We sampled each plot at least six times and excavated representative soil pits. Soils were sampled per horizon until the depth of 1 m. Soil analyses were performed on the fine earth fraction ($<2 \text{ mm}$). Stone contents (vol%) were estimated in the field. Bulk densities were measured using undisturbed soil cores (100 cm^3) after drying at 105°C. The carbon content (g carbon per kg soil) was determined for each horizon by analyzing air dried soil samples with a Vario El CN Analyser (Elementar, Hanau, Germany) in the laboratory in Palu, Indonesia. For each horizon the carbon stock (C_{stock}, in Mg ha^{-1}) was calculated using the equation $C_{stock} = C_{conc} \times BD \times d \times a \times CF_{stones}$ following [36], where C_{conc} represents the carbon content (in g kg^{-1}), BD is the soil bulk density of the respective horizon (in kg m^{-3}), d is the thickness of the horizon (in m), the related plot area a (one ha = 10,000 m^2), and a correction factor for the stone content of the soil samples (CFstones, (100–%stones)/100). The calculated soil carbon stocks per horizon were summed up to one meter depth and given as Mg ha^{-1} for each plot.

Quantification of Biodiversity

We assessed the species richness of trees, lianas and herbs, epiphytic liverworts from lower canopy trees, birds, butterflies, ants and beetles from lower canopy trees, dung beetles, bees, wasps, and their parasitoids for each 50×50 m^2 plot according to the methods described in [29] and briefly summarized here. In the largely aseasonal climate of our study region (mean monthly precipitation is >100 mm for each month; mean monthly temperatures vary <2°C over the year) no marked seasonal variations of species composition and abundance were detected over several years of field work [e.g. 26,37]. Accordingly, timing of the sampling should not have influenced main patterns of our results. *Trees:* All trees dbh ≥10 cm were mapped and individually numbered with aluminum tags, their dbh was measured, and their trunk height and total height were estimated. *Lianas and herbs:* In each study plot of 50×50 m^2 ten subplots of 2×2 m^2 each were randomly placed. Within these, all herb and liana species were inventoried, collected, and determined. *Epiphytic liverworts from lower canopy trees:* Two trees with a height up to 8 m, a dbh ranging of 20–60 cm, and comparable bark texture were selected in each study plot. Each tree was divided into zone 1 (treebase up to the first ramification), zone 2 (inner crown) and zone 3 (outer crown) according to modified Johansson zones for small trees. Within subplots of 200 cm^2, liverworts were sampled from each cardinal direction in all three zones. *Birds:* Each plot was visited on two mornings from 05:30 to 10:30 am. Birds were recorded visually and acoustically, and by systematic tape recordings. *Butterflies:* Butterflies were captured alive in traps baited with rotten mashed bananas in traps suspended from tree branches with strings about 1.5 m above the ground. *Ants and beetles from lower canopy trees:* Within each study plot, four trees were selected, which were of similar age and size. The insect fauna was sampled using canopy knockdown fogging, using a SwingFog TF35 to blow a fog of 1% pyrethroid insecticide (Permethrin) Killed arthropods were collected from a 4 m^2 sheet of white canvas placed directly under each tree. *Dung beetles:* Dung beetles were collected using baited pitfall traps baited with ca. 20 g of fresh cattle (*Bos taurus*) dung. *Bees, wasps, and their parasitoids:* Trap nests offer standardized nesting sites for above-ground nesting bees and wasps and can therefore be used to experimentally study these insects. They were constructed from PVC tubes with a length of 28 cm and a diameter of 14 cm. Internodes of the reed *Saccharum spontaneum* with varying diameter (3–25 mm) and a length of 20 cm were inserted into these tubes to provide nesting sites. Twelve trap nests (four in each stratum) were installed in three different heights from understorey and intermediate tree height to the canopy. Trap nests were checked every month and bee and wasp larvae were reared for later identification.

Correlations of Carbon Stocks and Biodiversity

To arrive at a comprehensive measure of biodiversity for each plot, we combined all groups. In order to weight all groups similarly, we first standardized the richness values of each group by setting the maximum plot count at 100% and all other counts respectively. We then averaged the standardized values for all 12 groups for each plot. This approach has previously been used to obtain generalized biodiversity patterns when numerous taxa have been sampled [38,39]. We then calculated simple, linear determination values (R^2) to relate carbon stocks to biodiversity. This was done for the entire carbon stocks as well as separately for above- and below-ground (soil + root) carbon stocks. Because relationships may be driven by the marked contrasts between natural forests and agroforests, both for biodiversity [29] and carbon stocks [40], in a second step we only included the 11 agroforestry plots in the analyses. Finally, because forest-based biodiversity is considered to be the most threatened in tropical forest ecosystems [5,28], we repeated all analyses by only including those species recorded in the natural forests of the study region [29]. In order to also be able to assess group-specific patterns, we repeated the above analyses for all groups independently (Figures S1, S2, S3). All statistical analyses have been done using the statistical platform R [41].

Results

In natural forests, carbon stocks were on average over twice as high as in the agroforests (Figure 1). The above-ground vegetation in natural forests held on average 54% of the total carbon stocks, with the root and soil components each contributing 14% and 32%, respectively. In agroforests, in contrast, the soil component on average included about 66% of the carbon stocks, followed by the above-ground vegetation (26%) and roots (7%). Perhaps the most striking result was that soil carbon stocks did not differ significantly between natural forests and agroforests (t-test, t = 0.68, P = 0.51), on average declining only from 87 Mg C ha^{-1} to 80 Mg C ha^{-1}, although a single agroforest plot had a value of 40 Mg C ha^{-1}.

The relationship of the species richness of all species to carbon stocks showed no or only marginally significant patterns, both when all plots and only the agroforest plots were considered (Figure 2, Figures S1, S2, S3). In contrast, when we only considered the forest-related species, we obtained highly significant relationships between species richness and carbon stocks when all plots were included. When we restricted this analysis to the agroforest plots, relationships were weaker but still significant for total and above-ground carbon stocks.

When we analyzed the species groups individually, only 1–3 groups showed significant positive or negative relationships to total carbon stocks, particularly trees, bryophytes and dung beetles (positive) as well as herbs, wasps, and their parasitoids (negative) (Figures S1, S2, S3). When we restricted the same analyses to the agroforestry plots, only a single relationship (lianas) was significant. When we restricted the analysis to forest species, no less than 23 of 36 (64%) of the relationships were significant. When the forest plots were excluded, median r-values decreased to 0.4–0.5 and only 8 (22%) relationships remained significant. In all cases except the analyses with all species across all plots, R^2 values were slightly higher when considering above-ground carbon stocks than below-ground stocks, with overall values intermediate.

Because trees are directly managed by the local farmers aiming to manipulate the shading level of cacao plantations, we explored the relationships of trees and biodiversity in more detail. Tree species richness was significantly positively correlated to the species

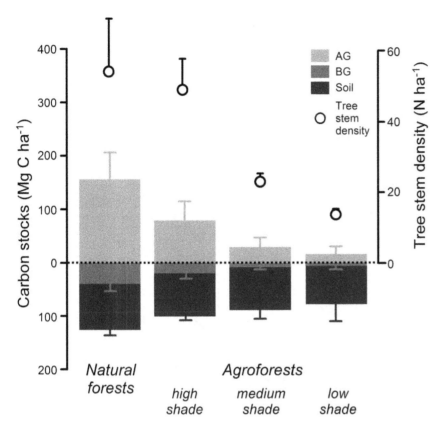

Figure 1. Carbon stocks in natural rainforests and cacao agroforests of varying tree density and shade levels. Shade levels were defined as: high shade: cacao agroforests with diverse, natural shade trees, retained after thinning of the previous forest cover, underplanted with cacao trees and few fruit trees; medium shade: cacao agroforests with shade tree stands dominated by various species of planted fruit and timber trees; low shade: cacao agroforests with a low diversity of planted shade trees, predominantly non-indigenous, nitrogen-fixing leguminous trees and a few native fruit tree species. Columns show mean carbon stocks (+1 SD) in the above-ground (AG) and below-ground (BG) plant components as well as in the soil. Also shown is the mean stem density (+1 SD) of trees with diameters ≥10 cm at breast height.

richness of only four of the other eleven study groups (Table 1). However, when we only considered the richness of tree species from the natural forests, the regression values between tree species richness and richness of the other groups were significantly higher, both across all study plots and in the agroforest plots only.

Discussion

Overall, we found that the relationship between carbon stocks and biodiversity was fairly weak and most pronounced when considering only forest-based biodiversity as well as when contrasting forest plots with agroforest plots. One may argue that the fact that carbon storage in the above-ground biomass of the agroforestry systems (being less bio-diverse) is less than that in the natural forest (harbouring higher biodiversity) are to be expected due to the large differences in stand structure. However, we also found that there is no simple, linear relationship between forest structure and biodiversity. While the simplistic assumption that more tree biomass automatically leads to higher biodiversity only holds true when we contrast natural forest with agroforestry systems, no such simple relationship is evident in different types of the latter. We therefore conclude that the conservation of carbon stocks and in particular of the forest-related biodiversity mainly depends on the preservation of natural forest habitats. Reduction of canopy tree density and cover within agroforestry systems leads to substantial carbon losses of around 50 Mg C ha^{-1}, but only to limited and taxon-specific biodiversity losses. Our study thus

suggests that remuneration schemes aimed at preserving or increasing carbon stocks in tropical forest regions should focus on maintaining natural forest ecosystems, in support of current political initiatives for Reducing Emissions from Deforestation and forest Degradation (REDD) [12,13,14].

On the other hand, carbon storage of the three most carbon-rich agroforest systems was about 60% of that of the three natural forest plots (181 *versus* 284 Mg C ha^{-1}), suggesting that 1.6 ha of optimally managed agroforest could contribute to the conservation of carbon stocks as much as 1 ha of natural forest. As for all observational studies, this result is only valid if the studied natural forest plots were typical for the whole study area and therefore representative for those forest stands that have been converted prior to the recent farmer agroforest management, which was the case in our study (Steffan-Dewenter et al. 2007). In particular, we found that land use change towards agroforests did not lead to excessive losses of the long-term soil carbon stocks, unlike observed in ploughed arable land-use systems, where up to 25–30% of the soil carbon is commonly lost in the tropics [40]. Agroforests may, in terms of soil organic carbon, thus be closer to secondary forests, which on average have 9% less soil carbon than primary forests [40]. We suspect that the reason for the limited loss of carbon stocks in the soil is that in our study region the creation of agroforests usually is not achieved via total removal of tree cover that leads to strong erosion and decomposition, but rather through the partial removal of natural trees and gradual replacement by other tree species [26], thus preserving much of the root systems

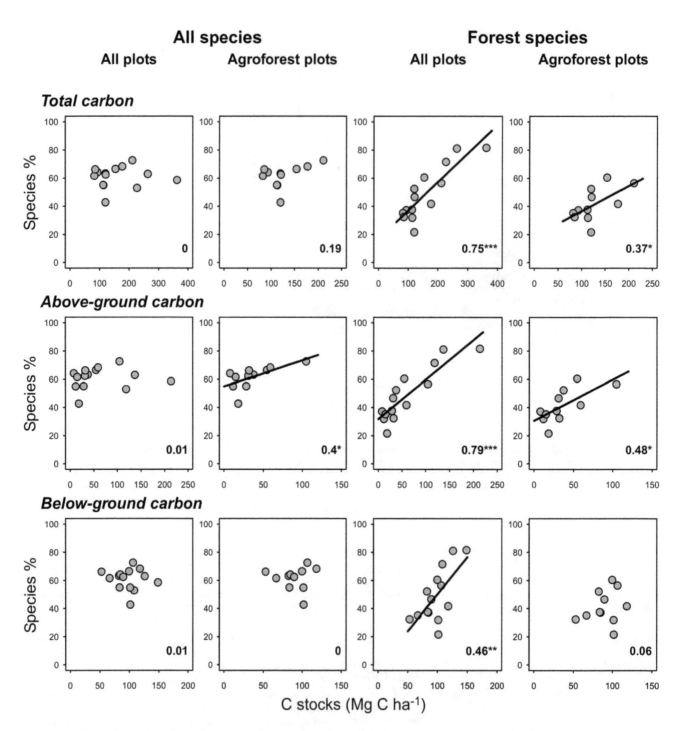

Figure 2. Relationships of species richness to carbon stocks. Relationships of species richness to carbon stocks, separated for all species and only those species recorded in the natural forest (forest species), for total, above-ground and below-ground carbon stocks as well as for all 14 study plots and only the 11 agroforest plots. To summarize the species richness patterns of the 12 focal plant and animal groups, richness values were all standardized to 100% relative to the highest plot values of each group and then averaged across all taxa. All individual relationships are shown in Figs. S1–3. Numbers in each graph are coefficients of determination (R^2-values), trend lines are shown for significant relationships only. *p<0.05, **p<0.01, ***p<0.001.

and preventing erosion and decomposition. From the point of view of carbon stock management, such a gradual transition is therefore preferable to the wholesale removal of natural tree cover and replacement by agroforest trees. However, since the carbon flux to the soil via leaf and root litter is markedly lower in agroforestry systems than in the natural forest [42], the soil carbon stocks might become lower in the long run.

In agroforests, we failed to detect a close relationship between carbon stocks and the species richness of most taxa. Thus, management strategies to maximise carbon storage, as supported by the Kyoto Protocol, do not automatically enhance biodiversity

Table 1. Results of Wilcoxon's tests for matched pairs, comparing (a) the linear correlation (r-)values of the species richness of various study groups against tree species richness with (b) the r-values of the species richness of the same study groups against the species richness of natural forest trees, either in all study plots (upper half of the table) or only in the agroforest study plots (lower half).

	All study groups (N = 11)		Only animal groups (N = 8)	
	W	**P**	**W**	**P**
All plots (N = 14)				
All species	22	n.s.	12	n.s.
Forest-adapted species	10	<0.05	7	<0.05
Only agroforest plots (N = 11)				
All species	31	n.s.	14	n.s.
Forest-adapted species	11	<0.05	0	<0.01

W = test statistic, P = probability.

in agroforests. This raises the need to identify promising solutions to optimise both carbon and biodiversity management in these economically and ecologically important agricultural systems. We explored this option by contrasting the carbon-biodiversity relationships when considering native versus non-native trees and found that biodiversity was more closely linked to the former. This suggests that carbon-biodiversity win-win solutions can be achieved when not all natural trees are removed during forest conversion. This may be further optimized by focussing on the identity of the tree species through shade-tree management that combines shade levels allowing for both high yield and low cacao stress with a selection of diverse shade-tree species from natural forests that enhance the biodiversity of other taxa [43]. This, and possibly other similar relationships involving functional traits of the shade trees such as fruit quantity and quality, opens promising perspectives for optimised joint carbon-biodiversity management strategies in agroforests.

Supporting Information

Figure S1 Species richness of selected organisms in relation to total carbon stocks in the 14 plots. Species richness (number of species per plot) of 12 groups of organisms in relation to total carbon stocks in 14 plots of natural forest and cacao agroforests. Large black circles denote natural forest plots, small circles agroforests of varying tree density (white: 0–79 trees >20 cm dbh/ha; medium: 80–159 trees/ha; dark: 160–240 trees/ ha) with blue symbols showing total species richness and red symbols richness of species also recorded in the natural forest. Coefficients of determination values (R^2 values) including all plots are given in normal font and for significant relationships are illustrated by continuous lines, values only including the agroforests are given in italics and illustrated by dashed lines. *p<0.05, **p<0.01, ***p<0.001.

Figure S2 Species richness in the study plots in relation to below-ground carbon stocks. Species richness of 12 groups of organisms in relation to below-ground (soil + root) carbon stocks in 14 plots of natural forest and cacao agroforests. Symbols as in Fig. S1.

Figure S3 Species richness in the study plots in relation to above-ground carbon stocks. Species richness of 12 groups of organisms in relation to above-ground carbon stocks in 14 plots of natural forest and cacao agroforests. Symbols as in Fig. S1.

Acknowledgments

We are very grateful to the Lore Lindu National Park administration for the permission to conduct the study. We furthermore thank P. Mann, Arifin, D. Stietenroth, A. Malik, W. Lorenz, S. Tarigan, P. Höhn, F. Orend, and all plantation owners for their help in this work.

Author Contributions

Conceived and designed the experiments: CL GG TT. Performed the experiments: SA MB DB SRG SK GM RP SS CHS SGS IS-D SST. Analyzed the data: MK DH HFJ JK GM. Contributed reagents/materials/analysis tools: GG SK TT MK. Wrote the paper: MK DH HFJ JK.

References

1. Houghton RA (2005) Aboveground forest biomass and the global carbon balance. Global Change Biol 11: 945–958.
2. Lal R (2005) Forest soils and carbon sequestration. For Ecol Manage 220: 242–258.
3. IPCC Guidelines for National Greenhouse Gas Inventories (2006) [IPCC website. Available: http://www.ipcc-nggip.iges.or.jp/public/2006gl/index.html. Accessed 2012 Sept 19.]. Institute for Global Environmental Strategies for the Intergovernmental Panel on Climate Change.
4. Houghton RA (2008) Carbon Flux to the Atmosphere from Land-use Changes: 1850–2005. Carbon Dioxide Information Analysis Center, OakRidge National Loratory, U.S. Department of Energy.
5. Sala OE, Chapin III FS, Armesto JJ, Berlow R, Bloomfield J, et al. (2000) Global biodiversity scenarios for the year 2100. Science 287: 1770–1774.
6. Lewis SL, Phillips OL, Baker TR, Lloyd J, Malhi Y, et al. (2009) Increasing carbon storage in intact African tropical forests. Nature 457: 1003–1006.
7. Sodhi NS, Ehrlich PR (2010) Conservation Biology for All. Oxford University Press, Oxford.
8. Bekessy S, Wintle BA (2008) Using carbon investment to grow the biodiversity bank. Conserv Biol 22: 510–513.
9. Ebeling J, Yasué M (2008) Generating carbon finance through avoided deforestation and its potential to create climatic, conservation and human development benefits. Phil Trans R Soc B 363: 1917–1924.
10. Venter O, Meijaard E, Possingham H, Dennis R, Sheil D, et al. (2009) Carbon payments as a safeguard for threatened tropical mammals. Conserv Lett 2: 123–129.
11. Campbell BM (2009) Beyond Copenhagen: REDD+, agriculture, adaptation strategies and poverty. Global Environ Change 19: 397–399.
12. ATBC/GTÖ Marburg declaration (2009) The urgent need to maximize biodiversity conservation in forest carbon-trading. A joint communiqué of the Association for Tropical Biology and Conservation (ATBC) and the Society for Tropical Ecology (GTÖ) during their joint annual meeting in Marburg, Germany, 26–29 July 2009. (Marburg University website. Available: http:// www.uni-marburg.de/aktuelles/news/2009b/0735.pdf. Accessed 2012 Sept 19.).
13. Grainger A, Boucher DH, Frumhoff PC, Laurance WF, Lovejoy T, et al. (2009) Biodiversity and REDD at Copenhagen. Current Biol 19: 974–976.
14. Ghazoul J, Buttler RA, Mateo-Vega J, Pin Koh L (2010) REDD: a reckoning of environmental and development implications. Trends Ecol Evol 25: 396–402.
15. Fischer J, Brosi B, Daily G, Ehrlich P, Goldman R, et al. (2008) Should agricultural policies encourage land sparing or wildlife-friendly farming? Front Ecol Environ 6: 380–385.
16. Strassburg BBN, Kelly A, Balmford A, Davies RG, Gibbs HK, et al. (2010) Global congruence of carbon storage and biodiversity in terrestrial ecosystems. Conserv Lett 3: 98–105.
17. Miles L, Kapos V (2008) Reducing greenhouse gas emissions from deforestation and forest degradation: global land-use implications. Science 320: 1454–1455.
18. Dent DH, Wright SJ (2009) The future of tropical species in secondary forests. A quantitative review. Biol Conserv 142: 2833–2843.
19. Hardner JJ, Frumhoff PC, Goetze DC (2000) Prospects for mitigating carbon, conserving biodiversity, and promoting socioeconomic development objectives

through the clean development mechanism. Mitig Adapt Strat Glob Change 5: 60–81.

20. Matthews S, O'Connor R, Plantinga AJ (2002) Quantifying the impact on biodiversity of policies for carbon sequestration in forests. Ecol Econ 40: 71–87.

21. Huston MA, Marland G (2003) Carbon management and biodiversity. J Environ Manage 67: 77–86.

22. Zomer RJ, Trabucco A, Coe R, Place F (2009) Trees on Farm: Analysis of Global Extent and Geographical Patterns of Agroforestry. ICRAF Working Paper no. 89. World Agroforestry Centre, Nairobi, Kenya.

23. FAOSTAT (2009) Statistical databases [FAOSTAT website. Available: http://faostat.fao.org/default.aspx. Accessed 2011 Jun 12.].

24. Clough Y, Faust H, Tscharntke T (2009) Cacao boom and bust: sustainability of agroforests and opportunities for biodiversity conservation. Conserv Lett 2: 197–205.

25. Perfecto I, Armbrecht I, Philpott SM, Soto-Pinto L, Dietsch TV (2007) Shaded coffee and the stability of rainforest margins in northern Latin America. In: Tscharntke T, et al., editors. Stability of Tropical Rainforest Margins: Linking Ecological, Economic and Social Constraints of Land Use and Conservation. Springer Verlag, Berlin, 227–264.

26. Steffan-Dewenter I, Kessler M, Barkmann J, Bos M, Buchori D, et al. (2007) Tradeoffs between income, biodiversity, and ecosystem function during rainforest conversion and agroforestry intensification. Proc Natl Acad Sci U S A 104: 4973–4978.

27. Clough Y, Dwi Putra D, Pitopang R, Tscharntke T (2009) Local and landscape factors determine functional bird diversity in Indonesian cacao agroforestry. Biol Conserv 142: 1032–1041.

28. Sodhi NS, Brook BW, Bradshaw CJA (2007) Tropical Conservation Biology. Blackwell Publishing, Oxford.

29. Kessler M, Abrahamczyk S, Bos M, Buchori D, Putra DD, et al. (2009) Alpha and beta diversity of plants and animals along a tropical land-use gradient. Ecol Appl 19: 2142–2156.

30. Hertel D, Moser G, Culmsee H, Erasmi S, Horna V, et al. (2009) Below- and above-ground biomass and carbon partitioning in a palaeotropical natural forest – an in-depth study from Central Sulawesi (Indonesia). For Ecol Manage 258: 1904–1912.

31. Chave J, Andalo C, Brown S, Cairns MA, Chambers JQ, et al. (2005) Tree allometry and improved estimation of carbon stocks and balance in tropical forests. Oecologia 145: 87–99.

32. Clark DA, Brown S, Kicklighter DW, Chambers JQ, Thomlinson JR, et al. (2001) Measuring net primary production in forests: concepts and field methods. Ecol Appl 11: 356–370.

33. Mokany K, Raison J, Prokushkin AS (2006) Critical analysis of root: shoot ratios in terrestrial biomes. Global Change Biol 12: 84–96.

34. Smiley GL, Kroschel J (2008) Temporal change in carbon stocks of cocoa-*Gliricidia* agroforests in Central Sulawesi, Indonesia. Agrofor Syst 73: 219–231.

35. Rahayu S, Lusiana B, van Noordwijk M (2005) Aboveground carbon stock assessment for various land use systems in Nunukan District, Kalimantan. In: Lusiana B, et al., editors. Carbon Stocks in Nunukan: a spatial monitoring and modeling approach. Carbon Monitoring Team of Forest Resource Management and Carbon Sequestration (FORMACS) Project, Bogor, Indonesia. World Agroforestry Centre - ICRAF, SEA Regional Office, 21–33.

36. Baritz R, Seufert G, Montanarella L, Van Ranst E (2010) Carbon concentrations and stocks in forest soils of Europe. For Ecol Manage 260: 262–277.

37. Clough Y, Barkmann J, Juhrbandt J, Kessler M, Wanger TC, et al. (2011) Combining high biodiversity with high yields in tropical agroforests. Proc Natl Acad Sci U S A 108: 8311–8316.

38. Westphal C, Bommarco R, Carré G, Lamborn E, Morison N, et al. (2008) Measuring bee diversity in different European habitats and biogeographical regions. Ecol Monogr 78: 653–671.

39. Kessler M, Abrahamczyk S, Bos M, Buchori D, Putra DD, et al. (2011) Cost-effectiveness of plant and animal biodiversity indicators in tropical forest and agroforest habitats. J Appl Ecol 38: 330–339.

40. Don A, Schuhmacher J, Freibauer A (2011) Impact of tropical land-use change on soil organic carbon stocks – a meta-analysis. Global Change Biol 17: 1658–1670.

41. R Development Core Team (2011) R: A language and environment for statistical computing. R Foundation for Statistical Computing, Vienna, Austria. ISBN 3-900051-07-0 [The R project website. Available: http://www.R-project.org/. Assessed 2012 Sept 21.]

42. Hertel D, Harteveld MA, Leuschner C (2009) Conversion of a tropical forest into agroforest alters the fine root-related carbon flux to the soil. Soil Biol Biochem 41: 481–490.

43. Tscharntke T, Clough Y, Bhagwat SA, Buchori D, Faust H, et al. (2011) Ecological principles of multifunctional shade-tree management in cacao agroforestry landscapes. J Appl Ecol 48: 619–629.

16

Remote Sensing Analysis of Vegetation Recovery following Short-Interval Fires in Southern California Shrublands

Ran Meng[1]*, Philip E. Dennison[1], Carla M. D'Antonio[2], Max A. Moritz[3]

1 Department of Geography, University of Utah, Salt Lake City, Utah, United States of America, **2** Department of Ecology, Evolution and Marine Biology, University of California Santa Barbara, Santa Barbara, California, United States of America, **3** Department of Environmental Science, Policy, and Management, University of California, Berkeley, California, United States of America

Abstract

Increased fire frequency has been shown to promote alien plant invasions in the western United States, resulting in persistent vegetation type change. Short interval fires are widely considered to be detrimental to reestablishment of shrub species in southern California chaparral, facilitating the invasion of exotic annuals and producing "type conversion". However, supporting evidence for type conversion has largely been at local, site scales and over short post-fire time scales. Type conversion has not been shown to be persistent or widespread in chaparral, and past range improvement studies present evidence that chaparral type conversion may be difficult and a relatively rare phenomenon across the landscape. With the aid of remote sensing data covering coastal southern California and a historical wildfire dataset, the effects of short interval fires (<8 years) on chaparral recovery were evaluated by comparing areas that burned twice to adjacent areas burned only once. Twelve pairs of once- and twice-burned areas were compared using normalized burn ratio (NBR) distributions. Correlations between measures of recovery and explanatory factors (fire history, climate and elevation) were analyzed by linear regression. Reduced vegetation cover was found in some lower elevation areas that were burned twice in short interval fires, where non-sprouting species are more common. However, extensive type conversion of chaparral to grassland was not evident in this study. Most variables, with the exception of elevation, were moderately or poorly correlated with differences in vegetation recovery.

Editor: Ben Bond-Lamberty, DOE Pacific Northwest National Laboratory, United States of America

Funding: Funding for this research was provided by California Energy Commission grant 500-10-045. The funders had no role in study design, data collection and analysis, decision to publish, or preparation of the manuscript.

Competing Interests: The authors have declared that no competing interests exist.

* Email: Ran.Meng@utah.edu

Introduction

Fire is an important ecological process with dynamic interactions between vegetation, biogeochemical cycles, and human activity [1–3]. Dry summer climates make Mediterranean ecosystems, like southern California chaparral, particularly prone to wildfire [4,5]. Southern California has experienced multiple, massive fires over the past two decades. For example, in October 2003, 300,277 ha (742,000 acres) were burned within one week in the Cedar, Otay and Paradise fires. Four years later many of the same areas were burned in the Harris, Witch and Poomacha fires. Repeat, short-interval fires may push ecosystems into new states, and recently there has been much discussion about disturbance regime thresholds beyond which ecosystem characteristics change dramatically due to a loss of resilience of the vegetation [6]. It has been suggested that in ecosystems dominated by fire adapted species, either very long or very short fire return intervals will drive long term compositional change [7,8]. Given widespread predictions of increasing fire frequency with ongoing climate change [9,10], it is important to explore whether altered fire return intervals can drive long term vegetation type change (often referred to as "type conversion"), and which environmental variables predict where such vegetation type change is likely to occur.

Chaparral, commonly found in the mountain areas of coastal California, Baja California and foothills of the Sierra Nevada [11], is dominated by evergreen, sclerophyllous, shrub species that have the ability to reestablish themselves following severe fires [12,13]. Chaparral species have different regeneration modes following fire, and generally can be grouped into three types: obligate seeders, obligate resprouters and facultative seeders [14]. Obligate seeders can germinate from fire-protected seed banks, obligate resprouters can resprout from fire resistant structures, and a few facultative species combine the two regeneration strategies. Hanes [12] observed that the abundance of obligate seeding species in southern California chaparral tended to decrease with the elevation, likely due to the co-variation between elevation and precipitation.

Landscape factors affecting chaparral regeneration include fire severity, fire history, indigenous life forms, site-specific water availability, latitude and topography [12,14–17]. Temporal factors (e.g. post-fire precipitation in wet season, post-fire temperature in winter) may also impact succession patterns in chaparral shrublands [12,14,15]. In general, the growth of chaparral shrubs is fast in the early post-fire years and slows when they become large enough to shade and cover sub-shrub and herbaceous plants that may persist [11]. Within five to seven years after fire, chaparral shrubs may regain dense cover [11], but only after two or three decades does their canopy recover to pre-fire status [12].

Although chaparral communities contain fire-adapted species, alterations in characteristics of fire regimes can make these communities vulnerable to alien plant invasion by reducing the abundance of vulnerable species [13,18]. Short fire return intervals can prevent obligate seeding species from reaching maturity and producing seed [19–24]. Zedler et al. [21] for example, showed that three years after a short (1 year) return interval fire in southern California shrublands, there was a drastic decline in density of an obligate seeding shrub species and large reductions in a resprouting species. Likewise, Keeley and Brennan [13] found that native chaparral woody species recovered well in 4 years after a first fire, but declined after a second fire that occurred with a short interval (3 years or 4 years), with obligate seeding shrubs being the most affected. In contrast, annual species increased and exotic annuals outnumbered native ones at the same sites after repeat fires [13]. Lippitt et al. [17] surveyed the post-fire change in chaparral stands after short interval fires (<5 years, 1977–2003) with the help of remote sensing imagery and a historical wildfire dataset in San Diego County, California, and they concluded that alteration and type conversion could occur under short fire return intervals. While these previous studies examined relatively small areas and short periods following multiple fires, they suggest that short fire return intervals can cause substantial changes in site-scale species composition. In addition to short fire return intervals, other factors, such as grazing and atmospheric nitrogen pollution, may contribute to alien plant invasions in semi-arid shrublands [25,26].

Extensive "range improvement" practices for increasing the production of herbaceous forage cover for livestock have affected the structure and composition of the western United States shrubland communities since the middle of 20th century [27,28]. These practices include removal and control of woody species, seeding with forage species, and grazing management [28–30]. Various techniques, including controlled burning, herbicide, and mechanical treatments, have been implemented to force type conversion from a community of woody vegetation to a stable and persistent community of herbaceous vegetation, or to at least postpone succession [27,31]. It has been shown that suitable soil, climate, and topography are necessary for successful type conversion for range use [27,28]. While periodic repeat burning has been used to reduce shrubby vegetation in diverse locations, in California it is considered the least effective method for removing woody vegetation for range improvement [27]. Despite extreme treatments and follow-up activities, chaparral can be very persistent, and the intended long term type conversion with the objective of range improvement can be difficult to achieve [28,31]. Thus, contrary to the hypothesis that repeat short interval fires will cause type conversion, past range improvement studies present an alternative hypothesis that type conversion may be difficult and should be a relatively rare phenomenon across the landscape.

To date, most studies of potential type conversion of chaparral by repeat fire have been implemented at the local-site scale over short time intervals (e.g. Keeley and Brennan [13]). Patterns demonstrated at these small scales may not be persistent over time or across larger regional scales. Due to the large-scale spatial and temporal variability of wildfire disturbance and post-fire response in terrestrial ecosystems, various studies have utilized remote sensing data to monitor and study post-fire vegetation response and recovery trajectories [17,32–36]. Remote sensing techniques

offer the opportunity to study fire effects and recovery dynamics across large areas, providing measurements at the landscape (local-to-regional) scale. To evaluate the effects of short fire return intervals on chaparral recovery across coastal southern California, we used Landsat-5 Thematic Mapper (TM) images and a remotely sensed vegetation index. The effects from two fires in rapid succession (fire return intervals of less than 8 years) were determined by comparing twelve areas that burned twice to adjacent control areas burned only once. We used our results to examine two alternative hypotheses: 1. Two fires separated by a short interval should adversely impact the recovery of obligate seeding species following the second fire, thus reducing chaparral cover in comparison to an adjacent area burned only once; 2. Persistent type conversion of chaparral shrubland to grassland is rare and spatially limited, even after short return interval fires (<8 years). Correlations between vegetation recovery and climate, fire history and elevation variables were also examined.

Methodology

Study region

The study area is located in coastal southern California within the combined area of three Landsat-5 TM scenes (path 40, row 37; path 41, row 36; path 42, row 36; Figure 1). Patchy mosaics of grassland, woodland, coastal sage scrub, and chaparral are the dominant vegetation types in the study area [37]. The Mediterranean climate is characterized by cool-wet winters (December to February) and dry-hot summers (June to September). Large fires occur predominantly from mid-summer until November, corresponding to low live fuel moisture [38] and the occurrence of Santa Ana winds [39]. While historic fire return intervals are poorly known, it is generally believed that they varied from 20–65 years depending on aspect, elevation and distance to coast [40,41].

Data

Landsat-5 TM imagery with a 30 m spatial resolution covering the study area in 1985 and 2010 was obtained from the USGS GLOVIS website (http://glovis.usgs.gov; Table 1). September dates were targeted for all imagery to ensure phenological comparability. ACORN (Atmosphere Correction Now; http://www.imspec.com) was selected to implement reflectance retrieval for the 1985 scene in each path-row, due to lower cloud cover for these scenes. The 2010 scenes were then calibrated to the 1985 scenes using pseudo-invariant features found by iMAD (iteratively re-weighted Multivariate Alteration Detection) [42].

Because pre-fire chaparral areas were targeted in this analysis, a map of vegetation type coincident with or prior to 1985 was needed. The 1977 CALVEG (Classification and Assessment with Landsat of Visible Ecological Groupings) —a statewide classification system developed by US Forest Service— was used to subset the Landsat imagery. The 1977 CALVEG map was created based on 1:250,000 scale Landsat Multispectral Scanner (MSS) imagery acquired between 1977 and 1979. Through identifying distinctions among canopy reflectance values of Landsat MSS imagery, field verification and professional guidance, vegetation type "series" based on dominant overstory species were mapped. The Minimum Mapping Unit (MMU) of the 1977 CALVEG is 162 ha (400 acres), mainly due to the limitations of image resolution (http://portal.gis.ca.gov/). MMU refers to the smallest size or dimensions for an entity to be mapped as a discrete feature for a given map scale [43].

Historical fire perimeters between 1985 and 2010 were derived from California Department of Forestry and Fire Protection FRAP (Fire and Resource Assessment Program; http://frap.cdf.ca.gov/)

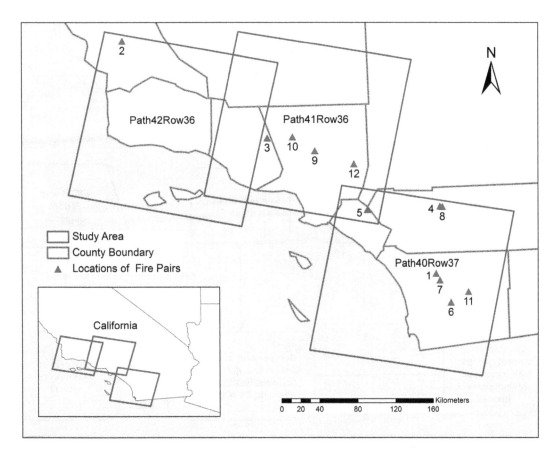

Figure 1. Study area. Study area covering coastal southern California. Path and row are indicated in the center of each Landsat TM scene. Locations and ID numbers of 12 example fire pairs are indicated on the map.

and US Forest Service/US Geological Survey MTBS (Monitoring Trends in Burn Severity; http://www.mtbs.gov/) datasets. Updated annually, the CALFIRE FRAP dataset represents the most complete GIS-format record of fire perimeters in California dating back to the early 1800s. This dataset provides extensive information for the period 1985–2010. The CALFIRE FRAP MMU is not consistent with the MMU for the CALVEG dataset, and varies by location and through time. During the period 1950–2001, the CALFIRE FRAP dataset included US Forest Service fires larger than 4 ha (10 acres) and CALFIRE fires larger than 121 ha (300 acres). Bureau of Land Management and National Park Service data were added starting in 2002, and the MMU for CALFIRE fires dropped to 20 ha (50 acres) for shrubland and 4 ha (10 acres) for forest. The variable MMU for this dataset is an

acknowledged weakness, but the CALFIRE FRAP dataset should still capture most of the area burned by wildfires in southern California during the study period. MTBS data were used to supplement the CALFIRE FRAP dataset. MTBS aims to map perimeters and burn severity of fires larger than 405 ha (1,000 acres) across the western United States using 30 m Landsat TM data [44].

Climate data—monthly mean precipitation and minimum temperature—between 1985 and 2010 were downloaded from the PRISM (Parameter-elevation Regressions on Independent Slopes Model) website (http://www.prism.oregonstate.edu/). PRISM makes use of point measurements of precipitation and temperature to generate continuous digital grid estimates of climatic data with a 4 km spatial resolution [45]. The DEM

Table 1. Dates and estimated cloud cover for each Landsat image used in the study.

Path	Row	Date	Estimated Cloud Cover (%)
40	37	14 Sep 1985	0
40	37	19 Sep 2010	23
41	36	21 Sep 1985	0
41	36	26 Sep 2010	0
42	36	12 Sep 1985	0
42	36	01 Sep 2010	33

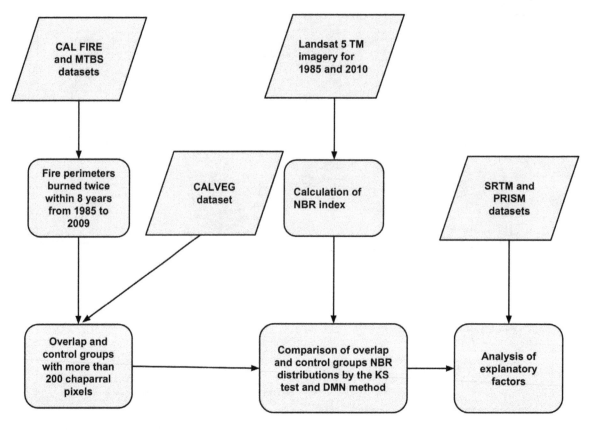

Figure 2. Flowchart of the analysis procedure.

(Digital Elevation Model) used in this study was derived from the SRTM (Shuttle Radar Topography Mission; http://www2.jpl.nasa.gov/srtm/) with a 30 m spatial resolution consistent with the Landsat-5 TM imagery. Climatic and topographic variables derived from PRISM and SRTM were used in linear regression as explanatory factors of vegetation recovery following short-interval fires.

Analysis procedure

Figure 2 illustrates the analysis procedure we used. Landsat-5 TM data were used for calculating NBR (Normalized Burn Ratio-described below) values for both 2010 and 1985 images. Conditional historical fire perimeters were extracted from the fire history dataset, in order to determine areas that burned twice (overlap) from October 1985 to December 2009 with a short fire return interval (defined here as <8 years) during this period, and similar areas that had only burned once in the latter of the two fires (control). Differences in the NBR distributions between overlap and control groups were used to compare vegetation recovery; conditions affecting the localized vegetation vigor or phenology in the control and overlap areas are assumed to be similar. Lastly, variables that could potentially influence vegetation recovery, including elevation, post-fire temperature, and post-fire precipitation, were compared to changes in NBR distributions using scatter plots and linear regression. A total 1804 fire cases were recorded in the CALFIRE FRAP datasets between 1984 and 2009 within the study area. Before analyzing the effects of explanatory factors on potential type conversion, global Moran's I was calculated to determine whether the change in DMN values (Table 2-described below) were spatially independent [46].

Moran's I measures the spatial autocorrelation based on feature locations and associated variable values.

Calculation of NBR. Various remote sensing instruments and techniques have been used for evaluating wildfire and its impacts [2]. It has been shown that bands 3, 4, 5, and 7 from Landsat-5 TM data have the largest responses to burn severity [47,48]. NBR uses a combination of band 4 (near infrared or NIR) and band 7 (shortwave infrared or SWIR) that provides the best distinction between burned and unburned areas. NBR is calculated as:

$$NBR = (R_{NIR} - R_{SWIR})/(R_{NIR} + R_{SWIR}) \quad \text{(Equation 1)}$$

where R_{NIR} and R_{SWIR} represent surface reflectance for TM band 4 (760–900 nm) and band 7 (2,080–2,350 nm). NBR is functionally identical to NDII (Normalized Difference Infrared Index), a SWIR-based vegetation index correlated with vegetation water content [49]. Changes in NBR and its relative form (relative differenced Normalized Burn Ratio (RdNBR)) derived from Landsat data have been widely demonstrated to correlate with field measurements of burn severity [48,50]. In this study, NBR is used as an indicator of vegetation cover, where an increase in NBR value is correlated with higher vegetation cover [48]. Lopez Garcia and Caselles [51] found that NBR not only can be used for the discrimination of burned areas, but was also capable of monitoring post-fire regeneration over burned areas in a Mediterranean ecosystem. In a more recent study, NBR was also found to support longer detectable periods of fire effects than other common band ratios, such as NDVI (Normalized Difference Vegetation Index) and EVI (Enhanced Vegetation Index) [52].

Table 2. Descriptions of indices used in this study.

Indices	Calculations	Indications
D_{1985}	The maximum difference between overlap and control groups in NBR cumulative distributions from the 1985 image	Difference in vegetation cover between a control area and a corresponding overlap area in 1985
DMN_{year}	Median (Control(NBR_{year})) − Median (Overlap(NBR_{year}))	Higher (positive)/lower (negative) vegetation cover within a control area relative to a corresponding overlap area in a specific year
Change in DMN	$DMN_{2010} - DMN_{1985}$	Increase (positive)/decrease (negative) in vegetation cover within a control area relative to a corresponding overlap area after repeat burn

NBR was calculated using Equation 1 [48] for all 1985 and 2010 Landsat-5 TM images.

Comparison of fire overlap and control groups. Fires that burned the same area twice within a period of eight years were selected for analysis. This time frame was chosen based on time required for obligate seeding shrub species to begin producing seed [8,53,54]. All successive fires with short return intervals (<8 years) within the study area, from October 1985 to December 2009, were selected. Some apparently incorrect fire perimeters in the CALFIRE dataset were deleted from the final result. In order to remove any edge effects, all fire perimeters were buffered by 3 pixels (90 m). The perimeters of successive fires from CALFIRE FRAP were replaced by MTBS where fire size exceeded the size necessary for inclusion in MTBS. MTBS burn severity was then used to exclude unburned islands inside of fire perimeters. Finally, any areas burned in the 8 years preceding the 1985 Landsat image, as indicated in the CALFIRE FRAP dataset, were excluded to ensure unlikely change in vegetation type since the 1977 CALVEG classification.

For each fire pair, NBR values of pixels classified as chaparral in the 1977 CALVEG classification and within the fire overlap area were derived from 1985 and 2010 imagery to constitute groups called "overlap". Similarly, NBR values within the latest (second) fire area, but excluding the overlap area, were also derived to constitute groups called "control". The overlap group contains NBR values of chaparral pixels burned twice, and the control group contains NBR values of chaparral pixels burned once since 1985 by the most recent fire of the two fires that burned the "overlap" group. The MMU inconsistency of CALFIRE FRAP dataset in the study period, as well as the difference in MMU between 1977 CALVEG classification and CALFIRE FRAP dataset, added uncertainty to the comparison of fire overlap and control groups.

Figure 3 shows an example of qualified fire perimeters within the Landsat-5 TM imagery before and after the repeat burn. Targeted chaparral pixels within fire perimeters were not required to be spatial contiguous, due to the multiple masks applied to the fire perimeters. A minimum pixel cluster size was not used to exclude isolated pixels from the analysis, since further analysis was based on the distribution of all pixels belonging to the control or overlap groups.

Distributions of NBR values in the overlap and control pairs with more than 200 chaparral pixels each were compared in two ways: the KS (Kolmogorov-Smirnov) test and a DMN (Difference Median NBR) method. The threshold of 200 pixels was established based on obtaining a population sufficient for calculating a distribution of values for comparison, producing 39 potential overlap and control pairs. In order to make sure that the

difference between overlap and control distributions in 2010 was likely due to repeat fires, and not differences in pre-fire vegetation cover, only the fire pairs with similar 1985 pre-fire NBR distributions were further compared. The KS test is a non-parametric test for comparing differences between two distributions. The D value returned by the KS test measures the maximum difference in two cumulative distributions [55]. In this study, the D value is the maximum difference between two cumulative distributions of NBR values in the pixels derived from control and overlap pairs (Table 2). Small D values calculated for 1985 indicated that pre-fire vegetation cover distributions were similar for the overlap and control areas. As D_{1985} values increased, pre-fire vegetation cover distributions were more distinct.

A threshold D_{1985} value of 0.16 was empirically established to identify pre-fire NBR distributions that were similar enough for post-fire comparison. This threshold was determined based on visual inspection of NBR distributions, and on correlations between D_{1985} and the difference in median NBR values, a measure described below. A local maximum in this correlation was found for a D_{1985} value of 0.16. The D_{1985} constraint produced 12 pairs of control and overlap areas with similar pre-fire distributions of NBR values and at least 200 chaparral pixels in both groups.

DMN was calculated using the equation in Table 2, taking the difference between median NBR values of control and overlap distributions. DMN is a measure of difference between control and overlap distributions, but unlike D value, DMN by definition is based on the center of the distribution and can be positive or negative. A positive DMN value indicates a higher median vegetation cover in a control area relative to an overlap area; a negative value indicates a higher median vegetation cover in an overlap area relative to a control area. DMN values should be strongly positive in 2010 relative to 1985 if type conversion occurred at the landscape scale, due to reduced vegetation cover in areas burned twice.

Analysis of explanatory factors. The potential explanatory factors listed in Table 3 were compared to changes in DMN values from pre-fire (1985) to post-fire (2010) using simple linear regression. For fire history factors, scatter plots were made to show the relationships between time interval between the two fires, time since second fire to 2010 (e.g. time for recovery), and change in DMN values. For each fire pair, the mean elevation, mean precipitation in the wet season (winter and spring) immediately following fire, and mean average January minimum temperature one year following fire were calculated from the PRISM and SRTM datasets (Table 3). Mean elevation, mean precipitation in the wet season (winter and spring), and mean average January

Figure 3. Fire areas and 1985 and 2010 Landsat images. Landsat-5 TM images covering areas burned by the Cedar fire in 2003 and the Witch fire in 2007. a. 1985 Landsat-5 TM, bands 5, 4, and 3 displayed as red, green, and blue, respectively. b. 2010 Landsat-5 TM, bands 5, 4, and 3 displayed as red, green, and blue, respectively.

minimum temperature were compared to change in DMN to analyze potential trends.

Validation of the DMN comparison. One hundred points were generated randomly in the both overlap and control areas of each qualified fire pair. High resolution imagery in Google Earth was used to inspect the vegetation cover at each point. The used Google Earth image dates ranged from 2009–2013, in order to use an image date that covered both the control and overlap areas in each pair and provide the highest contrast between evergreen chaparral canopies and senesced herbaceous vegetation or drought deciduous shrubs. Each randomly generated point was labeled as having evergreen shrub canopy or non-evergreen shrub canopy (e.g. grass, drought-deciduous shrub, soil). The percentages of evergreen shrub canopy pixels within the control and overlap groups were calculated, and differences between the control and overlap groups were compared to DMN_{2010}.

Results

Comparisons

Twelve of 39 fire pairs were sufficiently similar in the pre-fire (1985) distribution of NBR values (Table 4) to be considered adequate comparisons. The global Moran's I result (Moran's I index: 0.104; z-score: 1.07; p-value: 0.28) on the change in DMN indicates that values are spatially independent, likely due to the reduced degree of pixel contiguity caused by complex overlapping boundaries of chaparral polygons and burned areas (Figure 3).

All 12 control-overlap pairs were further compared using their NBR cumulative distributions. Control-overlap pairs generally had similar distributions in 1985 and 2010, but there were differences in the distributions between specific pairs. For example, for the 1999 Pine fire and 2006 Esperanza fire, the NBR distribution for the overlap area shifts to slightly higher median NBR than the control area in 2010 (Figure 4a). By contrast, for the 2003 Cedar

Table 3. Explanatory variables used in simple linear regression analysis.

Explanatory Variables	Descriptions
Time since second fire to 2010	Time since the occurrence of second fire in the short interval fire pair to 2010 (years)
Time interval between two fires	Interval between successive fires in fire pair (years)
Mean precipitation in winter season	Mean precipitation within area burned by second fire in December, January and February immediately following fire (mm)
Mean precipitation in spring season	Mean precipitation within area burned by second fire in March, April and May immediately following fire (mm)
Mean elevation	Mean elevation within area burned by second fire (m)
Mean average January minimum temperature	Mean average minimum temperature within area burned by second fire, in January following second fire (C)

fire and 2007 Witch fire, the difference between control and overlap distributions becomes smaller after being burned twice (Figure 4b). For the 1996 Highway58 fire and 2003 Parkhill fire, the overlap and control group distributions were more similar in 1985, but in 2010 there is a larger difference between them, with the overlap group shifted to lower NBR (lower vegetation cover) than the control group (Figure 4c). Table 4 shows that most of fire pairs did not undergo substantial changes in DMN. Six of twelve cases, representing a total area of 3515 ha (52% of the burned area evaluated), have a positive change in DMN consistent with our hypothesis that repeat short interval fire should cause type conversion. Six of twelve cases have a negative change in DMN, and the most extreme changes in DMN have negative values.

Validation results

Difference in percent evergreen shrub canopy cover between control and overlap areas, calculated from Google Earth imagery, was compared to DMN_{2010} (Figure 5). The two measures have a significant, positive correlation (Pearson's correlation: 0.63, p-value: 0.03), indicating that larger differences in evergreen shrub cover between control and overlap areas are correlated with larger differences in DMN calculated from 30 m Landsat-5 TM data.

Fire history factors

Time interval between two fires and time since second fire to 2010 were plotted against change in DMN (Figure 6a; Figure 6b). There was no apparent trend between time interval between fires or recovery time and change in DMN (Table 5). Shorter fire return intervals did not produce higher DMN values, which would be consistent with type conversion due to elimination of obligate seeding species.

Climatic and topographic factors

Correlation analysis of change in DMN with mean post-fire precipitation in wet season and mean average minimum temperature in January for each fire pair revealed no trends and is thus not shown here (Table 5). Elevation, by contrast, was most strongly correlated with the change in DMN between 1985 to 2010: as elevation increased, the change in DMN values from 1985 to 2010 tended to be more negative (Figure 7; Table 5).

Discussion

This remote sensing-based approach allows evaluation of the effects of short fire return interval on chaparral recovery on broader spatial and temporal scales than previous site-scale

Table 4. Comparison of results for fire pairs with similar cumulative distributions of 1985 NBR values[a].

ID Number	Path Row	Year1 Fire1	Year2 Fire2	D_{1985}	Change in DMN (2010–1985)	Overlap Area (ha)
1	p40r37	2003 Paradise	2007 Poomacha	0.135	0.045	1031
2	p42r36	1996 Highway58	2003 Parkhill	0.078	0.037	40
3	p40r37	2003 Piru	2007 Ranch	0.065	0.025	79
4	p40r37	2001 Silent	2006 Esperanza	0.140	0.017	22
5	p40r37	2002 Green	2006 Sierra	0.128	0.016	214
6	p40r37	2003 Cedar	2007 Witch	0.056	0.009	2130
7	p40r37	2003 Paradise	2007 Witch	0.057	−0.001	1518
8	p40r37	1999 Pine	2006 Esperanza	0.066	−0.005	210
9	p41r36	2007 North	2009 Station	0.086	−0.018	60
10	p41r36	2002 Copper	2007 Buckweed	0.060	−0.019	1368
11	p40r37	1999 Banner	2002 Pines	0.110	−0.051	23
12	p41r36	1996 Bichota	2002 Curve	0.094	−0.095	25

ID number refers to fire pairs shown in Figure 1.
[a]Sorted by decreasing change in DMN.

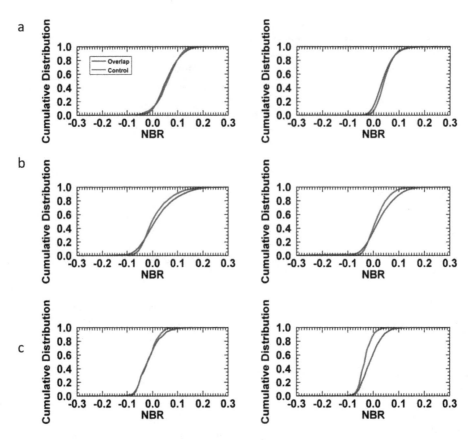

Figure 4. Pre-fire (1985) and post-fire (2010) cumulative distributions. Pre-fire (1985) and post-fire (2010) cumulative distributions of NBR values from different fire pairs. a. 1999 Pine fire and 2006 Esperanza fire b. 2003 Cedar fire and 2007 Witch fire. c. 1996 Highway58 fire and 2003 Parkhill fire.

studies. This study evaluated change at a relatively coarse spatial resolution (30 m), and thus our method could not detect changes in species composition within chaparral. Positive changes in DMN, indicative of possible reduction in vegetation cover, were found for 50% (6 out of 12) of the sites examined. Only two of these six overlap/control areas (16% of the burned area evaluated) at lower elevation showed large changes in DMN that are consistent with type conversion. In short, our evidence does not

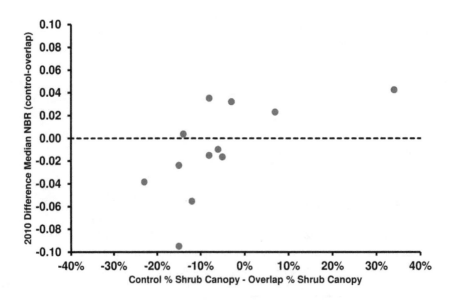

Figure 5. Post-fire difference in percent evergreen shrub canopy versus DMN$_{2010}$. Post-fire difference in percent evergreen shrub canopy, derived from Google Earth imagery for control and overlap areas, versus DMN$_{2010}$.

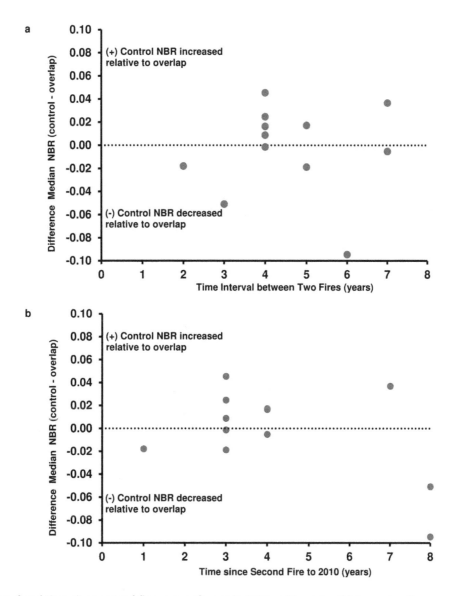

Figure 6. Time interval and time since second fire versus change in DMN. a. Time interval between two fires versus change in DMN from 1985 to 2010. **b.** Time since second fire to 2010 versus change in DMN from 1985 to 2010.

support that extensive type conversion has occurred on the landscape scale due to recent short-interval fires. Hence our results are consistent with the hypothesis that it is difficult to convert existing chaparral to grassland [28,31].

Replacement of specific shrub species with other shrub species may result in the same vegetation type and similar vegetation cover. This type of species-level change cannot be ruled out based on our landscape scale analysis. Also, use of median NBR as a

Table 5. Simple linear regression results for explanatory variables.

Variables	Pearson's r	R-squared	$P_r^a (>|t|^b)$
Time since second fire to 2010 (years)	−0.49	0.24	0.106
Time interval between two fires (years)	0.02	0.0004	0.963
Mean precipitation in winter season (mm)	−0.02	0.0004	0.963
Mean precipitation in spring season (mm)	−0.02	0.0004	0.963
Mean elevation (m)	−0.80	0.64	0.0019*
Mean average January minimum temperature (C)	0.02	0.0004	0.963

* P_r <0.01; [a] Probability of a standard normal variable; [b] Value of t distribution.

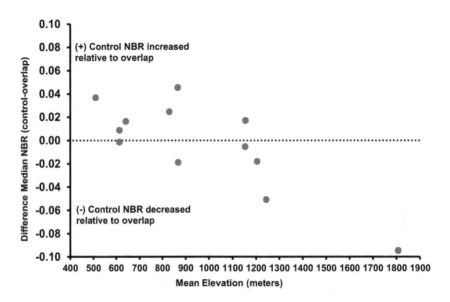

Figure 7. Mean elevation versus change in DMN.

metric does not exclude changes in the tails of NBR distributions that may be attributable to type conversion. However, distribution plots (Figure 4) did not indicate any systematic changes in the qualitative shape of NBR distributions. In addition, our methodology excluded cumulative, small scale type conversions that are beyond the capacity of Landsat-based analysis, and did not examine areas burned more than twice during the study period. It is also possible that locations most sensitive to long-term type conversion have already been converted many decades ago, so it may be difficult to discern which shrub species or topographic locations have suffered the most substantial losses.

Elevation has been deemed as one of the strongest predictors of post-fire regrowth patterns in chaparral [14,16]. In this study, simple linear regression also suggested that elevation explained 64% of the variability in change in DMN and produced the strongest correlation (−0.80). As precipitation increases with elevation, non-sprouting species become less common [12]. As multiple burns can reduce non-sprouting species more significantly [13,21], overlap areas at higher elevation could be expected to be less susceptible to type conversion than overlap areas at lower elevation. The invasion of alien plants in chaparral has been also attributed to the proximity of chaparral to coastal sage scrub and herbaceous communities at lower elevations with gentle slopes [17,21,56]. It is also likely that low elevation sites were more vulnerable to human activities, such as urbanization and increased ignition frequency.

Contrary to previous studies, effects of time interval between two fires and time since second fire were not found to be significant for explaining variability in post-fire recovery [17]. The apparently strong influence of elevation on differences in recovery between once- and twice-burned areas may have obscured more subtle differences in recovery caused by time interval or time since second fire. Also, reliance on distributions calculated from large areas prevented analysis of other factors that may influence recovery, including topographic slope and aspect, and soil type. Post-fire recovery of chaparral is likely to be sensitive to soil moisture, and moisture availability can be affected by topographic exposure [17]. Fire intensity and duration also were not incorporated to explain the variability in change in DMN. Nevertheless, impacts of fire intensity and duration on vegetation

recovery are complicated in chaparral; for some species, fire inhibits seeding recruitment or delays resprouting recruitment, but in other species fire enhances plant recovery [14].

Fire-adaptive traits enable chaparral species to reestablish themselves rapidly even following severe stand-replacing fire, referred to as auto-succession or a direct regeneration process first defined by Hanes [12]. Auto-succession assumes chaparral species are highly adapted to disturbance, and the phenomenon of vegetation type conversion (losing plant species) after fire is rare [57]; nevertheless, an extreme change in plant composition may occur when fire disturbance exceeds its natural range of variation [58]. In our study, most burned areas with short fire return interval demonstrated similar post-fire response to control areas with longer fire return intervals (> 8 year). Compounding effects of multiple disturbances may still trigger type conversion in chaparral, but this process may be rare and spatial-limited.

While multispectral remote sensing is insensitive to compositional shifts within shrubland vegetation, it may be able to complement future site-scale studies of fire recovery and be used to understand fire recovery processes occurring across large regions. Past studies using hyperspectral and very high spatial resolution remote sensing have demonstrated that dominant shrubland or forest species as well as main post-fire regeneration classes can be mapped [59–61]. Advanced image analysis techniques (i.e. object-oriented image analysis) may also potentially improve the accuracy of post-fire related information [62]. Current and proposed satellite missions providing improved fire detection and monitoring [63,64] and mapping of vegetation functional types [65] may allow further enhancements to monitoring of vegetation response to wildfire at continental-to-global scales.

Conclusions and Future Work

Using a historical wildfire dataset and remote sensing imagery, we have analyzed differences in vegetation cover following short interval fires across a broad expanse of southern California chaparral shrublands. Changes in the distributions of NBR values from qualified overlap/control fire pairs did not indicate extensive type conversion of shrubland to grassland, suggesting that type conversion of shrubland to grassland may be a spatially limited and regionally rare phenomenon. Sites with large, positive changes in DMN values most consistent with type conversion were at lower elevation in proximity to communities of coastal sage scrub, in accordance with previous results found by Lippitt *et al.* [17]. Simple linear regression indicated that only elevation was strongly correlated to observed variation in vegetation recovery across the once and twice burned pairs. At lower elevation, non-sprouting species are more common and thus the vegetation is likely more vulnerable to type conversion [12,14,17].

While loss of native shrublands is an important conservation issue, our findings suggest that recent localized studies demonstrating shrub declines should be resurveyed through time. Due to variability in chaparral recovery from short return interval fires, conservation strategies and management plans may need to take into account elevation and other site characteristics. Under future climate change scenarios in the western United States, a small change in climate could possibly lead to drastic and divergent shifts in fire activity and ability of chaparral species to recover from fire, with the potential of reaching "tipping points" of type conversion in Mediterranean-type ecosystems [66]. Prospective management plans should be prepared to deal with this situation. In addition, more studies are needed to investigate disturbance-chaparral interactions by quantifying the relative contributions of varying fire frequency on the composition of chaparral species.

Author Contributions

Conceived and designed the experiments: RM PED CMD MAM. Performed the experiments: RM PED. Analyzed the data: RM PED. Wrote the paper: RM PED CMD MAM.

References

1. Sugihara NG, van Wagtendonk JW, Fites-Kaufman J (2006) Fire as ecological process. In: Sugihara NG, van Wagtendonk JW, Shaffer KE, Fites-Kaufman J, Andrea ET, editors. Fire in California's ecosystems. Berkeley Los Angeles London: University of California Press. pp. 58–74.

2. Lentile LB, Holden ZA, Smith AMS, Falkowski MJ, Hudak AT, et al. (2006) Remote sensing techniques to assess active fire characteristics and post-fire effects. International Journal of Wildland Fire 15: 319–345.

3. Bowman DMJS, Balch JK, Artaxo P, Bond WJ, Carlson JM, et al. (2009) Fire in the earth system. Science 324: 481–484.

4. Keeley JE, Fotheringham CJ, Moritz MA (2004) Lessons from the October 2003 wildfires in southern California. Journal of Forestry 102: 26–31.

5. Keeley JE, Safford H, Fotheringham CJ, Franklin J, Moritz MA (2009) The 2007 southern California wildfires: Lessons in complexity. Journal of Forestry 107: 287–296.

6. Moritz MA, Hurteau MD, Suding KN, D'Antonio CM (2013) Bounded ranges of variation as a framework for future conservation and fire management. Annals of the New York Academy of Sciences 1286: 92–107.

7. Franklin J, Coulter CL, Rey SJ (2004) Change over 70 years in a southern California chaparral community related to fire history. Journal of Vegetation Science 15: 701–710.

8. Jacobsen AL, Davis SD, Fabritius SL (2004) Fire frequency impacts non-sprouting chaparral shrubs in the Santa Monica Mountains of southern California. In: Arianoutsou M, Papanastasis VP, editors. Proceedings 10th MEDECOS Conference. Rotterdam: Millpress Press. pp. 1–5.

9. Dale VH, Joyce LA, McNulty S, Neilson RP, Ayres MP, et al. (2001) Climate change and forest disturbances. BioScience 51: 723–734.

10. Krawchuk M, Moritz MA (2012) Fire and climate change in California - changes in the distribution and frequency of fire in climates of the future and recent past (1911–2099). California Energy Commission. Available: http://www.energy.ca.gov/2012publications/CEC-500-2012-026/CEC-500-2012-026.pdf. Accessed 2014 July 25.

11. Quinn RD, Keeley SC (2006) Introduction to California chaparral. Berkeley and Los Angeles, California: University of California Press.

12. Hanes TL (1971) Succession after fire in the chaparral of southern California. Ecological Monographs 41: 27–52.

13. Keeley JE, Brennan TJ (2012) Fire-driven alien invasion in a fire-adapted ecosystem. Oecologia 169: 1043–1052.

14. Keeley JE, Davis FW (2007) Chaparral. In: Barbour MG, Keeler-wolf T, Schoenherr AA, editors. Terrestrial vegetation of California. Berkeley and Los Angeles, California: University of California Press. pp. 339–362.

15. Keeley JE, Fotheringham CJ, Baer-Keeley M (2005) Factors affecting plant diversity during post-fire recovery and succession of mediterranean-climate shrublands in California, USA. Diversity and Distributions 11: 525–537.

16. Keeley JE, Keeley SC (1981) Post-Fire regeneration of southern California chaparral. American Journal of Botany 68: 524–530.

17. Lippitt CL, Stow DA, O'Leary JF, Franklin J (2013) Influence of short-interval fire occurrence on post-fire recovery of fire-prone shrublands in California, USA. International Journal of Wildland Fire 22: 184–193.

18. Keeley JE, Fotheringham CJ (2003) Impact of past, present, and future fire regimes on North American Mediterranean shrublands. In: Veblen TT, Baker WL, Montenegro G, Swetnam TW, editors. Fire and climatic change in temperate ecosystems of the Western Americas. New York: Springer Press. pp. 218–262.

19. Cooper WS (1922) Broad-sclerophyll vegetation of California: an ecological study of the chaparral and its related communities. No. 319. Carnegie Institution of Washington Press.

20. Wells PV (1962) Vegetation in relation to geological substratum and fire in the San Luis Obispo Quadrangle, California. Ecological Monographs 32: 79–103.

21. Zedler PH, Gautier CR, McMaster GS (1983) Vegetation change in response to extreme events: the effect of a short interval between fires in California chaparral and coastal scrub. Ecology 64: 809–818.

22. Callaway RM, Davis FW (1993) Vegetation dynamics, fire, and the physical environment in coastal central California. Ecology 74: 1567–1578.

23. Keeley JE (2002) Native American impacts on fire regimes of the California coastal ranges. Journal of Biogeography 29: 303–320.

24. Talluto MV, Suding KN (2008) Historical change in coastal sage scrub in southern California, USA in relation to fire frequency and air pollution. Landscape Ecology 23: 803–815.

25. Padgett PE, Allen EB (1999) Differential responses to nitrogen fertilization in native shrubs and exotic annuals common to mediterranean coastal sage scrub of California. Plant Ecology 144: 93–101.

26. Keeley JE, Lubin D, Fotheringham CJ (2003) Fire and grazing impacts on plant diversity and alien plant invasions in the southern Sierra Nevada. Ecological Applications 13: 1355–1374.

27. Burcham LT (1955) Recent trends in range improvement on California foothill ranges. Journal of Range Management 8: 121–125.

28. Bentley JR (1967) Conversion of Chaparral areas to grassland: techniques used in California. Agriculture handbook/Forest service USDA. Available: http://www.fs.fed.us/psw/publications/documents/misc/ah328.pdf?. Accessed 2014 July 25.

29. Williams WA, Love RM, Conrad JP (1956) Range improvement in California by seeding annual clovers, fertilization and grazing management. Journal of Range Management 9: 28–33.

30. Love RM, Berry LJ, Street JE, Osterli VP, Spurr AR, et al. (1960) Planned range improvement programs are beneficial. California Agriculture 14: 2–4.

31. Fuhrmann KN, Crews TE (2001) Long-term effects of vegetation treatments in the chaparral transition zone. Rangelands 23: 13–16.

32. McMichael CE, Hope AS, Roberts DA, Anaya MR (2004) Post-fire recovery of leaf area index in California chaparral: a remote sensing-chronosequence approach. International Journal of Remote Sensing 25: 4743–4760.

33. Röder A, Hill J, Duguy B, Alloza JA, Vallejo R (2008) Using long time series of Landsat data to monitor fire events and post-fire dynamics and identify driving factors: a case study in the Ayora region (eastern Spain). Remote Sensing of Environment 112: 259–273.

34. van Leeuwen WJD, Casady GM, Neary DG, Bautista S, Alloza JA, et al. (2010) Monitoring post-wildfire vegetation response with remotely sensed time-series data in Spain, USA and Israel. International Journal of Wildland Fire 19: 75–93.

35. Kinoshita AM, Hogue TS (2011) Spatial and temporal controls on post-fire hydrologic recovery in Southern California watersheds. CATENA 87: 240–252.

36. Henry MC, Hope AS (1998) Monitoring post-burn recovery of chaparral vegetation in southern California using multi-temporal satellite data. International Journal of Remote Sensing 19: 3097–3107.

37. Keeley JE (2006) South coast bioregion. In: Sugihara NG, Wagtendonk JWv, Shaffer KE, Fites-Kaufman J, Andrea ET, editors. Fire in California's ecosystems. Berkeley Los Angeles London: University of California Press. pp. 350–390.

38. Dennison PE, Moritz MA (2009) Critical live fuel moisture in chaparral ecosystems: A threshold for fire activity and its relationship to antecedent precipitation. International Journal of Wildland Fire 18: 1021–1027.

39. Schroeder MJ, Glovinsky M, Henricks VF, Hood FC, Hull MK (1964) Synoptic weather types associated with critical fire weather. U.S. Forest Service, Pacific Southwest Forest and Range Experiment Station Press.

40. Mensing SA, Michaelsen J, Byrne R (1999) A 560-year record of Santa Ana fires reconstructed from charcoal deposited in the Santa Barbara Basin, California. Quaternary Research 51: 295–305.

41. Keeley JE, Fotheringham CJ (2001) Historic fire regime in southern California shrublands. Conservation Biology 15: 1536–1548.

42. Canty MJ, Nielsen AA (2008) Automatic radiometric normalization of multitemporal satellite imagery with the iteratively re-weighted MAD transformation. Remote Sensing of Environment 112: 1025–1036.

43. Saura S (2002) Effects of minimum mapping unit on land cover data spatial configuration and composition. International Journal of Remote Sensing 23: 4853–4880.

44. Eidenshink J, Schwind B, Brewer K, Zhu Z, Quayle B, et al. (2007) A project for monitoring trends in burn severity. Fire Ecology 3: 3–21.

45. Daly C, Neilson RP, Phillips DL (1994) A statistical-topographic model for mapping climatological precipitation over mountainous terrain. Journal of Applied Meteorology 33: 140–158.

46. Moran PAP (1950) Notes on continuous stochastic phenomena. Biometrika 37: 17–23.

47. White JD, Ryan KC, Key CC, Running SW (1996) Remote sensing of forest fire severity and vegetation recovery. International Journal of Wildland Fire 6: 125–136.

48. Key CH, Benson NC (2006) Landscape assessment: sampling and analysis methods. Rocky Mountain Research Station General Technical Report RMRS-GTR-164-CD. Available: http://www.fs.fed.us/rm/pubs/rmrs_gtr164/rmrs_gtr164_13_land_assess.pdf. Accessed 2014 July 10.

49. Hunt JER, Rock BN (1989) Detection of changes in leaf water content using Near- and Middle-Infrared reflectances. Remote Sensing of Environment 30: 43–54.

50. Miller JD, Thode AE (2007) Quantifying burn severity in a heterogeneous landscape with a relative version of the delta Normalized Burn Ratio (dNBR). Remote Sensing of Environment 109: 66–80.

51. Lopez Garcia MJ, Caselles V (1991) Mapping burns and natural reforestation using Thematic Mapper data. Geocarto International 6: 31–37.

52. Chen X, Vogelmann JE, Rollins M, Ohlen D, Key CH, et al. (2011) Detecting post-fire burn severity and vegetation recovery using multitemporal remote sensing spectral indices and field-collected composite burn index data in a ponderosa pine forest. International Journal of Remote Sensing 32: 7905–7927.

53. Zammit CA, Zedler PH (1993) Size structure and seed production in even-aged populations of Ceanothus greggii in mixed chaparral. Journal of Ecology 81: 499–511.

54. Odion D, Tyler C (2002) Are long fire-free periods needed to maintain the endangered, fire-recruiting shrub Arctostaphylos morroensis (Ericaceae). Conservation Ecology. Available: http://www.consecol.org/vol6/iss2/art4/. Accessed 2014 July 10.

55. Kirkman TW (1996) Statistics to use. Available: http://www.physics.csbsju.edu/stats/. Accessed 2014 July 10.

56. Merriam KE, Keeley JE, Beyers JL (2006) Fuel breaks affect nonnative species abundance in Californian plant communities. Ecological Applications 16: 515–527.

57. Rodrigo A, Retana J, Picó FX (2004) Direct regeneration is not the only response of Mediterranean forests to large fires. Ecology 85: 716–729.

58. Dale VH, Joyce LA, McNulty S, Neilson RP (2000) The interplay between climate change, forests, and disturbances. Science of the Total Environment 262: 201–204.

59. Mitri GH, Gitas IZ (2013) Mapping post-fire forest regeneration and vegetation recovery using a combination of very high spatial resolution and hyperspectral satellite imagery. International Journal of Applied Earth Observation and Geoinformation 20: 60–66.

60. Huesca M, Merino-de-Miguel S, González-Alonso F, Martínez S, Miguel Cuevas J, et al. (2013) Using AHS hyper-spectral images to study forest vegetation recovery after a fire. International Journal of Remote Sensing 34: 4025–4048.

61. Dennison PE, Roberts DA (2003) The effects of vegetation phenology on endmember selection and species mapping in southern California chaparral. Remote Sensing of Environment 87: 295–309.

62. Benz UC, Hofmann P, Willhauck G, Lingenfelder I, Heynen M (2004) Multi-resolution, object-oriented fuzzy analysis of remote sensing data for GIS-ready information. ISPRS Journal of Photogrammetry and Remote Sensing 58: 239–258.

63. Pennypacker CR, Jakubowski MK, Kelly M, Lampton M, Schmidt C, et al. (2013) FUEGO—Fire Urgency Estimator in Geosynchronous Orbit—a proposed early-warning fire detection system. Remote Sensing 5: 5173–5192.

64. Schroeder W, Oliva P, Giglio L, Csiszar IA (2014) The New VIIRS 375m active fire detection data product: algorithm description and initial assessment. Remote Sensing of Environment 143: 85–96.

65. Green RO, Asner G, Ungar S, Knox R, NASA (2008) NASA mission to measure global plant physiology and functional types. IEEE Aerospace Conference. Available: http://ieeexplore.ieee.org/xpls/abs_all.jsp?arnumber=4526244&tag=1. Accessed 2014 July 16.

66. Batllori E, Parisien MA, Krawchuk MA, Moritz MA (2013) Climate change-induced shifts in fire for Mediterranean ecosystems. Global Ecology and Biogeography 22: 1118–1129.

17

Pulse Increase of Soil N_2O Emission in Response to N Addition in a Temperate Forest on Mt Changbai, Northeast China

Edith Bai[1]*, Wei Li[1,2], Shanlong Li[1,2], Jianfei Sun[1,2], Bo Peng[1,2], Weiwei Dai[1], Ping Jiang[1], Shijie Han[1]

1 State Key Laboratory of Forest and Soil Ecology, Institute of Applied Ecology, Chinese Academy of Sciences, Shenyang, China, 2 College of Resources and Environment, University of Chinese Academy of Sciences, Beijing, China

Abstract

Nitrogen (N) deposition has increased significantly globally since the industrial revolution. Previous studies on the response of gaseous emissions to N deposition have shown controversial results, pointing to the system-specific effect of N addition. Here we conducted an N addition experiment in a temperate natural forest in northeastern China to test how potential changes in N deposition alter soil N_2O emission and its sources from nitrification and denitrification. Soil N_2O emission was measured using closed chamber method and a separate incubation experiment using acetylene inhibition method was carried out to determine denitrification fluxes and the contribution of nitrification and denitrification to N_2O emissions between Jul. and Oct. 2012. An NH_4NO_3 addition of 50 kg N/ha/yr significantly increased N_2O and N_2 emissions, but their "pulse emission" induced by N addition only lasted for two weeks. Mean nitrification-derived N_2O to denitrification-derived N_2O ratio was 0.56 in control plots, indicating higher contribution of denitrification to N_2O emissions in the study area, and this ratio was not influenced by N addition. The N_2O to (N_2+N_2O) ratio was 0.41–0.55 in control plots and was reduced by N addition at one sampling time point. Based on this short term experiment, we propose that N_2O and denitrification rate might increase with increasing N deposition at least by the same fold in the future, which would deteriorate global warming problems.

Editor: Dafeng Hui, Tennessee State University, United States of America

Funding: This work was supported by the Major State Basic Research Development Program of China (973 Program) (2014CB954400 and 2011CB403202) and National Science Foundation of China (31100326). The funders had no role in study design, data collection and analysis, decision to publish, or preparation of the manuscript.

Competing Interests: The authors have declared that no competing interests exist.

* Email: baie@iae.ac.cn

Introduction

Anthropogenic nitrogen (N) has approximately doubled the input of reactive N to the Earth's land surface since the industrial revolution [1,2]. There are three potential fates of anthropogenic N entering terrestrial ecosystem: loss via nitrification and denitrification to the atmosphere as gases, loss via leaching into the aquatic ecosystems, and storage in plants and soils [3]. The gaseous products include NO_x, N_2O, N_2 and NH_3, among which N_2O is a potent greenhouse gas [4] and ozone-depleting agent [5] and has been a major concern for environmental scientists. Although previous studies suggested that many ecosystems have high N retention capacity without evidence of adverse effects of increasing N deposition [6], some other studies found a big increase of N_2O following increase of N inputs [7–9]. Therefore, the response of gaseous emissions to N deposition may be system-specific and depends on the N status [10] and environmental factors of the ecosystem [11].

Soil N_2O flux has been found to be affected by ammonium and nitrate availability [12], water filled pore space (WFPS) [13], soil temperature [14], soil moisture[15], soil pH [16], and carbon (C)

availability [17]. When increasing N deposition increases inorganic N availability for nitrification and denitrification, it is expected to increase N_2O emission [18]. However, inorganic N availability may not increase in response to increasing N deposition due to plant uptake and microbial immobilization [6]. In addition, soil pH has been found to decrease in response to increasing N deposition, which may lower N_2O emissions [19]. Decomposition of light fractions of soil C with low density (<1.7 g cm^{-3}) has been found to accelerate under long term N addition [20], which may affect C availability for nitrification and denitrification and thereby N_2O emission.

N_2O is derived from both nitrification and denitrification and it is important to distinguish these processes in order to better understand factors controlling the emission of N_2O [21]. The contribution of the two processes is mainly affected by oxygen availability and forms and availability of inorganic N [22,23]. Therefore, increasing N deposition may impact the ratio of nitrification derived to denitrification derived N_2O. However, very limited data exist on this ratio or the denitrification rate in temperate forest ecosystems [3,24,25].

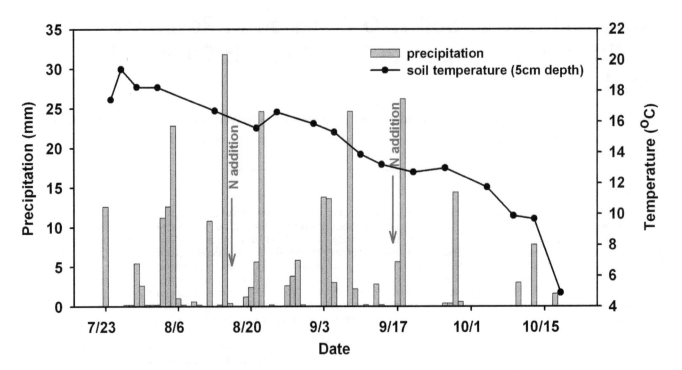

Figure 1. Variations of precipitation and soil temperature (5 cm depth) in the study area during the sampling periods.

This study was carried out in a broad-leaved Korean pine (*Pinus koraiensis* Siebold & Zucc) mixed forest, which is the dominant natural forest type in northeastern China. The system was previously thought to be N-limited but may be experiencing N saturation currently [26]. Hence, it is important to assess how increasing N deposition may affect the N cycling of this forest type. The objectives of our study were: 1) to measure soil N_2O fluxes, ratio of nitrification derived to denitrification derived N_2O, and denitrification rate, for the first time in the study area; 2) to examine the treatment effect of N addition on these fluxes; 3) to investigate the influences of environmental factors on these variables. The hypotheses of this study are: 1) N addition would increase soil N_2O emissions and denitrification rate; 2) soil temperature, moisture, WFPS, pH, and soil texture play an important role in soil N_2O emissions and denitrification.

Materials and Methods

The study was carried out within the research forest of Changbai Forest Ecosystem Research Station (CBFERS) established in 1979 (127°38' E, 41°42' N). No specific permits were

required for the described field studies. Research activities within that research forest do not need any specific permissions from any government levels, but need to inform Professor Han (Director CBFERS, co-author of the present paper). The field studies did not involve endangered or protected species.

Study area

The study region is characterized by a typical temperate climate, with long and cold winters, and warm summers. Mean annual temperature is 3.6 °C, with the highest temperature in July, and the lowest in January. Mean annual precipitation is 745 mm, mainly falling between May and September. The growing season is from Jun. to Sep., with mean temperature at 16.7 °C. Natural Korean pine and broad-leaved mixed forest is distributed from approximately 750 to 1100 m asl. The soil is a dark brown soil developed from volcanic ash (Albic Luvisol).

Experimental design

Present N deposition rate in China is 12.89–63.53 kg N ha^{-1} yr^{-1}, with higher values in eastern China [27]. It is predicted that in 2050, many areas in eastern China will have a deposition rate of

Table 1. Soil characteristics of treatment plots.

	pH	Clay (%)	Bulk Density (g/cm³)	WFPS (%)	Soil T (°C)	Soil moisture (kg H₂O/kg dry soil)	NH₄⁺ (mg N/kg soil)	NO₃⁻ (mg N/kg soil)	Inorganic N (mg N/kg soil)
Control	5.48±0.33[a]	31.67±4.40[a]	0.581±0.013[a]	70.48±15.88[a]	14.25±3.67[a]	94.81±21.99[a]	12.50±9.79[a]	19.11±23.87[a]	31.62±22.39[a]
N-treated	5.47±0.09[a]	30.98±1.11[a]	0.587±0.113[a]	76.07±20.15[a]	14.25±3.67[a]	101.90±26.94[a]	10.40±8.93[a]	22.56±26.26[b]	32.96±22.47[b]

Soils were collected 17 times from Jul. to Oct. 2012. Values are reported as mean ± standard deviation. WFPS: water-filled pore space. T: temperature. Sampling time was used as the within subject factor and N treatment was used as the between subject factor in repeated measures ANOVA. Different letter represents significant difference (p = 0.05) caused N treatment.

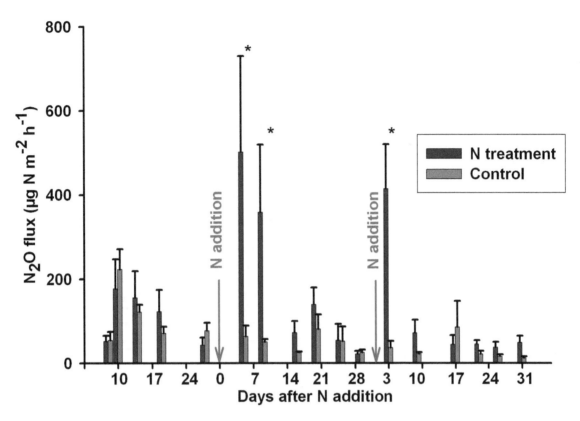

Figure 2. Effects of N addition on N₂O fluxes (n = 3). Arrows represent N addition on that date. Error bars are standard errors. * represents significant effect at p = 0.05.

50 kg N ha^{-1} yr^{-1} [1,28]. Current N deposition rate in the study area is 23 kg N ha^{-1} yr^{-1} [26], and we chose a N addition rate of 50 kg N ha^{-1} yr^{-1} in the treatment as a reasonable estimate of future change. Six sampling plots (25 m×50 m) were established: three replicates for both control and N addition treatment. Plots were separated by 10 m buffer strips. Vegetation was homogeneous based on visual observation.

Starting in 2008, NH$_4$NO$_3$ solution has been evenly sprayed for N addition treatment monthly from May to September at a rate of 10 kg N ha^{-1} (starting on May 17th), giving an annual N addition rate of 50 kg N ha^{-1} yr^{-1}. The same volume (40 L) of water was sprayed on the control plots.

Gas sampling and analysis

Soil N$_2$O emission was measured using closed chamber method [29]. A steel-made chamber (0.5 m×0.5 m×0.5 m) with a septum for gas sampling was placed on the base every time gases were sampled. The bases were installed into the soil one month before initial gas sampling to avoid disturbance on soils. Water was used to seal the connection between the chamber and the base. The chamber was covered with thermal insulation cotton to avoid temperature increase in the chamber. Fans were equipped in the chamber to increase air circulation. Gas sample was collected from each chamber every 30 mins for 2 hours and stored in air bags for measurement.

We sampled for three N addition periods in year 2012, 5–6 times for each period. Samples were taken once every 5–6 days. Given that NH$_4$NO$_3$ was sprayed on the day 17th each month, gas samples were collected in three periods: Period one (Jul. 17th–Aug. 16th): 7, 9, 13, 18, and 27 days after N addition; Period two (Aug. 17th–Sep. 16th): 4, 8, 15, 19, 24, and 28 days after N addition; Period three (Sep. 17th to Oct. 18th): 3, 9, 17, 22, 26, and 31 days after N addition. Samples were taken between 9:00 and 11:00 hours and between 15:00 and 17:00 hours because during these two time slots the flux can represent daily average [30]. N$_2$O concentration was measured by gas chromatography (HP 5890-II). The temperature of the chromatographic column was 55°C and the carrying gas was high purity N$_2$.

Acetylene inhibition method [31,32] was used to determine denitrification rates and to differentiate sources of N$_2$O based on the following principle: high concentration of acetylene (10 kPa) could inhibit both nitrification and the reduction of N$_2$O to N$_2$ by denitrifiers, so that the otherwise emitted N$_2$ remained as N$_2$O. When the acetylene concentration was low (10–100Pa), only nitrification was inhibited [33]. Emitted N$_2$O could be from three sources: N$_2$O$_{den}$ (N$_2$O produced by denitrification); N$_2$O$_{N2}$(additional N$_2$O produced by denitrification, instead of normal N$_2$ production); N$_2$O$_{nit}$ (N$_2$O produced by nitrification). In each sampling plot, soils were incubated $in\ situ$ in three tightly sealed bottles injected with different concentrations of C$_2$H$_2$ (purified by H$_2$SO$_4$ and distilled water): 0 (nothing inhibited), 60 Pa (nitrification inhibited), and 10 kPa (nitrification and the reduction of N$_2$O to N$_2$ inhibited). Correspondingly, measured N$_2$O of the three incubation bottles were N$_2$O$_{nit}$+N$_2$O$_{den}$, N$_2$O$_{den}$, and N$_2$O$_{den}$+N$_2$O$_{N2}$, respectively. Therefore, nitrification-derived N$_2$O, denitrification-derived N$_2$O, and denitrification-derived N$_2$

Table 2. Pearson's correlation coefficients between N₂O fluxes and soil characteristics.

Parameter	N₂O μg N/m²/h	Soil M (kg H₂O/kg dry soil)	WFPS (%)	NH₄⁺ (mg N/kg soil)	NO₃⁻ (mg N/kg soil)	Inorganic N (mg N/kg soil)	pH	Clay (%)	Soil T (°C)
N₂O	1	-0.108	-0.028	-0.103	0.225*	0.210*	0.050	-0.072	0.253*
Soil M		1	0.797*	-0.023	0.079	0.079	0.099	0.169	-0.113
WFPS			1	-0.021	0.144	0.153	0.085	0.023	-0.109
NH₄⁺				1	-0.475*	-0.116	-0.015	-0.027	0.016
NO₃⁻					1	0.929*	-0.042	-0.014	0.460*
Inorganic N						1	-0.054	-0.027	0.526*
pH							1	0.727*	0.000
Clay								1	0.000

WFPS: water-filled pore space. T: temperature. M: moisture. * represents significant correlation at p = 0.05.

can be calculated from these three measured N₂O fluxes. Denitrification rate is equal to $N_2O_{den}+N_2O_{N2}$.

60 Pa C_2H_2 was used for low concentration treatment as recommended in previous experiment in the study area [34]. This concentration has also been used in other experiments [35], although 10 Pa is more widely used.

Although the acetylene inhibition technique has not always been satisfactory due to incomplete or non-specific inhibition [36], at present it is still a widely-used method for obtaining a reasonable estimate of denitrification rate and ratio of nitrification to denitrification derived N₂O [3,37–40]. We used an *in situ* incubation method to reduce some potential bias [41].

Three sampling points were selected within each sampling plot. Nine soil cores (5 cm diameter×10 cm depth) were sampled by PVC tubes at each point. There were 25 holes (0.6 cm diameter) on each PVC tube to ensure acetylene circulation. At each sampling point, every three soil cores were assigned to one of the three incubation bottles. Therefore, we had three soil cores in each incubation bottle at each plot. Before the incubation, the headspace air was replaced by purified C_2H_2. The bottles were then buried *in situ* and 60 ml headspace samples were taken after 24 hours incubation and analyzed for N₂O concentration by gas chromatography (HP 5890-II). Gas samples were collected on Aug. 18th, Aug. 30th, Sep. 10th, Sep. 16th, Sep. 23rd, Sep. 30th, Oct. 7th, and Oct. 15th, 2012.

Soil sampling and analysis

Air temperature and soil temperature at 5 cm depth were recorded *in situ* at each sampling time of both closed chamber experiment and acetylene inhibition experiment using TP3001 thermometer (Boyang, China) (Fig. 1). Precipitation was recorded at the local weather station in the Changbai Mountain Nature Reserve (Fig. 1). Bulk density (BD) in each sampling plot (n = 3) was determined by the core method [42].

One soil sample (0–10 cm) in each plot was collected at each sampling time to determine pH, soil moisture content, and soil NH_4^+ and NO_3^-. Standard oven-drying method was used to measure soil gravimetric water content, which was expressed as kg H₂O/kg dry soil. WFPS was calculated based on Eqn.1:

$$WFPS = \text{soil gravimetric water content} \times BD/PS \quad (Eqn.1)$$

where PS is soil pore space and is calculated from Eqn.2 [43]:

$$PS = (1\text{-}BD/2.65) \times 100 \quad (Eqn.2)$$

20 g fresh soil was passed through 2 mm screen and extracted by 50 mL 2 M KCl. The solution was shaken for 1 hour and filtered through rapid filtering paper. Soil inorganic N in extracts was determined by Futura continuous flow analysis system (Alliance, France).

Soil pH was measured on a 1:2.5 soil solution in de-ionized water using a pH meter. Soil texture was determined by the pipette-sedimentation method [44].

Statistical analysis

Repeated measures ANOVA was used to examine the effects of N addition using N treatment method as the between subject factor and sampling time as the within subject factor. All analyses were carried out in SPSS [45].

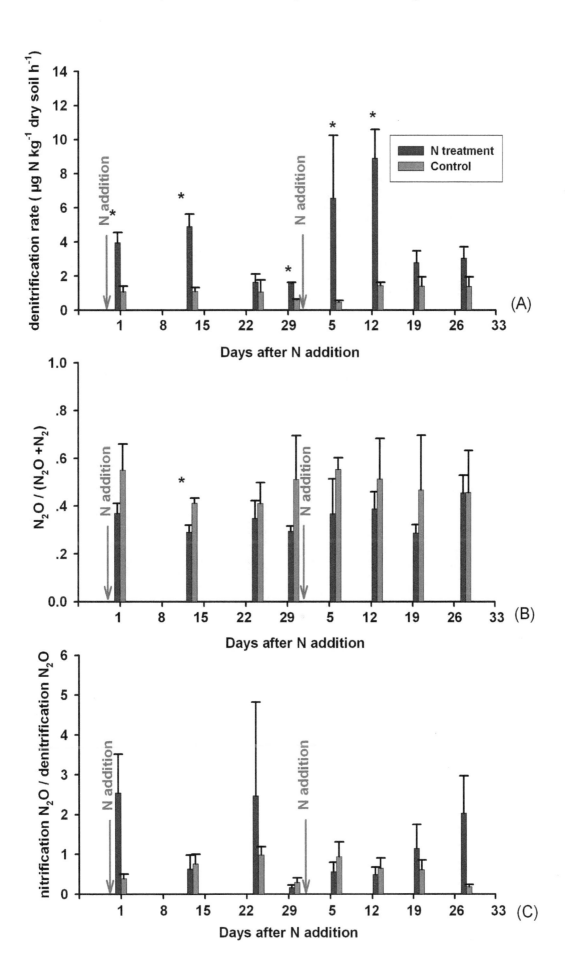

Figure 3. Effect of N addition on denitrification rate (A), N₂O: (N₂O+N₂) ratio (B), and nitrification-derived N₂O: denitrification-derived N₂O ratio (N₂Oₙᵢₜ:N₂Oₚₑₙ) (C) (n = 3). Arrows represent N addition on that date. * represents significant effect at p = 0.05.

Results

N₂O fluxes

Soil characteristics of the sampling plots were tested 17 times from Jul. to Oct. 2012 and mean results were reported in Table 1 (values were reported as mean ± standard deviation). Mean soil pH was 5.48 ± 0.33 for control plot and 5.47 ± 0.09 for N-treated plot. Soil clay content was $31.67\pm4.40\%$ and $30.98\pm1.11\%$ for control and N-treated plot, respectively. Soil bulk density was also similar between control and N-treated plots (0.581 ± 0.013 g cm^{-3} and 0.587 ± 0.113 g cm^{-3} respectively). Soil moisture and WFPS were not significantly different between N-treated plots (101.90 ± 26.94 kg H₂O kg^{-1} dry soil and $76.07\pm20.15\%$ respectively) and control plots (94.81 ± 21.99 kg H₂O kg^{-1} dry soil and $70.48\pm15.88\%$ respectively). Mean soil NO₃⁻ concentration was significantly increased in N-treated plots (22.56 ± 26.26 mg N kg^{-1} soil) compared to control plots (19.11 ± 23.87 mg N kg^{-1} soil) based on repeated measures ANOVA, but mean soil NH₄⁺ concentration was not changed (Table 1).

N-treated plots had significantly higher N₂O fluxes right after N addition for period 2 and 3 (Fig. 2). For example, four days after N addition in period 2, mean N₂O emission in N-treated plots reached 501.74 μg N m^{-2} h^{-1}, which was significantly higher than that in control plots (63.31 μg N m^{-2} h^{-1}). This difference between N-treated and control plots declined with time and became minimal on the 15th day after N addition. Similarly, for the next sampling period, mean N₂O emission in N-treated plots was 414.07 μg N m^{-2} h^{-1} three days after N addition, compared with 36.88 μg N m^{-2} h^{-1} in control plots. Only 17 days later, N₂O flux was similar in N-treated (44.13 μg N m^{-2} h^{-1}) and control (85.38 μg N m^{-2} h^{-1}) plots.

Pearson's correlation analysis showed that N₂O emission was positively correlated with soil NO₃⁻ (r = 0.225), soil total inorganic N (r = 0.210), and soil temperature (r = 0.253) (Table 2). N₂O emission first increased with WFPS then decreased, and the highest N₂O flux rate was observed at 80% WFPS. Soil pH was positively correlated with soil clay content (r = 0.727) (Table 2).

Sources of N₂O and denitrification flux

One day after N addition in August, denitrification rate was significantly higher in N-treated plots (3.95 μg N kg soil^{-1} h^{-1}) than in control plots (0.37 μg N kg soil^{-1} h^{-1}) (Fig. 3a). The effect lasted till the end of the treatment cycle. The same phenomenon was observed for the next sampling period, except that the effect of N addition was not statistically significant on the last two sampling dates (Fig. 3a). The N₂O to (N₂+N₂O) ratios in N-treated plots (0.29–0.45) were all lower than those in control plots (0.41–0.55), although the effect was not statistically significant (Fig. 3b). N-additions on Aug. 17th and Sep. 17th did not affect this pattern (Fig. 3b). We did not find a general pattern for the temporal variation of the N₂Oₙᵢₜ: N₂Oₚₑₙ ratio or a significant effect of N addition on the ratio (Fig. 3c). The N₂Oₙᵢₜ: N₂Oₚₑₙ ratio in N-treated plots had a wider range (0.17–4.47) than that in control plots (0.19–0.94).

Correlation analysis indicated both denitrification rate and N₂ flux had a significantly positive correlation with WFPS (r = 0.344 and 0.353 respectively). Neither the N₂O to (N₂+N₂O) ratio nor the N₂Oₙᵢₜ: N₂Oₚₑₙ ratio correlated with WFPS or any other measured soil variables (Table 3).

Discussion

N₂O emission and its sources

N₂O emission fluxes observed in the control plots were between 0.77 and 382.75 μg N m^{-2} h^{-1}, which mostly is within the range of soil N₂O emission in a 40-year-old pine forest ecosystems in the USA (3–424 μg N m^{-2} h^{-1}) [46]. Our results suggested N addition (50 kg N ha^{-1} yr^{-1}) could increase N₂O fluxes to up to 501 μg N m^{-2} h^{-1} and the monthly sum of N₂O emission in N-treated plots were five-times higher than that in control plots (Fig. 2). The positive correlation between N input rate and the magnitude of N₂O emissions has been found in previous studies [9,11,47]. Previously, Bowden et al. [48] reported N addition at 50–150 kg N/ha/yr only increased N₂O emission from 0.2–1.1 to 0.7–5.2 μg N m^{-2} h^{-1} in the pine forest plantation of Harvard Forest, US, which is much lower than our results. However, our results are consistent with those observations that showed a big increase of N₂O following increase of N inputs [7–9]. Our speculation on the discrepancy is that Harvard Forest in 1990s was N-limited and thereby could retain added N efficiently, while our forest in Changbai Mountain is experiencing "kinetic N saturation" currently [49], with lower N sinks than the N input rate.

N addition effect on N₂O emissions diminished quickly with time after N addition, because added NH₄⁺ and NO₃⁻ were also subject to plant uptake, microbial immobilization, and leaching (Fig. 2). We found the increase of N₂O emission after N addition only lasted for two weeks. Scheer et al. [7] also observed similar "N₂O emission pulses" which lasted for 7 days after N-fertilizer application in combination with irrigation events in a cotton field. This temporal variation suggests more frequent sampling is necessary to capture the dynamics of the emission pulse following N addition.

Precipitation and soil texture, which determines WFPS, may play an important role in soil N₂O emission. Positive correlations between WFPS and N₂O were usually observed because the contribution of denitrification-derived N₂O increases with WFPS [50,51]. Other potential reasons include: microbes were water-limited [52]; rain may disrupt physical aggregate and increase soil organic matter exposure for mineralization and denitrification [53]; and water may alleviate diffusional constraints [54,55]. However, when WFPS is very high (>80%), those factors were not limiting and N₂O fluxes may decline due to their further reduction to N₂ [56,57]. Because our mean WFPS mainly varied between 55% and 100% (Table 1), N₂O emissions increased with WFPS first and then decreased, and the highest N₂O flux rate was observed at WFPS = 80%. We speculate this was because N₂O from both nitrification and denitrification was high at this water and oxygen level.

We found N₂O fluxes were positively correlated with soil temperature (Table 3), which is consistent with many previous findings [58–60]. Both nitrification-derived N₂O [61] and denitrification-derived N₂O [62,63] have been found to be higher at lower pH in acid forest soils. However, nitrification-derived N₂O has more often been found to increase with pH [19]. Our N addition treatment did not affect soil pH (Table 1) and N₂O fluxes were not related to soil pH in our study (Table 2).

Table 3. Pearson's correlation coefficients between soil characteristics and denitrification rate, N_2 emission rate, $N_2O/(N_2O+N_2)$, and nitrification-derived N_2O: denitrification-derived N_2O ($N_2O_{nit}:N_2O_{den}$) ratios.

	Denitrification µg N/kg soil/h	N_2 µg N/kg soil/h	$N_2O/(N_2O+N_2)$	$N_2O_{nit}:N_2O_{den}$	WFPS (%)	NO_3^- (mg N/kg soil)	Inorganic N (mg N/kg soil)	pH	Clay (%)	Soil T (°C)	Soil M (kg H_2O/kg dry soil)
Denitrification	1	0.965*	-0.306*	-0.035	0.344*	0.202	-0.001	0.018	-0.092	0.030	0.273
N_2		1	-0.379*	-0.013	0.353*	0.191	0.006	0.054	-0.098	0.049	0.144
$N_2O/(N_2O+N_2)$			1	-0.173	-0.054	-0.131	-0.088	-0.107	-0.050	-0.045	-0.122
$N_2O_{nit}:N_2O_{den}$				1	-0.054	-0.061	0.045	0.008	0.094	0.029	0.196
WFPS					1	0.249	-0.028	0.131	0.013	-0.237	0.672*
NO_3^-						1	0.659*	0.019	-0.115	0.136	0.137
Inorganic N							1	0.072	-0.017	0.523*	0.022
pH								1	0.727*	0.000	0.105
Clay									1	0.000	0.177
Soil T										1	-0.225

WFPS: water-filled pore space. T: temperature. M: moisture. * represents significant correlation at p = 0.05.

The N_2O_{nit}: N_2O_{den} ratio was not affected by N addition (Table 3). The N_2O_{nit}: N_2O_{den} ratio in N-treated plots had a larger variation (0.17–4.47) than that in control plots (0.19–0.94). Mean N_2O_{nit}: N_2O_{den} ratio was 0.56 and 1.46 in control and N-treated plots respectively, which is lower than what Papen & Butterback-Bahl [47] found (2.33) in spruce and beech forest ecosystem in Germany. The WFPS in our area was high (Table 1), therefore more N_2O was emitted from denitrification. The N_2O_{nit}: N_2O_{den} ratio was not significantly correlated with any soil factors we examined (Table 3), which suggested that the effects of soil moisture, temperature, WFPS, and texture on N_2O_{nit}: N_2O_{den} ratio may be non-linear, or negligible.

Denitrification fluxes and the N_2O to (N_2+N_2O) ratio

Denitrification in the control plots were between 0.126 and 5.460 µg N kg^{-1} soil h^{-1} with a mean at 1.581 µg N kg^{-1} soil h^{-1} (Fig. 3). Bulk density of the study site was between 0.53 and 0.67 g cm^{-3} (Table 1) with a mean at 0.60 g cm^{-3}. Assuming denitrification keeps at the high flux rate during the growing season (May-September) and there is no denitrification in the other months [64], denitrification in the region was roughly estimated to be 0.346 g N m^{-2} yr^{-1}. This is close to the upper end of the modelled average value (0.156–0.368 g N m^{-2} yr^{-1}) in global temperate and boreal forest ecosystems by Bai et al. [65]. *In situ* denitrification data from temperate forest are very limited. Dannenmann et al. [66] reported a denitrification rate of 0.185–0.622 g N m^{-2} yr^{-1} in a mountainous beech forest using seasonally weighted means, although their direct up-scaling of plot means got much higher values (1.4–9.37 g N m^{-2} yr^{-1}). Wolf & Brumme [62] got 0.088–0.351 g N m^{-2} yr^{-1} *in situ* denitrification rates from beech forest floor and mineral soils. Our estimation is comparable to these above reported values.

N addition significantly increased denitrification rates to a mean of 5.539 µg N kg^{-1} soil h^{-1} (Table 3, Fig. 3a). With the current N deposition rate at 23 kg N ha^{-1} yr^{-1} [26], an N addition of 50 kg N ha^{-1} yr^{-1} (2.2 fold) has caused approximately a 2.5 fold increase in denitrification. Therefore, we predict that with the increase of N deposition, denitrification will increase correspondingly, at least by the same fold.

For the two gaseous products of denitrification, more N_2 is emitted after N addition compared with N_2O because the N_2O to (N_2+N_2O) ratio was lower in N-treated plots (Fig. 3b). The N_2O to (N_2+N_2O) ratios were between 0.41–0.55 and 0.29–0.45 in control and N-treated plots respectively. The N_2O to (N_2+N_2O) ratio has been used to estimate denitrification rate because N_2O was easier to measure [3,67]. Our data in control plots were close to the mean value in natural soils (0.43–0.56) compiled by Schlesinger [3]. Increased NO_3^- has been found to increase the N_2O to (N_2+N_2O) ratio because of the greater affinity of NO_3^- as terminal electron acceptor [68–70]. It is not clear why N addition slightly reduced the N_2O to (N_2+N_2O) ratio in our study, and further exploration is needed.

WFPS was positively correlated with N_2 emission, but not with N_2O fluxes (Table 2, Table 3). This is because the relationship between WFPS and N_2 fluxes is unidirectional, while bell-shaped relationship between WFPS and N_2O is more common [13].

In conclusion, N addition did not change soil pH or soil bulk density in the study area. Mainly due to the increase of substrate availability, N addition significantly increased N_2O and denitrification fluxes. However, the effect only lasted for two weeks. Based on this short term experiment, we propose that denitrification rate might increase at least by the same fold as the increase of N deposition in the future. N_2O may also increase with increasing N deposition in our studied forest, deteriorating global warming problems. Future studies on the underlying mechanism [71] behind increasing N_2O emissions following N addition are needed to better understand the processes contributing to increasing N_2O emissions.

Author Contributions

Performed the experiments: WL SL JS BP WD PJ. Analyzed the data: WL. Contributed reagents/materials/analysis tools: EB. Contributed to the writing of the manuscript: EB WL WD. Designed the experiment: EB SH.

References

1. Galloway JN, Dentener FJ, Capone DG, Boyer EW, Howarth RW, et al. (2004) Nitrogen cycles: past, present, and future. Biogeochemistry 70: 153–226.

2. Gruber N, Galloway JN (2008) An Earth-system perspective of the global nitrogen cycle. Nature 451: 293–296.

3. Schlesinger WH (2009) On the fate of anthropogenic nitrogen. Proc Natl Acad Sci USA 106: 203–208.

4. Ehhalt D, Prather M, Dentener F, Derwent R, Dlugokencky E, et al. (2001) Atmospheric chemistry and greenhouse gases. In: J . T . Houghton, Y . Ding, D. J . Griggs, M . Noguer, P. J. v. d . Linden, X . Dai, K . Maskell and C. A. Johnson, editors. IPCC Report 2001. Cambridge, UK: Cambridge University Press. pp. 241–280.

5. Ravishankara AR, Daniel JS, Portmann RW (2009) Nitrous oxide (N_2O): the dominant ozone-depleting substance emitted in the 21st century. Science 326: 123–125.

6. Bernal S, Hedin LO, Likens GE, Gerber S, Buso DC (2012) Complex response of the forest nitrogen cycle to climate change. Proc Natl Acad Sci USA 109: 3406–3411.

7. Scheer C, Wassmann R, Kienzler K, Ibragimov N, Eschanov R (2008) Nitrous oxide emissions from fertilized, irrigated cotton (Gossypium hirsutum L.) in the Aral Sea Basin, Uzbekistan: Influence of nitrogen applications and irrigation practices. Soil Biol Biochem 40: 290–301.

8. Zhang L, Song C, Wang D, Wang Y (2007) Effects of exogenous nitrogen on freshwater marsh plant growth and fluxes in Sanjiang Plain, Northeast China. Atmos Environ 41: 1080–1090.

9. Butterbach-Bahl K, Breuer L, Gasche R, Willibald G, Papen H (2002) Exchange of trace gases between soils and the atmosphere in Scots pine forest ecosystems of the northeastern German lowlands 1. Fluxes of N_2O, NO/NO_2 and CH_4 at forest sites with different N-deposition. Forest Ecol Manag 167: 123–134.

10. Hall SJ, Matson PA (2003) Nutrient status of tropical rain forests influences soil N dynamics after N additions. Ecol Monogr 73: 107–129.

11. Templer PH, Pinder RW, Goodale CL (2012) Effects of nitrogen deposition on greenhouse-gas fluxes for forests and grasslands of North America. Front Ecol Environ 10: 547–553.

12. Groffman PM, Tiedje JM (1989) Denitrification in north temperate forest soils— spatial and temporal patterns at the landscape and seasonal scales. Soil Biol Biochem 21: 613–620.

13. Davidson EA, Keller M, Erickson HE, Verchot LV, Veldkamp E (2000) Testing a conceptual model of soil emissions of nitrous and nitric Oxides. Bioscience 50: 667–680.

14. Heinen M (2006) Simplified denitrification models: Overview and properties. Geoderma 133: 444–463.

15. Schindlbacher A, Zechmeister-Boltenstern S, Butterbach-Bahl K (2004) Effects of soil moisture and temperature on NO, NO_2, and N_2O emissions from European forest soils. J Geophys Res - Atmos 109: D17302.

16. Stevens RJ, Laughlin RJ, Malone JP (1998) Soil pH affects the processes reducing nitrate to nitrous oxide and di-nitrogen. Soil Biol Biochem 30: 1119–1126.

17. Gregorich EG, Rochette P, Hopkins DW, McKim UF, St-Georges P (2006) Tillage-induced environmental conditions in soil and substrate limitation determine biogenic gas production. Soil Biol Biochem 38: 2614–2628.

18. Morse JL, Ardón M, Bernhardt ES (2012) Using environmental variables and soil processes to forecast denitrification potential and nitrous oxide fluxes in coastal plain wetlands across different land uses. J Geophys Res - Biogeo 117: G02023.

19. Baggs EM, Smales CL, Bateman EJ (2010) Changing pH shifts the microbial sourceas well as the magnitude of N_2O emission from soil. Biol Fert Soils 46: 793–805.

20. Neff JC, Townsend AR, Gleixner G, Lehman SJ, Turnbull J, et al. (2002) Variable effects of nitrogen additions on the stability and turnover of soil carbon. Nature 419: 915–917.

21. Kiese R, Butterbach-Bahl K (2002) N_2O and CO_2 emissions from three different tropical forest sites in the wet tropics of Queensland, Australia. Soil Biol Biochem 34: 975–987.

22. Bollmann A, Conrad R (1998) Influence of O_2 availability on NO and N_2O release by nitrification and denitrification in soils. Global Change Biol 4: 387–396.

23. Davidson EA, Matson PA, Vitousek PM, Riley R, Dunkin K, et al. (1993) Processes regulating soil emissions of NO and N_2O in a seasonally dry tropical forest. Ecology 74: 130–139.

24. Ambus P (1998) Nitrous oxide production by denitrification and nitrification in temperate forest, grassland and agricultural soils. Eur J Soil Sci 49: 495–502.

25. Wolf I, Brumme R (2002) Contribution of nitrification and denitrification sources for seasonal N_2O emissions in an acid German forest soil. Soil Biol Biochem 34: 741–744.

26. Wang C, Han S, Zhou Y, Yan C, Cheng X, et al. (2012) Responses of fine roots and soil N availability to short-term nitrogen fertilization in a broad-leaved Korean pine mixed forest in Northeastern China. PLoS ONE 7: e31042.

27. Lü C, Tian H (2007) Spatial and temporal patterns of nitrogen deposition in China: Synthesis of observational data. J Geophys Res - Atmos 112: D22S05.

28. Liu X, Zhang Y, Han W, Tang A, Shen J, et al. (2013) Enhanced nitrogen deposition over China. Nature 494: 459–462.

29. Hutchinson G, Mosier A (1981) Improved soil cover method for field measurement of nitrous oxide fluxes. Soil Sci Soc Am J 45: 311–316.

30. Alves BJR, Smith KA, Flores RA, Cardoso AS, Oliveira WRD, et al. (2012) Selection of the most suitable sampling time for static chambers for the estimation of daily mean N_2O flux from soils. Soil Biol Biochem 46: 129–135.

31. Tiedje J, Simkins S, Groffman P (1989) Perspectives on measurement of denitrification in the field including recommended protocols for acetylene based methods. Plant Soil 115: 261–284.

32. Ryden JC, Skinner JH, Nixon DJ (1987) Soil core incubation system for the field measurement of denitrification using acetylene-inhibition. Soil Biol Biochem 19: 753–757.

33. Klemedtsson L, Svensson B, Rosswall T (1987) Dinitrogen and nitrous oxide produced by denitrification and nitrification in soil with and without barley plants. Plant Soil 99: 303–319.

34. Xu H, Chen G-x, Li A-n, Han S-j, Huang G-h (2000) Nitrification and denitrification as sources of gaseous nitrogen emission from different forest soils in Changbai Mountain. J Forest Res 11: 177–182.

35. Deboer W, Tietema A, Gunneweik P, Laanbroek HJ (1992) The chemolithotrophic ammonium-oxidizing community in a nitrogen-saturated acid forest soil in relation to pH-dependent nitrifying activity. Soil Biol Biochem 24: 229–234.

36. Felber R, Conen F, Flechard CR, Neftel A (2012) Theoretical and practical limitations of the acetylene inhibition technique to determine total denitrification losses. Biogeosciences 9: 4125–4138.

37. Yamamoto A, Akiyama H, Naokawa T, Miyazaki Y, Honda Y, et al. (2014) Lime-nitrogen application affects nitrification, denitrification, and N_2O emission in an acidic tea soil. Biol Fert Soils 50: 53–62.

38. Zhong L, Du R, Ding K, Kang X, Li FY, et al. (2014) Effects of grazing on N_2O production potential and abundance of nitrifying and denitrifying microbial communities in meadow-steppe grassland in northern China. Soil Biol Biochem 69: 1–10.

39. Boulêtreau S, Salvo E, Lyautey E, Mastrorillo S, Garabetian F (2012) Temperature dependence of denitrification in phototrophic river biofilms. Sci Total Environ 416: 323–328.

40. Stief P, Poulsen M, Nielsen LP, Brix H, Schramm A (2009) Nitrous oxide emission by aquatic macrofauna. Proc Natl Acad Sci USA 106: 4296–4300.

41. Hatch DJ, Jarvis SC, Philipps L (1990) Field measurement of nitrogen mineralization using soil core incubation and acetylene inhibition of nitrification. Plant Soil 124: 97–107.

42. Burke W, Gabriels D, Bouma J (1986) Soil Structure Assessment. Rotterdam, Netherlands: A. A. Balkema Publishers. 92 p.

43. Parton WJ, Holland EA, Del Grosso SJ, Hartman MD, Martin RE, et al. (2001) Generalized model for NO_x and N_2O emissions from soils. J Geophys Res - Atmos 106: 17403–17419.

44. Gee GW, Bauder JW (1986) Particle-size analysis. In: A . Klute, editor editors. Methods of soil analysis, part I: physical and mineralogical methods. Madison, Wisconsin, USA: American Society of Agronomy, Inc. and Soil Science Society of America, Inc. pp. 383–411.

45. SPSS Inc. (Chicago, IL) SPSS for Windows, Version 11.5.

46. Goodroad LL, Keeney DR (1984) Nitrous oxide emission from forest, marsh, and prairie ecosystems. J Environ Qual 13: 448–452.

47. Papen H, Butterbach-Bahl K (1999) A 3-year continuous record of nitrogen trace gas fluxes from untreated and limed soil of a N-saturated spruce and beech forest ecosystem in Germany: 1. N2O emissions. J Geophys Res - Atmos 104: 18487–18503.

48. Bowden RD, Melillo JM, Steudler PA, Aber JD (1991) Effects of nitrogen additions on annual nitrous-oxide fluxes from temperate forest soils in the northeastern United-States. J Geophys Res 96: 9321–9328.

49. Lovett G, Goodale C (2011) A new conceptual model of nitrogen saturation based on experimental nitrogen addition to an oak forest. Ecosystems 14: 615–631.

50. Ruser R, Flessa H, Russow R, Schmidt G, Buegger F, et al. (2006) Emission of N_2O, N_2 and CO_2 from soil fertilized with nitrate: effect of compaction, soil moisture and rewetting. Soil Biol Biochem 38: 263–274.

51. Kachenchart B, Jones DL, Gajaseni N, Edwards-Jones G, Limsakul A (2012) Seasonal nitrous oxide emissions from different land uses and their controlling factors in a tropical riparian ecosystem. Agric, Ecosyst Environ 158: 15–30.

52. Fierer N, Schimel JP, Holden PA (2003) Influence of drying–rewetting frequency on soil bacterial community structure. Microbial Ecol 45: 63–71.

53. Goebel M-O, Bachmann J, Woche SK, Fischer WR (2005) Soil wettability, aggregate stability, and the decomposition of soil organic matter. Geoderma 128: 80–93.

54. Schjønning P, Thomsen IK, Moldrup P, Christensen BT (2003) Linking soil microbial activity to water- and air-phase contents and diffusivities. Soil Sci Soc Am J 67: 156–165.

55. Petersen SO, Schjønning P, Thomsen IK, Christensen BT (2008) Nitrous oxide evolution from structurally intact soil as influenced by tillage and soil water content. Soil Biol Biochem 40: 967–977.

56. Ciarlo E, Conti M, Bartoloni N, Rubio G (2007) The effect of moisture on nitrous oxide emissions from soil and the $N_2O/(N_2O+N_2)$ ratio under laboratory conditions. Biol Fert Soils 43: 675–681.

57. Parton WJ, Mosier AR, Ojima DS, Valentine DW, Schimel DS, et al. (1996) Generalized model for N_2 and N_2O production from nitrification and denitrification. Global Biogeochem Cy 10: 401–412.

58. Li K, Gong Y, Song W, Lv J, Chang Y, et al. (2012) No significant nitrous oxide emissions during spring thaw under grazing and nitrogen addition in an alpine grassland. Global Change Biol 18: 2546–2554.

59. Hénault C, Bizouard F, Laville P, Gabrielle B, Nicoullaud B, et al. (2005) Predicting *in situ* soil N_2O emission using NOE algorithm and soil database. Global Change Biol 11: 115–127.

60. Del Grosso SJ, Parton WJ (2012) Climate change increases soil nitrous oxide emissions. New Phytol 196: 327–328.

61. Martikainen PJ, de Boer W (1993) Nitrous oxide production and nitrification in acidic soil from a dutch coniferous forest. Soil Biol Biochem 25: 343–347.

62. Wolf I, Brumme R (2003) Dinitrogen and nitrous oxide formation in beech forest floor and mineral soils. Soil Sci Soc Am J 67: 1862–1868.

63. Weslien P, Klemedtsson AK, Borjesson G, Klemedtsson L (2009) Strong pH influence on N_2O and CH_4 fluxes from forested organic soils. Eur J Soil Sci 60: 311–320.

64. Mogge B, Kaiser E-A, Munch J-C (1999) Nitrous oxide emissions and denitrification N-losses from agricultural soils in the Bornhöved Lake region: influence of organic fertilizers and land-use. Soil Biol Biochem 31: 1245–1252.

65. Bai E, Houlton BZ, Wang YP (2012) Isotopic identification of nitrogen hotspots across natural terrestrial ecosystems. Biogeosciences 9: 3287–3304.

66. Dannenmann M, Butterbach-Bahl K, Gasche R, Willibald G, Papen H (2008) Dinitrogen emissions and the N_2:N_2O emission ratio of a Rendzic Leptosol as influenced by pH and forest thinning. Soil Biol Biochem 40: 2317–2323.

67. Seitzinger SP (1988) Denitrification in freshwater and coastal marine ecosystems: ecological and geochemical significance. Limnol Oceanogr 33: 702–724.

68. Weier KL, Doran JW, Power JF, Walters DT (1993) Denitrification and the dinitrogen/nitrous oxide ratio as affected by soil water, available carbon, and nitrate. Soil Sci Soc Am J 57: 66–72.

69. Cho CM, Burton DL, Chang C (1997) Kinetic formulation of oxygen consumption and denitrification processes in soil. Can J Soil Sci 77: 253–260.

70. Gillam KM, Zebarth BJ, Burton DL (2008) Nitrous oxide emissions from denitrification and the partitioning of gaseous losses as affected by nitrate and carbon addition and soil aeration. Can J Soil Sci 88: 133–143.

71. Butterbach-Bahl K, Baggs EM, Dannenmann M, Kiese R, Zechmeister-Boltenstern S (2013) Nitrous oxide emissions from soils: how well do we understand the processes and their controls? Philos T Roy Soc B 368: 20130122.

18

Negative Density Dependence Regulates Two Tree Species at Later Life Stage in a Temperate Forest

Tiefeng Piao[1,2], Jung Hwa Chun[2]*, Hee Moon Yang[2], Kwangil Cheon[2]

1 Institute of Soil and Water Conservation, Northwest A&F University, Yangling, China, **2** Division of Forest Ecology, Department of Forest Conservation, Korea Forest Research Institute, Seoul, Republic of Korea

Abstract

Numerous studies have demonstrated that tree survival is influenced by negative density dependence (NDD) and differences among species in shade tolerance could enhance coexistence via resource partitioning, but it is still unclear how NDD affects tree species with different shade-tolerance guilds at later life stages. In this study, we analyzed the spatial patterns for trees with *dbh* (diameter at breast height) ≥ 2 cm using the pair-correlation $g(r)$ function to test for NDD in a temperate forest in South Korea after removing the effects of habitat heterogeneity. The analyses were implemented for the most abundant shade-tolerant (*Chamaecyparis obtusa*) and shade-intolerant (*Quercus serrata*) species. We found NDD existed for both species at later life stages. We also found *Quercus serrata* experienced greater NDD compared with *Chamaecyparis obtusa*. This study indicates that NDD regulates the two abundant tree species at later life stages and it is important to consider variation in species' shade tolerance in NDD study.

Editor: Ben Bond-Lamberty, DOE Pacific Northwest National Laboratory, United States of America

Funding: This study was supported by the "Long-Term Monitoring and Assessment of Climate Change Impacts on Forest Ecosystem" project of Korea Forest Research Institute (KFRI). The funders had no role in study design, data collection and analysis, decision to publish, or preparation of the manuscript.

Competing Interests: The authors have declared that no competing interests exist.

* Email: chunjh69@forest.go.kr

Introduction

Tree populations are often thought to be regulated by negative density dependence (NDD) that can occur during several life stages, because higher conspecific density can impair performance due to stronger intraspecific competition for resources, more susceptibility to pathogens and easier detection by herbivores [1]. Therefore, NDD can result in lower growth and survival of local abundant species [2]. The supporting results include observational [2–8] and experimental studies [9–13] that take place in tropical [14–17], subtropical [18] and temperate forest [19–22]. Johnson et al. [23] even further concluded that NDD explains the latitudinal gradient of tree species richness.

A challenge for studies of NDD is caused by lagged effects [24] which are non-fatal effects on individuals during one life stage that influence the growth and mortality rate during the following life stage [25]. Mortality patterns in seedlings can normally be analyzed through direct observation, due to their high susceptibility to natural enemies and environmental stress. However, for established trees that have lower mortality rates, long time-scales observation is required to analyze mortality pattern because NDD may not be strong enough to induce immediate mortality, but instead cause limited growth over short time-scales [24]. This may explain why few studies have been implemented to test NDD on trees at later life stages [4,26–29]. Peters [4] found patterns consistent with NDD for saplings and trees of >75% of the species tested at sites in Pasoh, Malaysia, and BCI, Panama. Lan et al. [29] found saplings of 83.2% of species have a aggregated pattern, whereas adults of 96.2% species have a random distribution, implying that NDD can make the spatial pattern of tropical trees more regular with time. Evidence have shown that different

mechanisms shape plant communities at different life stages [30–32], therefore, it is important to evaluate NDD at later life stages. One robust way to indirectly examine NDD on the long history of a forest is to analyze the changes in spatial patterns of trees at different life stages [20,33,34]. This is possible because the regulating mechanisms that are operating in a forest should have left a detectable spatial signature [35]. If NDD is prevalent and works constantly during the whole life-cycle of trees, the following signatures should be left on the spatial pattern of trees: trees survive better with fewer conspecific neighbors and the clustering of conspecific trees declines with time [36,37]. However, the above predictions can be obscured by habitat heterogeneity [33], because the performance of individuals can be greatly influenced by the availability of environmental resources [38–40], and the functional traits that are involved in plant-enemy interactions might be altered by abiotic and biotic factors [41]. This change in performance with environment can affect a population's susceptibility to herbivory to better link NDD with habitat heterogeneity [18]. For example, one population may show NDD pattern if resource availability is low and reduces the ability of individuals to survive herbivore damage, whereas one population with ample resources may not show the same effect. Therefore, habitat heterogeneity must be considered in NDD study. However, this is made difficult by the fact that many environmental covariates are difficult to quantify [20]. Getzin et al. [33] used a simple method to solve this problem. By utilizing the spatial pattern of adult trees (*dbh* ≥ 10 cm), they factored out large-scale habitat heterogeneity and were able to detect NDD in western hemlock populations.

Species varies in the strength of NDD, arise from differences in morphological, physiological and allocational traits [42]. Shade-tolerant, slow-growing species may experience less NDD than shade-intolerant, fast-growing species [43,44]. One of the possible mechanisms behind the differences is that shade-tolerant species tend to have higher tissue density and carbohydrate storage for increasing tolerance to herbivores [43] and recover from damage compared with shade-intolerant species [45,46]. Previous studies have shown that seedling shade tolerance is negatively correlated with disease susceptibility [40] and depression of growth and survival [47]. Also, in a tropical forest study, Comita and Hubbell [7] found seedling survival tended to be positively correlated with species abundance before controlling for shade tolerance. After controlling for shade tolerance, they found a significant negative relationship between seedling survival and species' shade tolerance. Therefore, shade tolerance should be accounted for in NDD study. However, few NDD studies implemented at later life stages considered the variation in shade tolerance among species.

The aim of this study is to examine NDD at later life stage for the most abundant shade-tolerant and shade-intolerant species at Keumsan long-term ecological research plot (Keumsan LTER) which is located in South Korea. The study was implemented by testing whether the intensity of aggregation was decreased from saplings to juveniles. The large-scale habitat heterogeneity was accounted for by utilizing the spatial pattern of adult trees. We first determined whether saplings exhibit additional aggregated patterns relative to adults, then tested whether the additional aggregation decreases from saplings to juveniles (the definitions of "sapling", "juvenile" and "adult" are provided in the Methods). Finally, we compared whether the shade-intolerant species experiences stronger NDD compared with the shade-tolerant species across size classes.

Materials and Methods

Study site

Our 1-ha (100×100 m) study site, Keumsan LTER (34°30′N, 127°59′E), is located in the southern area of the Hallyeohaesang National park in Gyeongsangnam-do, South Korea. The area is located in a warm temperate forest zone. The annual mean temperature and precipitation were 14.0°C and 2,180.0 mm, respectively, and the mean monthly temperature ranged from −1.5°C (January) and 25.6°C (August) in 2011 [48]. The study site is a secondary forest which had suffered great damage during the Korean War (1950–1953). Thereafter, the forest has been well conserved and underwent secondary succession [49]. The Keumsan LTER is located at a 360–430 m hillside elevation with slopes ranging from 12°–28°, and is influenced by habitat heterogeneity such as edaphic gaps (e.g. gravels or wet drainage sites) [50]. The most abundant species are *Quercus serrata*, *Chamaecyparis obtusa*, *Styrax japonica*, *Acer pseudo-sieboldianum*, *Carpinus tschonoskii* and *Stewertia pseudo-camellia*. The Hallyeohaesang National Park is owned and managed by the state and its government and the location including our study area is not privately-owned. No specific permits were required for the described field studies. The field studies did not involve endangered or protected species.

Data collection

The Keumsan LTER was established in 2000. All woody stems ≥2 cm *dbh* were mapped, measured, identified to species, and tagged. The plot was recensused in 2006 and 2011. In the 2011 census, we documented 2,412 free-standing live individuals ≥2 cm *dbh* belonging to 20 families, 22 genera and 35 species. In this paper, we used data on live trees ≥2 cm *dbh* from the 2011 census for spatial point pattern analysis.

We selected *Chamaecyparis obtusa* (CHOB) and *Quercus serrata* (QUSE) as focal species. *Chamaecyparis obtusa*, a slow-growing, late successional species [51,52], was the most abundant shade-tolerant tree species at the site, and accounted for 32.5% of the total individuals and 11.3% of total basal area (Table 1); *Quercus serrata*, an early successional species [53], was the most abundant shade-intolerant tree species at the site, and accounted for 10.7% of the total individuals and 53.9% of total basal area (Table 1). Young trees of the two species were usually found under different light conditions: QUSE were often found in forest gaps, while CHOB were often found in shaded areas.

All individuals of the two species were divided into three size classes as an indication for life history stages: sapling, juvenile and adult (Note the terminology differ from their original meanings, they were used only for classifying different size classes of trees and facilitating spatial point pattern analyses). For CHOB: saplings with *dbh* ranging from 2 to 5 cm, juveniles with *dbh* ranging from 5 to 10 cm and adults with *dbh* ≥10 cm; for QUSE: saplings with *dbh* ranging from 2 to 20 cm, juveniles with *dbh* ranging from 20 to 30 cm and adults with *dbh* ≥30 cm. *Dbh* cut-offs were selected to ensure adequate sample sizes for the spatial point pattern analysis. Both species had more than 60 individuals in each size class (Table S1 and S2). Because using different cut-offs may lead to different results, we also used all other cut-offs (with an increment of 1 cm) (on the condition that more than 60 individuals are included in each life stage) (Table S1 and S2) to implement the analyses.

Data analysis

In recent studies, spatial point pattern analysis which compares patterns of trees in different size classes has been shown to be an effective approach for testing NDD on established trees [18,20,33,34,54], with the assumption that populations of established trees are in an equilibrium stage. We utilized the bivariate pair correlation $g(r)$ function [55,56] to implement the analyses. The $g(r)$ function is the probability density function of the broadly used Ripley's $K(r)$ function [57], which is calculated as:

$$g(\mathrm{r}) = \frac{dK(\mathrm{r})}{d(\mathrm{r})} \frac{1}{2\pi r}$$

$K(r)$ is the cumulative distribution function of the expected number of points of a pattern within the whole circle of a given radius r around a typical point of the pattern divided by the intensity λ (points per unit area) of the pattern. $K(r)$ is a cumulative distribution function where $K(r)$ is the expected number of other points of a pattern within the whole circle of a given radius r around a typical point of the pattern divided by the intensity λ of the pattern. In contrast, the $g(r)$ is a non-cumulative distribution function in which $g(r)$ is the expected density of other points in a ring of a given distance r around a focal point divided by the intensity λ of the pattern [55]. The $g(r)$ function can be used to estimate the strength of aggregation at specific scales. If $g(r) \gg 1$, there are more points at scale r than expected under a random distribution, which indicates an aggregated pattern at scale r. For the correction of edge effects, the translation correction method was used in the analysis [55].

We used the method of random-labeling null model within a case–control design [33] to study NDD on the two focal species from saplings to juveniles, where saplings and juveniles were used as cases (pattern i) and adults as controls (pattern j). In the case-control design, the control pattern is used to account for the large-scale habitat heterogeneity and intensity of seed rain [33,34]. This

Table 1. Species composition in the Keumsan LTER.

Species name	No. Ind. (%)	Basal area (m²) (%)	Shade tolerance
Chamaecyparis obtusa	784 (32.5)	3.68 (11.3)	Shade-tolerant
Quercus serrata	257 (10.7)	17.6 (53.9)	Shade-intolerant
Acer pseudo-sieboldianum	253 (10.5)	1.08 (3.3)	Shade-tolerant
Styrax japonicas	138 (5.7)	1.19 (3.6)	Mid-tolerant
Stewartia pseudo-camellia	125 (5.2)	0.62 (1.9)	Shade-tolerant
Carpinus cordata	97 (4)	0.42 (1.3)	Shade-intolerant
Sapium japonicum	98 (4)	0.26 (0.8)	Shade-tolerant
Cornus kousa	92 (3.8)	0.33 (1)	Mid-tolerant
Carpinus tschonoskii	93 (3.8)	2.43 (7.5)	Shade-tolerant
Acer palmatum	79 (3.3)	0.22 (0.7)	Shade-tolerant
Rhododendron schlippenbachii	77 (3.2)	0.06 (0.2)	Mid-tolerant
Fraxinus sieboldiana	51 (2.1)	0.11 (0.3)	Mid-tolerant
Others	268 (11.1)	4.62 (14.2)	-
Total	2412	32.62	-

The figures in brackets show the percentage of the total amounts. The table is sorted in descending order according to the number of individuals.

method is based on two assumptions: 1) small-scale patterns are usually attributed to plant–plant interactions, whereas large-scale patterns are usually attributed to habitat heterogeneity [58] and 2) adult trees have undergone excessive thinning over time due to habitat heterogeneity and most likely represent those which lived in sites most favorable for the species. Therefore, the large-scale pattern (e.g. at $r > 10$ m) of adults can be used as an indicator of environmental habitat preferences and can be used to control for the large-scale habitat heterogeneity [59,60]. Under the null hypothesis of random-labeling, labels (case or control) are assigned to points randomly, conditioning the observed locations of the points in the joined patterns of cases and controls [61]. If the null hypothesis is rejected, the case-control approach may identify specific factors (i.e. NDD) other than habitat heterogeneity that may influence the spatial pattern of trees, by comparing the differences between case patterns and control patterns of trees in different life stages [18,33]. The $g(r)$ functions are invariant under random thinning of trees, hence we would expect $g_{ii}(r) = g_{ij}(r) = g_{ij}(r) = g_{ji}(r)$. In this study, NDD is examined by using $g_{ii}(r) - g_{ij}(r)$ as test statistics [33,34,62].

According to the prediction of NDD, tree survival would be low if individuals are growing in high density patches of conspecifics. Therefore, if there is strong NDD in saplings, the surviving juveniles would be less aggregated than saplings. To test this hypothesis, we studied the difference in the degree of aggregation between saplings and juveniles. We used $A_i(r) = g_{ii}(r) - g_{ij}(r)$ as the test statistic to determine whether cases i show an additional pattern that is independent of the controls j. If $A_i(r) >> 0$, cases can be said to exhibit additional aggregated patterns relative to adults, irrespective of whether habitat heterogeneity is present or not [62,63]. The change in additional aggregation from saplings to juveniles at scale r can be expressed by the formula: $T(r) = A_{saplings}(r) - A_{juveniles}(r)$. For a species, if $A_{saplings}(r) >> 0$ and $T(r) >> 0$, we would infer that NDD existed in saplings. If this is the case, $T(r)$ can be used to indicate the strength of NDD at different scales.

All spatial point pattern analyses were implemented with the "spatstat" package of R [64]. We focused on the scale of 0–20 m for the analyses above, assuming that tree-tree interactions could

be efficiently indicated by this scale [65]. We performed 199 Monte Carlo simulations of the random labeling null model and used the 5th-lowest and 5th-highest values (i.e., extreme 0.25% simulated cases at either end) as simulation envelopes. However, this simulation inference yields an underestimated Type I error rate because the tests are performed at different concurrent scales [66]. We combined the simulation method with a goodness-of-fit test (GOF) [67]. Significant deviations were only determined for those data sets where the observed GOF's P value (P_{GOF}) were less than 0.025 [58,66].

Results

Diameter distributions

The diameter distribution of CHOB was strongly right-skewed with 89% trees less than 10 cm dbh (Fig. 1a); this is a pattern typical of late-successional species that have a large number of suppressed young trees. In contrast, the diameter distribution of QUSE was symmetrical with a lack of small trees (Fig. 1b); this is a pattern typical of early-successional species that establish as approximately even-aged cohorts in large forest gaps. For both CHOB and QUSE, dead trees were concentrated in the smaller diameter classes (Fig. 1). Young trees of QUSE had a much greater mortality rate compared with CHOB (Fig. 1). For trees that were

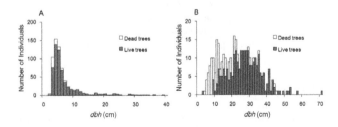

Figure 1. *Dbh* distributions of *Chamaecyparis obtusa* **(a)** and *Quercus serrata* **(b).** Live trees are those individuals that were alive in the 2011 census; dead trees are those individuals that were alive in the 2000 census but were found dead in the 2011 census.

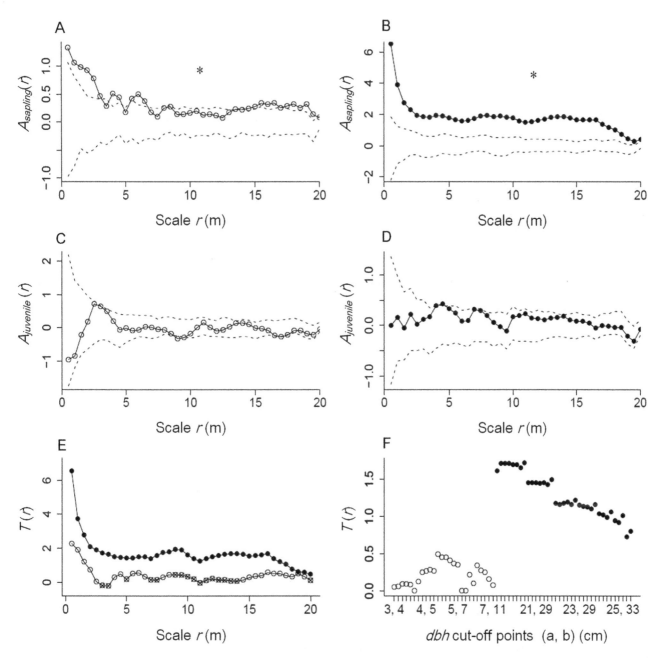

Figure 2. Changes of additional aggregation from saplings to juveniles. Open circles show the result of *Chamaecyparis obtusa* and filled circles show the result of *Quercus serrata*. The test statistic $A_j(r)$ ($A_{saplings}(r)$ or $A_{juveniles}(r)$) evaluates if there is an additional aggregation within cases (saplings or juveniles) that is independent from adults. $A_j(r) \gg 0$ means there is an additional aggregation within cases relative to adults. $A_j(r) \approx 0$ means patterns of cases and adults are created by the same stochastic process. The scale r is the radius around a focal tree of the pattern. Dotted lines were the 95% simulation envelopes constructed using the 5th-lowest and 5th-highest value of 199 Monte Carlo simulations of the null model. Comparison of the strength of NDD effect experienced by the two focal species (e) and Comparison of the strength of NDD experienced by the two focal species using all *dbh* cut-offs (f). *Dbh* cut-off points (a, b) are the cut-offs for defining life stages of trees: sapling (≥ 2 and $<a$), juvenile ($\geq a$ and $<b$) and adult ($\geq b$). Units of cut-off points are centimeters. The cut-off point lists can be found in Table S1 and S2. *: significant difference by the goodness-of-fit test ($P_{GOF} < 0.025$).

alive and <20 cm *dbh* in the year 2000 census, 50.1% of QUSE were dead in the year 2011 census; whereas, the mortality rate was only 10.8% for CHOB.

Negative density dependence

For both CHOB and QUSE, saplings showed additional aggregated patterns relative to adults, as their test statistic $A_{saplings}(r) \gg 0$ ($P_{GOF} = 0.005$ for CHOB and QUSE) (Fig. 2a,

b). For CHOB, significant deviations were found at the scale of 0–3, 4–4.5, 5.5–6.5, 8.5, 14–16.5, 17.5–19.5 m (Fig. 2a); for QUSE, significant deviations were found at all test scales (Fig. 2b). In juveniles, these aggregated patterns diminished for both species, as their test statistics demonstrate ($A_{juveniles}(r) \approx 0$) (Fig. 2c, d). These results indicate that both focal species were influenced by NDD at the sapling stage. For both species, the decrease in additional aggregation was the greatest at the smallest scale and had a trend of decreasing with increasing scales (Fig. 2e), indicating that NDD occurred predominantly at close distances among neighbors. QUSE experienced stronger NDD in the sapling stage as $T(r)$ is greater for QUSE compared with CHOB at all test scales (Fig. 2e). For all *dbh* cut-offs, the mean $T(r)$ values were greater for QUSE than CHOB (Fig. 2f, Table S1 and S2), indicating the above finding was not influenced by the selection of different *dbh* cut-offs.

Discussion

NDD has been shown to play an important role in the population dynamics of tree communities in numerous studies [1,10,31,68–70]. Our study examined the spatial patterns and tested NDD for two abundant tree species with different shade tolerance at later life stages.

Plants are most susceptible to natural enemies and abiotic stresses in early life stages, therefore, most demographic winnowing of propagules takes place at early life stages [71–73]. Thus, one might expect NDD to be subdued at later life stages. However, recent studies pointed out that NDD also exists at later life stages [4,15,18,22,54,74]. For example, Zhu et al. [18] studied 47 common tree species (*dbh* ≥ 1 cm) and found 39 of them exhibited conspecific density-dependent thinning in a subtropical forest by comparing the strength of aggregation from saplings to juveniles. Piao et al. [54] studied 15 common tree species (*dbh* ≥ 1 cm) in a temperate forest and found 11 of them showed conspecific density-dependent thinning, also by comparing the strength of aggregation from saplings to juveniles. Carson et al. [75] suggested that ecologists should focus on NDD occurring at later life stages in order to get a complete understanding of the effect in promoting species coexistence. In our later life stage study, we found NDD regulated both focal species. We also found NDD was the greatest at small scales and decreased with increasing scales for saplings. These results may be due to the decline in aggregation of saplings with increasing distance from parent trees, which had likely resulted from dispersal limitation [76]. Those findings are consistent with the distance-dependent model which predicts that survival is negatively correlated with the distance from the conspecific adult [77]. Distance dependence and NDD leads to a decreased probability that a species will replace itself locally and therefore promotes species coexistence [5,77].

Shade-tolerant tree species have been shown to have better defenses against natural enemies than shade-intolerant tree species [42,43]. Therefore, one could predict that shade-intolerant tree species experience greater NDD than shade-tolerant tree species. For example, Augspurger and Kelly [11] found that seedling disease susceptibility was negatively correlated with shade tolerance. Comita et al. [7] found that variations in species' shade tolerance could mask the negative relationship between seedling survival and abundance. Our result is consistent with the above prediction: we found that QUSE (a shade-intolerant and early-successional species) experienced greater NDD compared with CHOB (a shade-tolerant and late-successional species). Because this result can be caused by the great differences in the size class cut-offs between the two species, we used all possible size class cut-offs to implement the analyses and therefore our results are not likely caused by the specific size cut-offs selected. Our observation and previous studies suggest that attempting to quantify the contribution of NDD should take into account the variation in species' shade tolerance.

The stronger NDD experienced by QUSE may play an important role in the replacement of QUSE by other late-successional species such as CHOB, as self-thinning of early-successional trees could progressively open up the forest canopy and create open spaces that promote the survival of late-successional trees [20]. However, further studies should be implemented to examine whether the subsequent survival of late-successional trees such as CHOB are higher in those open spaces created by QUSE.

Conclusions

Our study tested NDD for two abundant species found in the Keumsan LTER. We found NDD on the two abundant tree species at later life stages. We also found the shade-intolerant species suffered stronger NDD compared with the shade-tolerant species. Our findings highlighted the importance of testing NDD at later life stage for improved understanding of species coexistence. We also highlighted the importance of considering species shade tolerance in NDD study.

Supporting Information

Table S1 List of *dbh* cut-off points for defining life stages of *Chamaecyparis obtusa*: sapling (≥2 and <a), juvenile (≥a and <b) and adult (≥b).

Table S2 List of *dbh* cut-off points for defining life stages of *Quercus serrata*: sapling (≥2 and <a), juvenile (≥a and <b) and adult (≥b).

Acknowledgments

We thank the editor and two anonymous reviewers for their constructive comments and suggestions to improve the quality of the paper.

Author Contributions

Conceived and designed the experiments: JHC. Performed the experiments: JHC TP KC. Analyzed the data: TP. Contributed reagents/materials/analysis tools: HMY KC. Wrote the paper: TP. Wrote part of the programming code: JHC.

References

1. Wright SJ (2002) Plant diversity in tropical forests: a review of mechanisms of species coexistence. Oecologia 130: 1–14.
2. Comita LS, Muller-Landau HC, Aguilar S, Hubbell SP (2010) Asymmetric density dependence shapes species abundances in a tropical tree community. Science (New York, NY) 329: 330–332.
3. Harms KE, Wright SJ, Calderón O, Hernández A, Herre EA (2000) Pervasive density-dependent recruitment enhances seedling diversity in a tropical forest. Nature 404: 493–495.
4. Peters HA (2003) Neighbour-regulated mortality: the influence of positive and negative density dependence on tree populations in species-rich tropical forests. Ecol Lett 6: 757–765.
5. Queenborough SA, Burslem DFRP, Garwood NC, Valencia R (2007) Neighborhood and community interactions determine the spatial pattern of tropical tree seedling survival. Ecology 88: 2248–2258.

6. Bai X, Queenborough SA, Wang X, Zhang J, Li B, et al. (2012) Effects of local biotic neighbors and habitat heterogeneity on tree and shrub seedling survival in an old-growth temperate forest. Oecologia 170: 755–765.
7. Comita LS, Hubbell SP (2009) Local neighborhood and species' shade tolerance influence survival in a diverse seedling bank. Ecology 90: 328–334.
8. Queenborough SA, Burslem DFRP, Garwood NC, Valencia R (2009) Taxonomic scale-dependence of habitat niche partitioning and biotic neighbourhood on survival of tropical tree seedlings. Proceedings Biological sciences/The Royal Society 276: 4197–4205.
9. Bell T, Freckleton RP, Lewis OT (2006) Plant pathogens drive density-dependent seedling mortality in a tropical tree. Ecol Lett 9: 569–574.
10. Liu X, Liang M, Etienne RS, Wang Y, Staehelin C, et al. (2012) Experimental evidence for a phylogenetic Janzen-Connell effect in a subtropical forest. Ecol Lett 15: 111–118.

11. Augspurger CK, Kelly CK (1984) Pathogen mortality of tropical tree seedlings: experimental studies of the effects of dispersal distance, seedling density, and light conditions Oecologia 61: 211–217.

12. Walsh RK, Bradley C, Apperson CS, Gould F (2012) An Experimental Field Study of Delayed Density Dependence in Natural Populations of Aedes albopictus. 7: e35959.

13. Mangan SA, Schnitzer SA, Herre EA, Mack KML, Valencia MC, et al. (2010) Negative plant-soil feedback predicts tree-species relative abundance in a tropical forest. Nature 466: 752–755.

14. Condit R, Hubbell SP, Foster RB (1992) Recruitment Near Conspecific Adults and the Maintenance of Tree and Shrub Diversity in a Neotropical Forest. Am Nat 140: 261–286.

15. Wills C, Condit R, Foster RB, Hubbell SP (1997) Strong density- and diversity-related effects help to maintain tree species diversity in a neotropical forest. Proc Natl Acad Sci U S A 94: 1252–1257.

16. Uriarte M, Canham CD, Thompson J, Zimmerman JK (2004) A neighborhood analysis of tree growth and survival in a hurricane-driven tropical forest. Ecol Monogr 74: 591–614.

17. Comita LS, Uriarte M, Thompson J, Jonckheere I, Canham CD, et al. (2009) Abiotic and biotic drivers of seedling survival in a hurricane-impacted tropical forest. J Ecol 97: 1346–1359.

18. Zhu Y, Mi X, Ren H, Ma K (2010) Density dependence is prevalent in a heterogeneous subtropical forest. Oikos 119: 109–119.

19. Kenkel NC (1988) Pattern of self-thinning in jack pine: testing the random mortality hypothesis. Ecology 69: 1017–1024.

20. He F, Duncan RP (2000) Density-dependent effects on tree survival in an old-growth Douglas fir forest. J Ecol 88: 676–688.

21. Lambers JHR, Clark JS, Beckage B (2002) Density-dependent mortality and the latitudinal gradient in species diversity. Nature 417: 732–735.

22. Zhang J, Hao Z, Sun I-F, Song B, Ye J, et al. (2009) Density dependence on tree survival in an old-growth temperate forest in northeastern China. Ann For Sci 66: 204–204.

23. Johnson DJ, Beaulieu WT, Bever JD, Clay K (2012) Conspecific negative density dependence and forest diversity. Science (New York, NY) 336: 904–907.

24. Ratikainen II, Gill JA, Gunnarsson TG, Sutherland WJ, Kokko H (2008) When density dependence is not instantaneous: theoretical developments and management implications. Ecol Lett 11: 184–198.

25. Webster MS, Marra PP, Haig SM, Bensch S, Holmes RT (2002) Links between worlds: unraveling migratory connectivity. Trends Ecol Evol 17: 76–83.

26. Nathan R, Safriel UN, Noy-Meir I, Schiller G (2000) Spatiotemporal variation in seed dispersal and recruitment near and far from Pinus halepensis trees. Ecology 81: 2156–2169.

27. Howe HF, Miriti MN (2004) When Seed Dispersal Matters. BioOne 54: 651–660.

28. Schupp EW, Jordano P (2011) The full path of Janzen-Connell effects: genetic tracking of seeds to adult plant recruitment. Mol Ecol 20: 3953–3955.

29. Lan GY, Zhu H, Cao M, Hu YH, Wang H, et al. (2009) Spatial dispersion patterns of trees in a tropical rainforest in Xishuangbanna, southwest China. Ecol Res 24: 1117–1124.

30. Comita LS, Condit R, Hubbell SP (2007) Developmental changes in habitat associations of tropical trees. J Ecol 95: 482–492.

31. Paine CET, Norden N, Chave J, Forget P-M, Fortunel C, et al. (2012) Phylogenetic density dependence and environmental filtering predict seedling mortality in a tropical forest. Ecol Lett 15: 34–41.

32. Grubb PJ (1977) The maintenance of species-richness in plant communities: the importance of the regeneration niche. BRCPS 52: 107–145.

33. Getzin S, Wiegand T, Wiegand K, He F (2008) Heterogeneity influences spatial patterns and demographics of a forest stands. J Ecol 96: 807–820.

34. Luo Z, Mi X, Chen X, Ye Z, Ding B (2012) Density dependence is not very prevalent in a heterogeneous subtropical forest. Oikos 121: 1239–1250.

35. Hubbell SP (2001) Local neighborhood effects on long-term survival of individual trees in a neotropical forest. Ecol Res 72: 35–875.

36. Hubbell SP (1979) Tree Dispersion, Abundance, and Diversity in a Tropical Dry Forest. Science 203: 1299–1309.

37. Moeur M (1997) Spatial models of competition and gap dynamics in old-growth plicata forests. For Ecol Manage 94: 175–186.

38. Beckage B, Clark JS (2003) Seedling survival and growth of three forest tree species: The role of spatial heterogeneity. Ecology 84: 1849–1861.

39. Harper JL (1977) Population biology of plants. London, UK: Academic Press.

40. Augspurger CK, Kelly CK (1984) Pathogen mortality of tropical tree seedlings: experimental studies of the effects of dispersal distance, seedling density, and light conditions. Oecologia 61: 211–217.

41. Bachelot B, Kobe RK (2013) Rare species advantage? Richness of damage types due to natural enemies increases with species abundance in a wet tropical forest. J Ecol 101: 846–856.

42. Kobe RK, Vriesendorp CF (2011) Conspecific density dependence in seedlings varies with species shade tolerance in a wet tropical forest. Ecol Lett 14: 503–510.

43. Coley PD, Barone JA (1996) Herbivory and plant defenses in tropical forests. Annual Review of Ecology, Evolution, and Systematics 27: 305–335.

44. Kitajima K, Poorter L (2010) Tissue-level leaf toughness, but not lamina thickness, predicts sapling leaf lifespan and shade tolerance of tropical tree species. New Phytol 168: 708–721.

45. Kobe RK (1997) Carbohydrate allocation to storage as a basis of interspecific variation in sapling survivorship and growth. Oikos 80: 226–233.

46. Myers JA, Kitajima K (2007) Carhohydrate storage enhances seedling shade and stress tolerance in a neotropical forest. J Ecol 95: 383–395.

47. McCarthy-Neumann S, Kobe RK (2008) Tolerance of soil pathogens co-varies with shade tolerance across species of tropical tree seedlings. Ecology 89: 1883–1892.

48. Korea Meteorological Administration (2012) Available: http://www.kma.go.kr.

49. Bae JS, Joo RW, Kim YS (2012) Forest transition in South Korea: reality, path and drivers. Land Use Policy 29: 198–207.

50. Kim C, Lim JH, Lee IK, Park BB, Chun JH (2013) Annual variations of litterfall production in a broadleaved deciduous forest at the Mt. Keumsan LTER site. Journal of Korean Forest Society 102: 210–215.

51. Fujii S, Kasuya N (2008) Fine root biomass and morphology of Pinus densiflora under competitive stress by Chamaecyparis obtusa. J For Res 13: 185–189.

52. Fujimaki R, McGonigle TP, Takeda H (2004) Soil micro-habitat effects on fine roots of Chamaecyparis obtusa Endl.: a field experiment using root ingrowth cores. Plant Soil 266: 325–332.

53. Seiwa K, Kikuzawa K, Kadowaki T, Akasaka S, Ueno N (2006) Shoot lifespan in relation to successional status in deciduous broad-leaved tree species in a temperate forest. New Phytol 169: 537–548.

54. Piao T, Comita LS, Jin G, Kim JH (2013) Density dependence across multiple life stages in a temperate old-growth forest of northeast China. Oecologia 172: 207–217.

55. Stoyan D, Stoyan H (1994) Fractals, random shapes, and point fields: methods of geometrical statistics. Wiley, Chichester.

56. Illian J, Penttinen A, Stoyan H, Stoyan D (2008) Statistical analysis and modelling of spatial point patterns: Wiley, London.

57. Ripley BD (1976) The second-order analysis of stationary point processes. J Appl Prob 13: 255–266. J Appl Probab 13: 255–266.

58. Wiegand T, Gunatilleke S, Gunatilleke N (2007) Species associations in a heterogeneous Srilankan dipterocarp forest. Am Nat 170: E77–95.

59. Stoyan D, Penttinen A (2000) Recent applications of point process methods in forestry statistics. Statistical Science 15: 61–78.

60. Condit R, Ashton PS, Baker P, Bunyavejchewin S, Gunatilleke S, et al. (2000) Spatial patterns in the distribution of tropical tree species. Science (New York, NY) 288: 1414–1418.

61. Wiegand T, Moloney KA (2004) Rings, circles and nullmodels for point pattern analysis in ecology. Oikos 104: 209–229.

62. Getzin S, Dean C, He F, Trofymow JA, Wiegand K, et al. (2006) Spatial patterns and competition of tree species in a Douglas-fir chronosequence on Vancouver Island. Ecography 29: 671–682.

63. Watson DM, Roshier DA, Wiegand T (2007) Spatial ecology of a parasitic shrub: patterns and predictions. Austral Ecol 32: 359–369.

64. R Development Core Team (2011) R: a language and environment for statistical computing. R foundation for statistical computing, Vienna. ISBN 3-900051-07-0. Available: http://www.R-project.org. Accessed 2011 April 13.

65. McCarthy-Neumann S, Kobe RK (2010) Conspecific plant–soil feedbacks reduce survivorship and growth of tropical tree seedlings. J Ecol 98: 396–407.

66. Loosmore NB, Ford ED (2006) Statistical inference using the G or K point pattern spatial statistics. Ecology 87: 1925–1931.

67. Diggle P (2003) Statistical analysis of spatial point patterns. Arnold, London.

68. Chesson P (2000) Mechanisms of maintenance of species diversity. Annu Rev Ecol Syst 31: 343–366.

69. Volkov I, Banavar JR, He F, Hubbell SP, Maritan A (2005) Density dependence explains tree species abundance and diversity in tropical forests. Nature 438: 658–661.

70. Yamazaki M, Iwamoto S, Seiwa K (2009) Distance- and density-dependent seedling mortality caused by several diseases in eight tree species co-occurring in a temperate forest. Plant Ecol 201: 181–196.

71. Terborgh J (2012) Enemies maintain hyperdiverse tropical forests. Am Nat 179: 303–314.

72. Connell JH (1971) On the role of natural enemies in preventing competitive exclusion in some marine animals and in rain forest trees. In: den Boer PJ, Gradwell GR, editors. Dynamics of populations: Centre for Agricultural Publications and Documentation, Wageningen. pp. 298–310.

73. Hyatt LA, Rosenberg MS, Howard TG, Bole G, Fang W, et al. (2003) The distance dependence prediction of the Janzen-Connell hypothesis: a meta-analysis. Oikos 103: 590–602.

74. Stoll P, Newbery DM (2005) Evidence of species-specific neighborhood effects in the Dipterocarpaceae of a Bornean rain forest. Ecology 86: 3048–3062.

75. Carson WP, Anderson J, Leigh E, Schnitzer SA (2008) Challenges Associated with Testing and Falsifying the Janzen-Connell Hypothesis: A Review and Critique. In: Carson WP, Schnitzer SA, editors. Tropical Forest Community Ecology: Blackwell Publishing, Oxford. pp. 210–241.

76. He F, Legendre P, LaFrankie JV (1997) Distribution patterns of tree species in a Malaysian tropical rain forest. Journal of Vegetation Science 8: 105–114.

77. Janzen DH (1970) Herbivores and the Number of Tree Species in Tropical Forests. Am Nat 104: 501–528.

19

Land Use and Land Cover Change Dynamics across the Brazilian Amazon: Insights from Extensive Time-Series Analysis of Remote Sensing Data

João M. B. Carreiras[1,2]*, Joshua Jones[3], Richard M. Lucas[3,4], Cristina Gabriel[1]

1 Tropical Research Institute (IICT), Lisboa, Portugal, 2 National Centre for Earth Observation (NCEO), Centre for Terrestrial Carbon Dynamics (CTCD), University of Sheffield, Sheffield, South Yorkshire, United Kingdom, 3 Department of Geography and Earth Sciences, Aberystwyth University, Aberystwyth, Ceredigion, United Kingdom, 4 Centre for Ecosystem Science, School of Biological, Earth and Environmental Sciences, University of New South Wales, Sydney, New South Wales, Australia

Abstract

Throughout the Amazon region, the age of forests regenerating on previously deforested land is determined, in part, by the periods of active land use prior to abandonment and the frequency of reclearance of regrowth, both of which can be quantified by comparing time-series of Landsat sensor data. Using these time-series of near annual data from 1973–2011 for an area north of Manaus (in Amazonas state), from 1984–2010 for south of Santarém (Pará state) and 1984–2011 near Machadinho d'Oeste (Rondônia state), the changes in the area of primary forest, non-forest and secondary forest were documented from which the age of regenerating forests, periods of active land use and the frequency of forest reclearance were derived. At Manaus, and at the end of the time-series, over 50% of regenerating forests were older than 16 years, whilst at Santarém and Machadinho d'Oeste, 57% and 41% of forests respectively were aged 6–15 years, with the remainder being mostly younger forests. These differences were attributed to the time since deforestation commenced but also the greater frequencies of reclearance of forests at the latter two sites with short periods of use in the intervening periods. The majority of clearance for agriculture was also found outside of protected areas. The study suggested that a) the history of clearance and land use should be taken into account when protecting deforested land for the purpose of restoring both tree species diversity and biomass through natural regeneration and b) a greater proportion of the forested landscape should be placed under protection, including areas of regrowth.

Editor: Bruno Hérault, Cirad, France

Funding: This study was supported by the "REGROWTH-BR - Remote sensing of regenerating tropical forests in Brazil: mapping and retrieving biophysical parameters" project (Ref. PTDC/AGR-CFL/114908/2009), funded by the Foundation for Science and Technology (FCT, http://www.fct.pt/index.phtml.en), Portugal. J.M.B. Carreiras' work was partially funded by NERC National Centre for Earth Observation (Ref: R8/H12/82). The funders had no role in study design, data collection and analysis, decision to publish, or preparation of the manuscript.

Competing Interests: The authors have declared that no competing interests exist.

* Email: j.carreiras@sheffield.ac.uk

Introduction

Humans are increasingly changing the state and dynamics of the Earth system, affecting processes within and between the biosphere, hydrosphere and atmosphere. Climate change is a recognized consequence of such changes and would be occurring at a faster rate if several ecosystem services at the global scale were absent [1]. For example, less than half of the total amount of carbon dioxide (CO_2) released to the atmosphere each year remains there because of removal by terrestrial and ocean carbon (C) sinks [2]. Losses of biodiversity from terrestrial ecosystems have also been substantial.

The role of terrestrial ecosystems in the global C cycle had been widely recognized in the literature (e.g., [3]), with particular recognition given to the pan-tropical belt [4,5]. In the past two decades, two main international conventions have sought to establish mechanisms aimed at stabilizing greenhouse gases (GHG) concentrations in the atmosphere (the United Nations Framework Convention on Climate Change; UNFCCC) and protecting biological diversity (Convention on Biological Diversity; CBD). Specifically, the alarmingly high rate of tropical land use and land

cover change (LULCC) [6] and resulting biodiversity loss with further severe consequences for ecosystem function and structure [7] has driven the UNFCCC to establish several investment mechanisms and market based C transactions. These are related to the enhancement of forest C stocks and the decrease of deforestation and forest degradation, while promoting sustainable development in developing countries (UNFCCC Non-Annex I countries). Such mechanisms include the UNFCCC Clean Development Mechanism (CDM) initiative and the post-Kyoto Protocol Reduced Emissions from Deforestation and Degradation (REDD+) program (e.g., [8]). These efforts are proving successful in many regions leading to greater conservation of the intact forests of the Neotropical, Afrotropical, Australasia and Indo-Malay biogeographical regions [9] and the carbon and biodiversity they contain. However, many of the deforested and degraded lands of the tropics are also capable of supporting forests and hence, in addition, there is potential to restore some of the ecosystem values (e.g., carbon amounts and biodiversity) lost through previous disturbances [10,11]. This capacity depends in part, however, on the history of forest clearance and land use that has occurred and hence a clearer understanding of the potential of

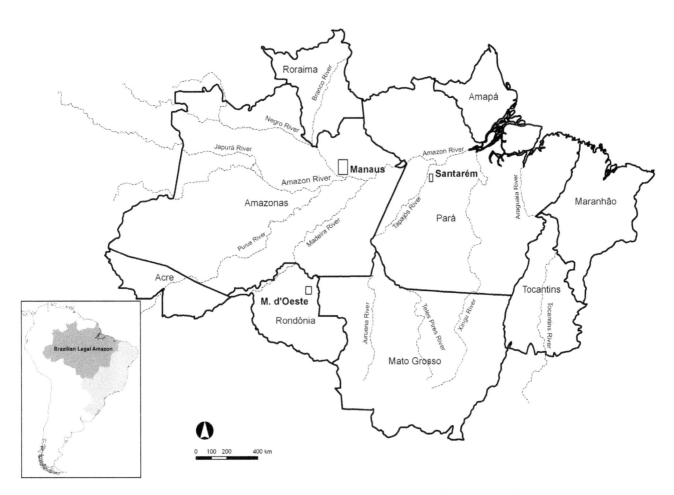

Figure 1. Location of the three study sites. Location of the three study sites: Manaus, Santarém and Machadinho d'Oeste (M. d'Oeste), and the insertion (bottom left) of the Brazilian Legal Amazon (dark grey) in Brazil (light grey) and South American continent. Also shown are the major rivers included in the Amazon basin.

deforested areas to recover ecosystems is needed. This study addresses these issues by focusing on sites in the Brazilian Legal Amazon (BLA) that have experienced deforestation and variable histories of subsequent land use.

The BLA covers an area of approximately 5,000,000 km^2 and consists primarily of primary tropical forest (rain and seasonal forest) [12]. Brazil's National Institute for Space Research (Instituto Nacional de Pesquisas Espaciais, INPE) has, since 1988, undertaken annual mapping of deforestation within the BLA using remote sensing data [13], which has been highly variable across the region. For example, an average deforestation rate of 21,050 km^2 yr^{-1} was reported between 1977 and 1988, which decreased subsequently to 11,030 km^2 yr^{-1} up to 1991; from 1991 to 2004 the deforestation rate increased to 27,772 km^2 yr^{-1} (reaching a record high of 29,059 km^2 yr^{-1} from 1994 to 1995) and then decreased to a record low of 4,571 km^2 yr^{-1} in 2012 [13]. Currently, much of the deforested area is under agricultural use (following disturbance mainly with slash-and-burn practices) but in many areas, the land has been abandoned and frequently supports forests at different stages of regeneration [14].

Several estimates of the deforested land occupied with regeneration have been generated, mainly from interpretation of remote sensing data, but also using transition modeling. Data from the National Oceanic and Atmospheric Administration (NOAA) Advanced Very High Resolution Radiometer (AVHRR) from 1988–1991 were used by Stone et al., 1994 [15] to generate a land

cover map of South America, and subsequently Schroeder and Winjum, 1995 [16] used this dataset to estimate the extent of forest regeneration in the BLA at ~151×10^3 km^2. Fearnside, 1996 [17], using a matrix of annual transition probabilities, estimated that approximately 48% (~195×10^3 km^2) of the deforested landscape in 1990 supported forest regeneration. Lucas et al., 2000 [18] using NOAA AVHRRR data estimated that ~158×10^3 km^2 supported some type of forest regeneration in the period 1991–1994. Cardille and Foley, 2003 [19] used a matrix of annual transition probabilities to estimate that 36% (~91×10^3 km^2) of the area deforested between 1980 and 1995 in the entire Amazon river drainage basin was in some stage of secondary succession forest. Carreiras et al., 2006 [20] exploited a time-series of 12 monthly composite images of the year 2000, derived from the *Satellite Pour l'Observation de la Terre* (SPOT-4) VEGETATION sensor and estimated that, in 2000, ~140×10^3 km^2 of land was occupied by regenerating forests. Neeff et al., 2006 [21] used large-area land cover maps derived from remote sensing datasets to generate estimates of regrowth extent within the BLA and concluded that regrowth increased from ~29×10^3 km^2 in 1978 to ~161×10^3 km^2 in 2002. Whilst providing information on the extent of regenerating forests at certain points in time, the dynamics of regenerating forests and the land on which they occupy has only been attempted using remote sensing by a few studies (e.g., [22]). However, such information is needed across large areas as regenerating forests may represent one of the prime

Table 1. Landsat Multi-spectral Scanner (MSS), Thematic Mapper (TM) and Enhanced TM (ETM+) data available for the three Amazonian sites.

Manaus (Path 231, Row 62)		Santarém (Path 227, Row 62)		Machadinho d'Oeste (Path 231, Row 67)	
Date[1]	Sensor	Date[1]	Sensor	Date[1]	Sensor
19730707	MSS	19840824	TM	19840617	TM
19770731	MSS	19850726	TM	19860810	TM
19780822	MSS	19860729	TM	19870712	TM
19790703	MSS	19870716	TM	19890717	TM
19830709	MSS	19880803	TM	19900618	TM
19850604	TM	19890822	TM	19910925	TM
19880815	TM	19900809	TM	19940816	TM
19890802	TM	19910711	TM	19950803	TM
19910808	TM	19931020	TM	19960704	TM
19920607	TM	19951010	TM	19970723	TM
19941019	TM	19960825	TM	19980624	TM
19950920	TM	19970727	TM	19990729	TM
19960720	TM	19980815	TM	20010803	TM
19990713	TM	19990903	TM	20030724	TM
20010827	ETM+	20000905	TM	20050713	TM
20020830	ETM+	20010916	ETM+	20060716	TM
20030809	TM	20030829	TM	20070703	TM
20060716	TM	20050701	TM	20080806	TM
20070804	TM	20060805	TM	20090809	TM
20080806	TM	20070621	TM	20100625	TM
20090910	TM	20081130	TM	20110612	TM
20100727	TM	20090712	TM		
20110831	TM	20100629	TM		

[1]Date format is yyyymmdd.

mechanisms by which biomass, biological diversity and ecosystem services can be restored [23–25].

The aim of this study was to establish the extent to which the age class distribution of regenerating forests was determined by the history of deforestation and subsequent use and management of the land, thereby giving insight into the potential for forest restoration and conservation into the future. This was achieved by comparing Landsat sensor data classified into the broad amalgamated categories of mature forest, non-forest (i.e., agricul-

ture, including stock pasture) and secondary forest over decadal periods and time-separated by a maximum of four years. Three study areas were chosen to perform the analysis: Manaus (Amazonas state), Santarém (Pará state) and Machadinho d'Oeste (Rondônia state). Several measures were used to compare the three areas since the inception of widespread agricultural practices in the region (1970s for Manaus and Santarém and 1980s for Machadinho d'Oeste) up to the present: Deforestation (conversion from mature forest to non-forest or from secondary forest to non-

Table 2. Error matrix obtained from the accuracy assessment of the 2007 land cover map of the Manaus site (MF – mature forest, NF – non-forest, SF – secondary forest).

Reference	Classified				
	MF	NF	SF	Total	Omission error
MF	186	4	10	200	0.070
NF	8	185	7	200	0.075
SF	11	2	187	200	0.065
Total	205	191	204	600	
Commission error	0.093	0.031	0.083		

The overall accuracy was 0.930.

Table 3. Error matrix obtained from the accuracy assessment of the 2010 land cover map of the Santarém site (MF – mature forest, NF – non-forest, SF – secondary forest).

| Reference | Classified | | | | |
	MF	NF	SF	Total	Omission error
MF	196	0	4	200	0.020
NF	0	198	2	200	0.010
SF	23	11	166	200	0.170
Total	219	209	172	600	
Commission error	0.105	0.053	0.035		

The overall accuracy was 0.93.

forest) and regrowth rates (conversion from non-forest to regeneration), the age of secondary forest, the period of active land use prior to abandonment to regeneration, and the frequency of clearance. The study builds on that of Prates-Clarke et al., 2009 [38] in that the time-series was extended from 2003 to 2011 for Manaus and Santarém and a new time-series was generated for Machadinho d'Oeste.

Study Areas

For all three sites (Manaus, Santarém and Machadinho d'Oeste; Figure 1), the natural vegetation prior to disturbance was primary (undisturbed) *terra firme* (i.e., non-flooded) forest. In each case, the canopy heights averaged between 25 and 35 m, with emergent trees exceeding 50 m at the more productive sites (e.g., Manaus). Species diversity is also high, often exceeding 225 species ha^{-1} [26], with an estimated mean above ground biomass of approximately 270 Mg ha^{-1} [27].

The first site, north of Manaus, the capital of Amazonas state, occupied an area of 5,042 km^2 ($2°33'11''$ S, $60°5'7''$ W, Figure 1) and included the Instituto Nacional de Pesquisas da Amazônia (INPA) and the Smithsonian Institution (SI) Biological Dynamics of Forest Fragments Project (BDFFP) research sites (established in 1979) [28], the Adolfo Ducke and Walter Egler forest reserves, and other state and federal environmental protection areas. Significant deforestation began in the area following construction in the early 1970s of the BR-174 highway connecting Manaus with Boa Vista in Roraima. Most deforestation activity occurred either side of the highway, with this fuelled primarily by agricultural expansion. Recognition of the opportunity to study the impacts of clearance and forest fragmentation on biodiversity as well as ecosystem

services spurred the preservation of fragments of various dimensions (typically less than 2 km^2) prior to felling of the surrounding tropical forest. Many of these fragments were, however, reconnected because of rapid regrowth of forests following abandonment of many clearances from the mid-1980s onwards. Up until the present, a substantial amount of research has been conducted, with these informing future management strategies across the Amazon [28].

The second site ($3°10'5''$ S, $54°55'42''$ W) covered an area of 1,118 km^2 and was located approximately 80 km to the south of Santarém, the second main city in the Pará state (Figure 1). This study area is within the Tapajós National Forest, between the Tapajós River and the BR-163 highway connecting Santarém to the state capital of Mato Grosso, Cuiabá. This conservation unit was created in 1974 and has been recognized as a model for sustainable forest management, including logging activities, at levels ranging from individuals to communities (e.g., [29,30]). Compared to Manaus, forests are less dense with a mixed arrangement of broadleaves and palm trees such as Açaí (*Euterpe oleracea*) and Babaçu (*Orbignya phalerata*) being commonplace [31].

The third site ($9°32'56''$ S and $62°6'27''$ W) occupied an area of 1,780 km^2 and was located mostly within the Machadinho d'Oeste municipality, Rondônia state. This municipality originated from the former Machadinho settlement project, deployed by the Brazilian federal government (through the *Instituto Nacional de Colonização e Reforma Agrária*, INCRA) in 1982 as part of the POLONORDESTE Program [32]. As such, the majority of its inhabitants are dependent on agriculture for subsistence [33]. Open rain forest is the dominant vegetation type in the municipality [32]. According to data from the Brazilian Ministry

Table 4. Error matrix obtained from the accuracy assessment of the 2010 land cover map of the Machadinho d'Oeste site (MF – mature forest, NF – non-forest, SF – secondary forest).

| Reference | Classified | | | | |
	MF	NF	SF	Total	Omission error
MF	200	0	0	200	0.000
NF	0	197	3	200	0.015
SF	19	37	144	200	0.280
Total	219	234	147	600	
Commission error	0.087	0.158	0.020		

The overall accuracy was 0.90.

Table 5. Area (ha) and relative incidence (RI, %) of perennial (PC) and tree (TC) crops with the mapped area of mature forest (MF), non-forest (NF) and secondary forest (SF).

| Classified | Manaus (2007) TC | | Machadinho d'Oeste (2010) PC | | TC | |
	Area (ha)	RI (%)	Area (ha)	RI (%)	Area (ha)	RI (%)
MF	139	0.0	31	0.1	0	0.0
NF	398	1.0	540	0.9	14	0.0
SF	4,237	4.4	1,323	2.1	3	0.0
Total	4,774		1,894		17	

Estimates were based on the land cover map closest to the date that was used to delineate crop types using very high resolution (VHR) imagery.

of the Environment (MMA), this site includes several conservation units, mainly extractive reserves, which were established in and after 1995.

The area of the Manaus site was approximately three to four times the size of the other two study areas for several reasons. Deforestation in Manaus started in the early 1970s and land conversion from primary forest to agriculture/pasture was much more scattered. By comparison, deforestation patterns were more concentrated at Santarém (either side of and in proximity to a highway) and Machadinho d'Oeste (a planned settlement). A larger area was also selected to encompass older deforestation within the BDFFP study area and more recent clearing towards the city of Manaus, thereby allowing a wide range of LULCC processes to be captured.

At all three sites, much of the history of deforestation and land use has been captured by optical (primarily Landsat) sensors during the dry season, when cloud cover is minimal. In Manaus, a moderately strong dry season occurs between June and October [28], in Santarém between May and October [34], and in Machadinho d'Oeste between April and November [32]. Rainfall is generally higher at Manaus (between 1,900 and 2,500 mm annually) and lower at Santarém and Machadinho d'Oeste with an annual average of 2000 mm. The mean annual temperatures at the sites range from 25°C to 26°C [32,34,35]. The distribution of forests types is determined in part by the topography and geology. The topography is moderately flat (up to 160 m elevation) north of Manaus and is divided by a large number of waterlines. Soils are nutrient-poor (i.e., Oxisols), with these being typical of many across the Amazon basin [36]. The area south of Santarém is also moderately flat, ranging from approximately 50 to 240 m, and consists of nutrient poor soils (i.e., Oxisols and Ultisols) [34,37]. The terrain is moderately undulated at Machadinho d'Oeste, ranging from approximately 90 to 370 m, and the dominant soil types are Alfisols, Oxisols, Ultisols, and Alluvial [32].

Data

For Manaus, Landsat Multi-spectral Scanner (MSS), Thematic Mapper (TM) and Enhanced Thematic Mapper Plus (ETM+) data were acquired between 1973 and 2011 (path 231, row 62; Table 1). The Landsat MSS data acquired between 1973 and 1983 were only available as hard-copy prints [38]. The remaining Landsat TM and ETM+ scenes (1985 onwards) were provided in digital format from both INPE and the United States Geological Service (USGS).

In the previous study of Prates-Clark et al., 2009 [38], scenes acquired between 1985 and 2003 were used to study forest

regeneration in this area. In this study, the time-series was extended to 2011 using six additional annual scenes, which were relatively cloud free (<20%), or cloud was distributed in only certain areas of the image. In total, the time-series extended over an approximate 40-year period from 1973 (when the deforested area was minimal) to 2011. For Santarém, Prates-Clark et al., 2009 [38] again used Landsat TM and ETM+ imagery acquired between 1984 and 2003 (path 227, row 62), with most scenes unaffected by substantive cloud cover. The time-series was extended using Landsat TM data acquired for each year from 2005 to 2010. Coverage for 2011 was actively sought but no data with minimal cloud coverage were available. Whilst most changes at Santarém were associated with deforestation for agriculture and pasture, extensive wildfires damaged the surrounding forests in 1992 and 1997. The last fire episode was associated with the 1997–1998 El Niño Southern Oscillation (ENSO) event. The time-series classification for Machadinho d'Oeste had not been undertaken previously and so 21 new time-series digital images acquired between 1984 and 2011 were obtained from the USGS. Gaps in the time-series ranged from between one (70%) to approximately three years (5%). Cloud cover was minimal on all but two dates, although the vast majority of the deforested area was visible within these.

Methods

4.1. Image pre-processing

Landsat images acquired by the USGS were orthorectified and calibrated to units of spectral radiance (W m^{-2} sr^{-1} μm^{-1}) and then calibrated to top of atmosphere (TOA) reflectance using calibration factors and equations provided by Chander et al., 2009 [39]. Images acquired by INPE were processed to TOA reflectance and geometrically corrected using Environment for Visualizing Images (ENVI) software (Exelis Visual Information Solutions, Boulder, CO, USA), a third order polynomial and a nearest neighbor transformation [38]. Each Landsat image was then subsetted to encompass the main area of deforestation and the intersection of all images in the time series.

4.2. Image classification

As with Prates-Clark et al., 2009 [38], each digital image within the time-series for each site was classified into mature forest (MF), non-forest (NF), and secondary forest (SF), with the second category including crops (herbaceous and woody) and pastures. Areas of open water were also mapped and a common mask was applied to all dates in the time-series. Several classification approaches were utilized with the intention of providing the best

A

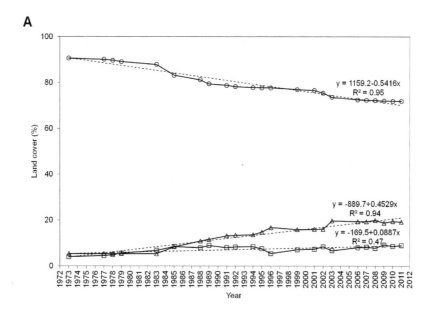

B

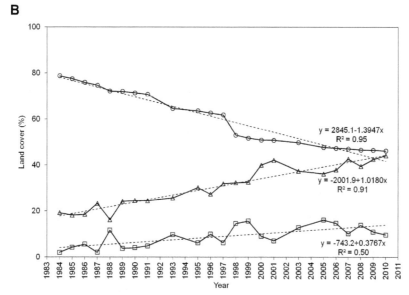

C

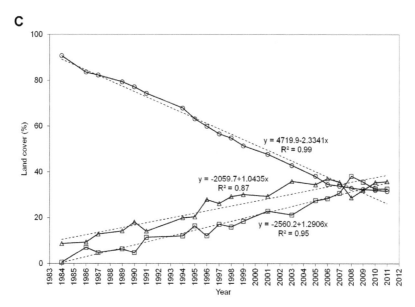

Figure 2. Land cover change (%) across the three selected sites. Land cover change (%) across the three selected sites; A) Manaus (1973–2011), B) Santarém (1984–2010), and C) Machadinho d'Oeste (1984–2011). Each dashed line represent the linear fit of the proportion of each land cover class as a function of the year; also showing are the corresponding equations and coefficient of determination (R^2).($\bigcirc$ - mature forest, $\square$ - non-forest, Δ- secondary forest).

classification of the desired categories. At Manaus, and for digital images acquired before 2006, supervised classifications based on the minimum distance or maximum likelihood algorithms were applied (e.g., [40]), with regions of interest representing the main cover types defined around a number of target areas described on the basis of field observations or through reference to very high resolution data. For the Landsat TM images acquired between 2006 and 2011, an object oriented classification (e.g., [41]) was followed and involved a decision-rule classification applied within the eCognition software (Trimble Geospatial Imaging, Munich, Germany). The rule base method used data from the available Landsat sensor bands as well as data layers derived from spectral indices, namely the Normalized Difference Vegetation Index (NDVI) and Normalized Difference Water Index (NDWI). The classification of MF was refined with a cloud-free mask of the MF area obtained through classification of the most-cloud-free and recent image in the time-series. Images acquired on the 31st August, 2011, 29th June, 2010 and 12th June 2011 were used to generate the MF mask, on the assumption that the land cover type had remained as MF for the whole of the time-series. If MF was covered by cloud in earlier images of the time-series, this was subsequently classified as MF. At Manaus, SF were older compared to the other sites and the structure of the upper canopy of the regenerating forests led to a spectral response in the near and shortwave infrared channels that was similar to MF; hence, reference to images earlier in the time-series was necessary [18,22,42]. At Santarém, the Landsat data acquired between 1984 and 2003 were classified using a fuzzy logic approach that was applied to the original Landsat bands as well as derived fractional images (shade/moisture, vegetation and soil). Each pixel of the Landsat data acquired between 2005 and 2010 was classified into one of the three main land cover classes (described above) using the random forests machine learning classification algorithm (e.g., [43]) and training data. This algorithm has a number of options (e.g., [43]) that were tested for each date to obtain the best possible classifier. As with Santarém, all Landsat TM dates for Macha-dinho d'Oeste were classified using the random forests algorithm. In all cases, the objective was to obtain the best classification of NF, MF and SF for each year of observation.

4.3. Post-classification

The comparison of the time-series of classified images (as MF, NF, or SF) identified several cases where some pixels classified as SF or NF in a given date were classified as MF in the following date. As this sequence is not plausible, in-house Interactive Data Language (IDL) code (Exelis Visual Information Solutions, Boulder, CO, USA) was written to identify these cases by comparing the classification over two consecutive dates. Where SF or NF areas were classified as such in one year and as MF in the next, the MF class was reallocated to SF. These areas were not reclassified as NF because it was assumed that spectral confusion between NF and MF was unlikely. This procedure was similar to that undertaken by Roberts et al., 2002 [44], who also considered "disallowed transitions" between cover types.

4.4. Deforestation and regrowth rates

Annual rates of change from MF (or SF) to NF (deforestation) and from NF to SF (regeneration) were calculated for consecutive

dates in the time-series using equations (1) and (2) [45]

$$r = \frac{1}{t_2 - t_1} \ln \frac{A_2}{A_1} \qquad (1)$$

$$R = \frac{A_1 - A_2}{t_2 - t_1} \qquad (2)$$

where r and R are the annual rate of change expressed in percentage per year and hectares per year respectively; A_1 and A_2 are the MF (or SF, in the case of deforestation occurring in regeneration areas) or NF cover areas (hectares) at time t_1 and t_2 respectively (time period). These rates were used to characterize the temporal evolution of change across the three sites, to detect events (primarily clearance) and describe general trends.

4.5. Age of secondary forest, period of active land use, and frequency of clearance

Algorithms were written and implemented in IDL to compare the classifications of MF, NF, and SF between dates. Subsequently, datasets relating to the history and dynamics of land use, namely the age of secondary forest (ASF), the period of active land use (PALU) prior to abandonment, and the frequency of clearance (FC) were generated for each site and for each year in which an image had been acquired. The ASF was estimated by summing the time (in years) that each pixel was occupied by SF since the last clearance event. However, when SF occurs in the first date of the time series, and it is not subsequently cleared, the ASF can only be considered as a minimum age as the exact date of land abandonment is not known. The PALU was defined as the difference (in years) between the time of initial forest clearance and the onset of regeneration. However, where reclearance of regenerating forest had occurred, the PALU was calculated by summing the period since the last reclearance event until the forest cover had re-established. The FC was estimated by summing the frequency of transitions from MF (or SF) to NF. For both the Manaus and Santarém study areas, fire scars were evident within some clearances by the lower near and shortwave infrared reflectance compared to the original vegetated surface [46]. However, many burned areas acquired a vegetated cover quite rapidly [47], and so only a partial fire history could be retained.

To better spatially represent and discuss the main results, classes of ASF, PALU, and FC were generated and used to assess their temporal evolution. Each metric was classified into three classes as to identify a lower, middle and upper interval range. However, these intervals do have some degree of subjectivity. ASF classes were defined as initial (≤ 5 years), intermediate (6–15 years) and advanced (≥ 16 years), adapted from Lucas et al., 2000 [18]. PALU classes were defined as short (≤ 2 years), medium (3–4 years) and long (≥ 5 years), with these representing various different crop or pasture cycles. FC classes were defined as low (1 time), medium (2 times) and high (≥ 3 times), representing several land use patterns in terms of the frequency of the deforestation -> agriculture/pasture -> abandonment -> re-growth temporal sequence.

Table 6. Land cover proportion (%) of mature forest (MF), non-forest (NF) and secondary forest (SF) in each year of the time-series at Manaus, Santarém and Machadinho d'Oeste.

Manaus				Santarém				Machadinho d'Oeste			
Year	MF	NF	SF	Year	MF	NF	SF	Year	MF	NF	SF
1973	91	4	5	1984	79	2	19	1984	91	1	9
1977	90	4	5	1985	78	4	18	1986	84	7	9
1978	90	5	5	1986	76	6	19	1987	82	5	13
1979	89	6	5	1987	75	2	23	1989	79	6	14
1983	88	7	5	1988	72	12	16	1990	77	5	18
1985	83	8	8	1989	72	4	24	1991	74	11	14
1988	81	8	11	1990	71	4	25	1994	68	12	20
1989	79	9	12	1991	71	5	24	1995	63	16	20
1991	79	8	13	1993	65	10	26	1996	60	12	28
1992	78	8	13	1995	64	6	30	1997	57	17	26
1994	78	8	14	1996	63	10	27	1998	55	16	29
1995	78	7	15	1997	62	6	32	1999	51	18	30
1996	78	5	17	1998	53	15	32	2001	48	23	29
1999	77	7	16	1999	52	16	33	2003	43	21	36
2001	77	7	16	2000	51	9	40	2005	38	27	34
2002	76	8	16	2001	51	7	42	2006	35	28	37
2003	74	7	20	2003	50	13	37	2007	34	31	35
2006	73	8	19	2005	48	16	36	2008	33	38	29
2007	72	8	19	2006	47	15	38	2009	32	36	32
2008	72	8	20	2007	47	10	43	2010	32	33	35
2009	72	9	19	2008	47	14	40	2011	32	33	36
2010	72	9	19	2009	46	11	43				
2011	72	9	19	2010	46	10	44				

Table 7. Summary of deforestation and regrowth rates between consecutive dates in the Manaus time-series (MF – mature forest; SF – secondary forest).

	MF annual rate of deforestation		SF annual rate of deforestation		Annual rate of deforestation		Annual rate of regrowth	
	($\%$ yr^{-1})	(ha yr^{-1})	($\%$ yr^{-1})	(ha yr^{-1})	($\%$ yr^{-1})	(ha yr^{-1})	($\%$ yr^{-1})	(ha yr^{-1})
Minimum	0.1	277	0.0	0	0.1	691	0.0	0
Maximum	2.9	13,451	27.7	8,453	3.8	17,936	133.9	24,738
Mean	0.7	3,123	6.4	3,231	1.3	6,355	34.1	6,017
Median	0.4	1,591	5.0	2,725	0.9	4,513	24.8	4,793
Inter-quartile range	0.4	2,172	5.1	3,006	1.3	5,881	31.9	6,156
Standard deviation	0.8	3,726	6.5	2,699	1.1	4,905	33.7	5,756

doi:10.1371/journal.pone.0104144.t007

Table 8. Summary of deforestation and regrowth rates between consecutive dates in the Santarém time-series (MF – mature forest; SF – secondary forest).

	MF annual rate of deforestation		SF annual rate of deforestation		Annual rate of deforestation		Annual rate of regrowth	
	($\%$ yr^{-1})	(ha yr^{-1})	($\%$ yr^{-1})	(ha yr^{-1})	($\%$ yr^{-1})	(ha yr^{-1})	($\%$ yr^{-1})	(ha yr^{-1})
Minimum	0.4	235	0.5	194	0.5	448	10.5	832
Maximum	15.6	11,046	82.3	11,352	17.7	16,952	261.5	12,257
Mean	2.1	1,598	16.7	3,189	4.9	4,787	78.5	4,675
Median	1.2	963	10.7	2,769	4.0	4,236	71.3	3,361
Inter-quartile range	1.3	1,085	16.6	2,350	3.0	2,831	79.6	4,031
Standard deviation	3.2	2,292	18.7	2,849	4.4	4,188	61.5	3,332

Table 9. Summary of deforestation and regrowth rates between consecutive dates in the Machadinho d'Oeste time-series (MF – mature forest; SF – secondary forest).

	MF annual rate of deforestation		SF annual rate of deforestation		Annual rate of deforestation		Annual rate of regrowth	
	(% yr⁻¹)	(ha yr⁻¹)	(% yr⁻¹)	(ha yr⁻¹)	(% yr⁻¹)	(ha yr⁻¹)	(% yr⁻¹)	(ha yr⁻¹)
Minimum	1.2	748	0.0	17	1.5	2,386	3.7	103
Maximum	9.9	9,223	88.1	16,711	17.7	18,202	84.2	17,389
Mean	3.9	4,142	19.5	5,289	7.2	9,286	25.8	6,804
Median	3.5	4,171	14.3	4,273	6.0	7,436	19.1	6,172
Inter-quartile range	3.4	4,057	13.5	7,114	6.7	9,571	18.0	4,609
Standard deviation	2.3	2,394	19.5	4,467	4.4	5,248	21.5	4,141

4.6. Accuracy assessment and area calibration of the time-series of classified images

The most recent very high resolution (VHR) imagery was used to carry out the accuracy assessment of the time-series classifications at each site. As VHR data were not available to perform the accuracy assessment for all the classified images in the time-series, the accuracy in the classification of the image that was closest in time was assumed to be similar for the remaining images. To quantify the accuracy of the 2010 classification for Santarém, reference was made to 5 m SPOT-5 panchromatic data from 2009 and 2011 whilst the 2007 and 2010 classifications for Manaus and Machadinho d'Oeste respectively were validated using GeoEye imagery available on Google Earth.

At each site, 200 random points located at the center of larger polygons (greater than 6 ha) were generated for each class. The points on the classification were compared with the VHR imagery from the same date. At Santarém, it was assumed that if the same land cover was present in both the 2011 and 2009 image, the land cover in 2010 would be the same. The accuracy assessment was reported as a standard error matrix, including the overall accuracy and omission and commission errors (e.g., [48]). From the same VHR imagery and past field studies of the study sites (e.g., [32,38]), woody agriculture crops had established on some of the deforested areas. At Manaus, a large area of tree crops (TC, mainly oil palm) was progressively planted in the north west of the site from the 1983 to 1989, whilst at Machadinho d'Oeste, perennial crops (PC, mainly coffee plantations) were more common. To assess how well the areas of MF, NF and SF were discriminated using the classification, these areas were delineated manually from the VHR imagery and used in the generation of the error matrices.

The most straightforward way of estimating total areas for each class was to count the number of pixels in the MF, NF and SF classes with respect to the ground truth data. This is called "naïve estimation" in the remote sensing literature [49]. As a matter of fact, areal estimates obtained this way are biased (e.g., [50]). Some post-classification methods have been used to improve these estimates, namely the so-called calibration techniques that are divided into classical and inverse methods. The information contained in the error matrix was used to correct for misclassification bias [50]. Walsh and Burk, 1993 [50] carried out a simulation study and concluded that the inverse method consistently performed better than the classical method and lead to unbiased estimates for total areas. The idea behind this calibration technique was to use the misclassification probabilities among classes to revise the proportions given by pixel counting (p_i) [51]. According to Tenenbein, 1972 [51] and Walsh and Burk, 1993 [50], the revised (calibrated) proportion of the total area occupied by class i (π_i) can be estimated with equation (3)

$$\pi_i = \sum_{j=1}^{k} \left[(f_j + n_j)/N \right] \left(n_{ij} / n_j \right) \tag{3}$$

where i represents the observed class, j the predicted class, k the number of classes, n the number of observations in the validation sample, N the combined number of observations in the validation and satellite datasets, f_j the $(N-n)$ in class j, n_{ij} the number of observations in class i that were classified as j, and $n_{.j}$ the sum of all training observations predicted as class j. This method was applied to calibrate the area estimates obtained from each date of classified images using the error matrices that were produced for the three sites.

Table 10. An assessment of the dynamics of land use at Manaus, Santarém and Machadinho d'Oeste (figures in brackets represent the number of years).

Indicators	Impact	Sites		
		Manaus	Santarém	Machadinho d'Oeste
a) Clearance of MF only with little or no deforestation over SF	High pressure for new land	1973–1985 (12)	Pre-1984	1984–1987 (3)
b) Area of MF cleared>SF cleared.	Pressure for new land with some contained re-use of the existing deforested area (through reclearance)	1985–1989 (4)	1986–1987 (1)	1987–1990 (3)
		2002–2003 (1)	1991–1993 (2)	1991–1996 (5)
			1997–1998 (1)	1997–1998 (1)
			2006–2007 (1)	2001–2003 (2)
			2008–2009 (1)	
c) Area of MF cleared<SF cleared	Greater re-use of existing deforested land but still requirement for more land	1989–2002 (13)	1984–1986 (2)	1990–1991 (1)
		2003–2011 (11)	1987–1991 (4)	1996–1997 (1)
			1993–1997 (4)	1998–2001 (3)
			1998–2006 (8)	2003–2011 (8)
			2007–2008 (1)	
			2009–2010 (1)	
d) Area of MF and SF cleared>area of SF regenerating		1973–1985 (12)	1984–1986 (2)	1984–1986 (2)
		1988–1989 (1)	1987–1988 (1)	1987–1989 (2)
		1991–1994 (3)	1989–1993 (4)	1990–1995 (5)
		1996–2002 (6)	1995–1996 (1)	1996–1997 (1)
		2003–2007 (4)	1997–1999 (2)	1998–2001 (3)
		2008–2009 (1)	2001–2005 (4)	2003–2008 (5)
		2010–2011 (1)	2007–2008 (1)	2010–2011 (1)
e) Reclearance of SF but no clearance of MF	Contained re-use of area already deforested	No years	No years	No years
f) Area of SF regeneration>MF and SF deforestation	Net abandonment of land to regenerating forests	1985–1988 (3)	1986–1987 (1)	1986–1987 (1)
		1989–1991 (2)	1988–1989 (1)	1989–1990 (1)
		1994–1996 (2)	1993–1995 (2)	1995–1996 (1)
		2002–2003 (1)	1996–1997 (1)	1997–1998 (1)
		2007–2008 (1)	1999–2001 (2)	2001–2003 (2)
		2009–2010 (1)	2005–2007 (2)	2008–2010 (2)
			2008–2010 (2)	

Results

5.1. Accuracy assessment of land cover classification

The error matrices resulting from the accuracy assessment over Manaus, Santarém, and Machadinho d'Oeste are presented in Table 2, 3 and 4 respectively. For all three sites, the overall accuracy was high (above 0.90), although some relatively high omission and commission errors were identified, especially in Santarém and Machadinho d'Oeste. For Manaus, the omission and commission errors were always below 10%. For Santarém, a higher omission error in the SF class was detected (17.0%), mainly due to misclassification as MF (11.5%), which reflects also in the high commission error in the MF class (10.5%). At Machadinho d'Oeste, a major source of error also comes from a high omission error in the SF class (28%), with this mainly being a consequence of misclassification as NF (18.5%) and to a lesser extent as MF (9.5%); a high commission error was also observed in the NF class (15.8%), with this being due to misclassification with SF.

In Manaus, TC (4,774 ha) was mostly incorrectly classified as SF (89%), illustrating the similarity in their spectral signatures, but accounted only for 4.4% of the SF area in 2007 (Table 5). However, most of the TC classified as SF in Manaus was associated with the oil palm plantation situated in the north west of the study site. The majority of TC in Machadinho d'Oeste was correctly classified as NF, although the overall area was very small (17 ha). Conversely, 70% of the PC (mainly coffee plantations) at Machadinho d'Oeste was misclassified as SF, although this only accounted for 2.1% the entire SF area in 2010.

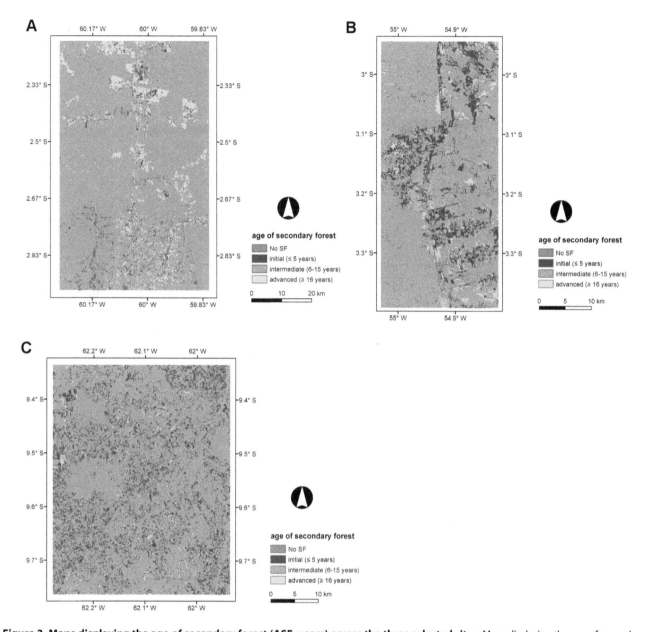

Figure 3. Maps displaying the age of secondary forest (ASF, years) across the three selected sites. Maps displaying the age of secondary forest (ASF) across the three selected sites for areas undergoing secondary forest (SF) in the last year of the corresponding time-series; A) Manaus (2011), B) Santarém (2010), and C) Machadinho d'Oeste (2011).

5.2. Land cover change

The evolution of land cover change in Manaus, Santarém and Machadinho d'Oeste is depicted in Figure 2A, 2B and 2C respectively and in Table 6, which shows that the most rapid change occurred in Machadinho d'Oeste, followed by Santarém and Manaus. For all sites, there was a strong correspondence between the land cover percentage and time (year) for the MF class, with the rate decreasing by an average of 0.54% in Manaus to 1.39% in Santarém and 2.33% in Machadinho d'Oeste. For all sites, a corresponding increase in the area deforested occurred, with the land cover alternating between NF and SF in all cases. The change rate in the NF class increased from 0.09% (Manaus) and 0.38% (Santarém) to 1.29% (Machadinho d'Oeste) whilst the SF change rate progressively increased from Manaus (0.45%) to Santarém (1.02%) and Machadinho d'Oeste (1.04%). The

relationships between the proportion (%) of MF, NF and SF and time (year) were significant at all sites (P<0.001).

5.3. Deforestation and regrowth rates

Tables depicting relative and absolute deforestation and regeneration rates between consecutive dates for Manaus, Santarém and Machadinho d'Oeste are presented as Supporting Information (Table S1, S2 and S3 respectively in File S1) and a summary is given in Tables 7, 8 and 9 respectively. The maximum relative annual rate of deforestation over MF was observed at Santarém, with a 15.6% yr^{-1} loss between 1997 and 1998, with this associated with the extensive wildfires in 1997, and the highest average annual rate of deforestation was in Machadinho d'Oeste (3.9% yr^{-1}). The maximum relative rates of SF clearance were in Machadinho d'Oeste (88.1% yr^{-1}, 1990–1991) and Santarém

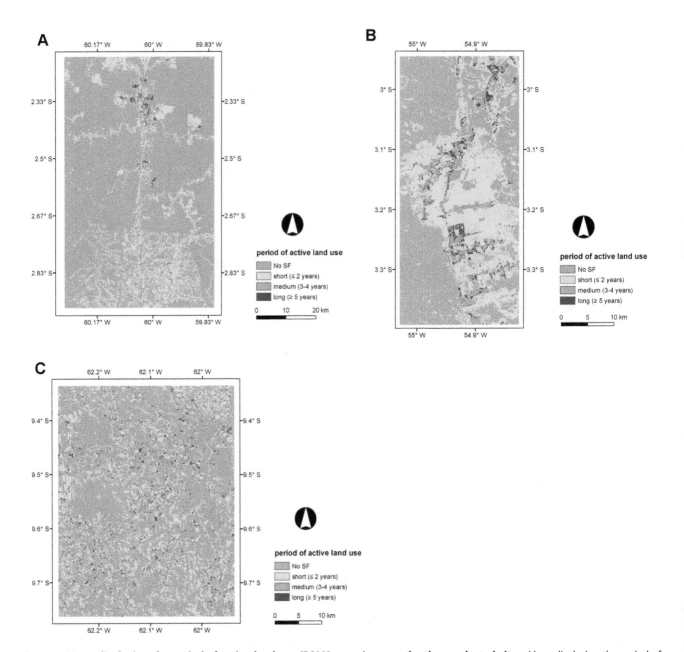

Figure 4. Maps displaying the period of active land use (PALU, years) across the three selected sites. Maps displaying the period of active land use (PALU, years) across the three selected sites for areas undergoing secondary forest (SF) in the last year of the corresponding time-series; A) Manaus (2011), B) Santarém (2010), and C) Machadinho d'Oeste (2011). An area of 667 ha (1.5%) with a PALU of zero in the Santarém site was included in the short PALU class (≤2 years) and corresponds to areas that were SF from 1984 to 2010.

(82.3% yr^{-1}, 1987–1988), with average rates being 19.5% and 16.7% for these sites respectively. The maximum and the highest average relative rates of regeneration were observed at Santarém, with 261.5% yr^{-1} (1997–1998) and 78.5% yr^{-1} respectively. The lowest relative rates of regrowth were at Machadinho d'Oeste, with these averaging 25.8% yr^{-1}. The average overall relative deforestation rates (combined MF and SF) were greatest for Machadinho d'Oeste (7.2% yr^{-1}), followed by Santarém (4.9% yr^{-1}) and Manaus (1.4% yr^{-1}). On average, deforestation rates over SF were higher than those over MF, indicating reclearance at all sites and particularly Machadinho d'Oeste (averaging 19.5% yr^{-1}) and Santarém (averaging 16.7% yr^{-1}).

Absolute areas (annual or based on periods of observation) are useful when comparing deforestation and regeneration rates between sites and the processes of change. Several key indicators are then listed in Table 10 together with the dates for each site to which these apply. For all sites, a trend from indicators a) to f) was observed with the majority being cleared initially, regenerating forests progressively establishing with these then recleared to varying degrees, and, of these, several are then abandoned to regrowth again.

5.4. Age of secondary forest (ASF), period of active land use (PALU), and frequency of clearance (FC)

Maps of the ASF, PALU, and FC for the last date of the time-series are presented in Figures 3, 4 and 5 for Manaus (2011), Santarém (2010) and Machadinho d'Oeste (2011) respectively and the proportion of areas with respect to these classes is indicated in

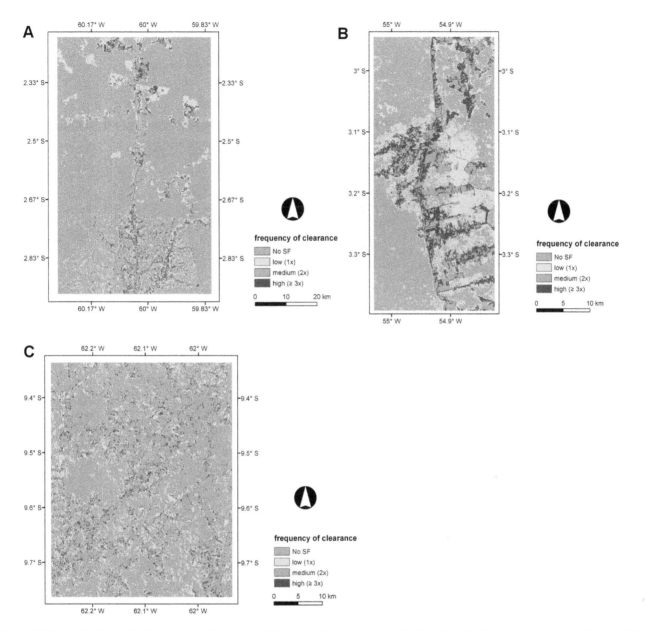

Figure 5. Maps displaying the frequency of clearance (FC) across the three selected sites. Maps displaying the frequency of clearance (FC) across the three selected sites for areas undergoing secondary forest (SF) in the last year of the corresponding time-series; A) Manaus (2011), B) Santarém (2010), and C) Machadinho d'Oeste (2011). In Manaus, Santarém and Machadinho d'Oeste an area of 103 ha (0.1%), 726 ha (1.6%), and 55 ha (0.1%), respectively, with a PALU of zero was included in the low FC class (1 time) and corresponded to areas that were non-forest (NF) in the first date of the time-series and no clearance has occurred in the secondary forest (SF) that persisted until the end of the time-series. In Santarém, the 726 ha also included areas that were already SF in 1984 that persisted until 2010.

Figure 6. At Manaus, regrowth forests were comparatively older, with 50% occupied by forests ≥16 years (Figure 6A). At Santarém, 57% of forests were aged 6–15 years, and at Machadinho d'Oeste, the majority of the area occupied by SF aged ≤5 years (46%). Hence, the three sites contain different distributions of age classes, with this attributable to differences in the time since deforestation but also the PALU and the frequency of clearance events. At all sites, the majority of land has been used actively for ≤2 years (Figure 6B), either following the initial clearance of the primary forest or subsequent reclearance events, but the deforestation had commenced at different times. At Manaus, the PALU was also ≤2 years for over 64% of the area and many of the forests had been able to regenerate undisturbed,

with some approaching 40 years in 2011, and had been cleared mainly on one (65%) occasion (Figure 6C). At Santarém, 88% of the area under regrowth had been used actively for ≤2 years prior to the last abandonment and the majority (53%) cleared only once. At Machadinho d'Oeste, the PALU was typically ≤2 years (75%), and again, forests were relatively young because most of the clearance occurred only on one (57%) occasion. Hence, different typologies of SF were observed at each site.

As a result of combining the 3-class PALU and FC maps for the last date of the time-series, a proxy for land use intensity in each site was generated (Table 11). At Manaus, of the 65% of land occupied by SF in 2011 that was subjected to only one clearing event, 41% had a short PALU but a significant proportion (16%

A

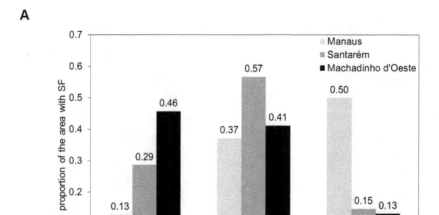

B

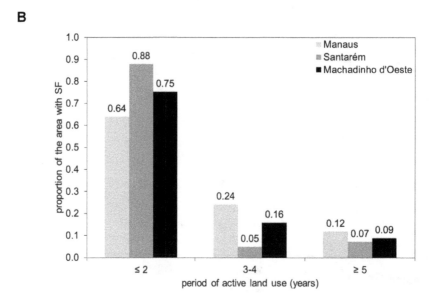

C

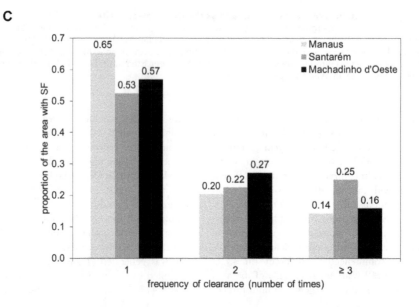

Figure 6. Classes of age of secondary forest, period of active land use, and frequency of clearance for the three sites. Classes of A) age of secondary forest (ASF), B) period of active land use (PALU), and C) frequency of clearance (FC) for the three sites, as a proportion of the area with secondary forest (SF) in the last date of the time-series. The proportion of the area with SF in the first class of PALU (≤2 years) in the Santarém site includes a small proportion (1.5%) corresponding to areas of SF persisting in all dates of the time-series. The proportion of the area with SF in the first class of FC (1x) in Manaus, Santarém and Machadinho d'Oeste includes a small percentage (0.1%, 1.6%, 0.1%, respectively) corresponding to those cases where non-forest (NF) was observed in the first date of the time-series and no clearance has occurred (i.e., the only transition was from NF to SF).

and 9%) had a medium and long PALU respectively. Santarém is paradigmatic, as almost all land with SF in 2010 that was cleared only once was subjected to a short PALU. Assuming that a high FC and short and medium PALU will represent a higher land use intensity, then Manaus had 13% of the area undergoing SF in 2011 in these conditions, while Santarém and Machadinho d'Oeste had 20% and 15% in 2010 and 2011 respectively.

To investigate whether these typologies were dependent on the last date of the time-series, the same analysis was performed for all the dates in the time-series. At Manaus, the proportion of the area with SF in the ≤5 year ASF class decreased irregularly with the onset of regeneration up to approximately 15% in the last date of the time-series (Figure 7A), with this pattern being mirrored for the 6–15 year ASF class, and the ≥16 year ASF class steadily increasing up to around 50% in 2011. At Santarém (Figure 7B), the proportion of the area with SF in the initial ASF class decreased to around 30% in 2010, again mirrored by an overall increase to about 60% in the 6–15 year ASF class. As with Manaus, the proportion of the area with SF in the advanced ASF class increased from 15 years following the onset of regeneration to cover approximately 15% in 2010. In Machadinho d'Oeste (Figure 7C), the proportion of SF in the initial ASF class systematically decreased from the onset of regeneration leveling at around 45% in 2011, and an opposite pattern was observed in the intermediate ASF class, with a systematic increase since the onset of regeneration up to approximately 40% in 2011. As in Manaus and Santarém, a steady increase in the proportion of area with regeneration in the advanced ASF class was observed, from 17 years after the onset of regeneration up to around 15% in 2011. By combining the initial and intermediate ASF classes, a more

consistent but asymptotic increase in the area of SF was observed, with this compensating for the interplay between these ASF classes, which occurs because of reclearance. The general trend across all sites is a decrease in the proportion of the area with SF in the initial ASF class up to the present, with a simultaneous increase in the proportion of the intermediate and advanced ASF classes. No asymptotic pattern is visible in the relationship between the proportion of the area with SF and the number of years following the onset of regeneration for this variable in the three selected sites.

At Manaus, the proportion of areas with a short PALU was initially high (100%) but, over time, this area decreased to about 40% but then increased subsequently to 60%, where it remained constant 32 years from the onset of regeneration (Figure 8A). This decrease was associated with an increase in the area with a medium and long PALU (14–16 years after the onset of SF regeneration). At Santarém (Figure 8B), the proportion with short PALU remained high, with this suggesting clearance of SF was common practice. The areas with medium and long PALU remained relatively low (<5–10%) but increased slightly towards the end of the time-series. At Machadinho d'Oeste, the temporal pattern was similar to that observed at Manaus; after 12–14 years, the proportion of short, medium and long PALU stabilized around 75%, 15% and 10% respectively.

The FC over areas undergoing SF also varied between sites (Figure 9). At Manaus, the SF had been cleared only once until about 14 years after the onset of SF, but this proportion decreased, stabilizing at approximately 65% in 2011. This occurred because SF began to be cleared again 15 years after the onset of regeneration and again after 19 years (i.e., 3 FC events), with these areas stabilizing at approximately 20% and 15% respectively. At

Table 11. Percentage of area with SF in the last date of the time-series, per combination of PALU and FC classes.

PALU (classes)	FC (classes)			Total
Manaus	**Low (1x)**	**Medium (2x)**	**High (≥3x)**	
Short (≤2 years)	41	13	10	64
Medium (3–4 years)	16	5	3	24
Long (≥5 years)	9	2	1	12
Total	66	20	14	
Santarém	**Low (1x)**	**Medium (2x)**	**High (≥3x)**	
Short (≤2 years)	51	20	17	88
Medium (3–4 years)	1	1	3	5
Long (≥5 years)	0	2	5	7
Total	52	23	25	
Machadinho d'Oeste	**Low (1x)**	**Medium (2x)**	**High (≥3x)**	
Short (≤2 years)	48	17	11	75
Medium (3–4 years)	5	7	4	16
Long (≥5 years)	4	4	1	9
Total	57	28	16	

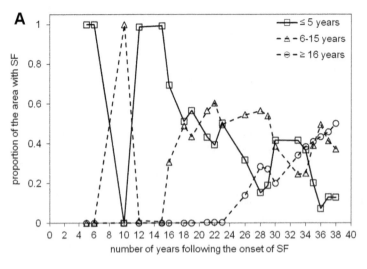

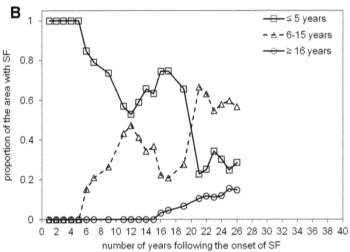

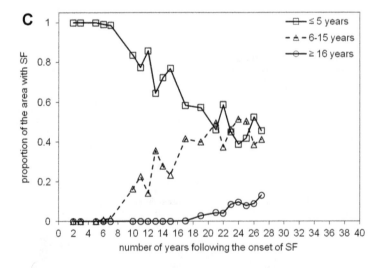

Figure 7. Proportion of the area with secondary forest (SF) as a function of the number of years following the onset of SF for the three classes of age of secondary forest. Proportion of the area with secondary forest (SF) as a function of the number of years following the onset of SF for the three classes of age of secondary forest (ASF) in A) Manaus, B) Santarém, and C) Machadinho d'Oeste: initial, ≤5 years; intermediate, 6–15 years; advanced, ≥16 years.

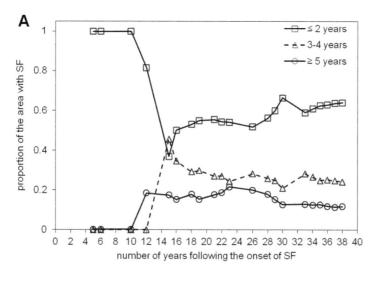

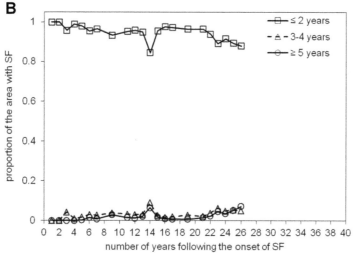

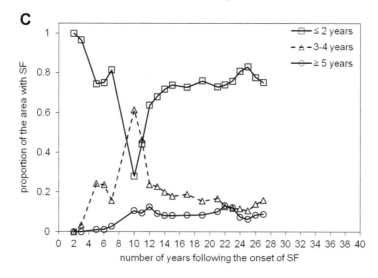

Figure 8. Proportion of the area with secondary forest (SF) as a function of the number of years following the onset of SF for the three classes of period of active land use. Proportion of the area with secondary forest (SF) as a function of the number of years following the onset of SF for the three classes of period of active land use (PALU) in A) Manaus, B) Santarém, and C) Machadinho d'Oeste: short, ≤2 years; medium, 3–4 years; long, ≥5 years.

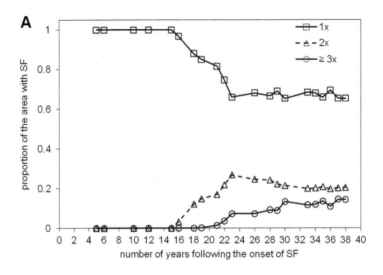

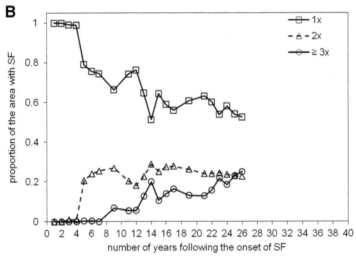

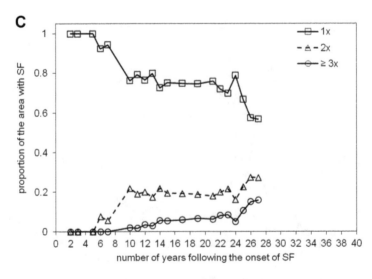

Figure 9. Proportion of the area with secondary forest (SF) as a function of the number of years following the onset of SF for the three classes of frequency of clearance. Proportion of the area with secondary forest (SF) as a function of the number of years following the onset of SF for the three classes of frequency of clearance (FC) in A) Manaus, B) Santarém, and C) Machadinho d'Oeste: low, 1 time; medium, 2 times; high, ≥3 times.

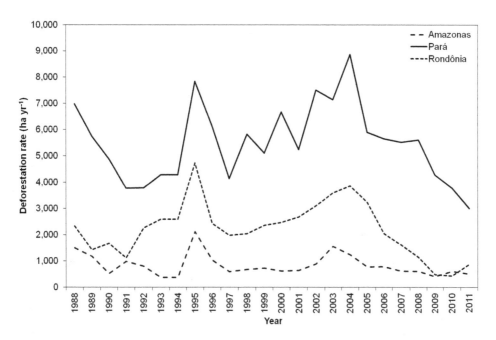

Figure 10. Deforestation rates in the Amazonas, Pará and Rondônia states according to INPE, 2013 [13]. Deforestation rates since 1988 (ha yr^{-1}) in the Amazonas, Pará and Rondônia states according to the data reported by INPE, 2013 [13].

Santarém, SF was cleared within 2–4 years from the onset of regeneration and continually thereafter. This was reflected in the increase in the proportion of SF areas cleared twice (after 4 years) and three times (after 7 years) with these both stabilizing at approximately 20%. At Machadinho d'Oeste, the proportion of areas with only a clearing event decreased to 55% 28 years after the onset of regeneration. The medium and high FC classes commenced after 4 years and 7 years respectively with the proportions of these remaining relatively constant (25% and 15%).

Discussion

6.1. Accuracy assessment of the land cover time-series

Several studies carried out in the Amazon have acknowledged some issues in discriminating SF from MF and, sometimes, NF. Lucas et al., 2000 [18] reported omission and commission errors ranging from 55–75% and 51–76%, respectively, associated with a greater confusion between NF and SF rather than between MF and SF. Carreiras et al., 2006 [20] reported an omission and commission error in the SF class of 66% and 43%, respectively, due to confusion between the SF, MF and NF classes. These authors concluded that misclassification among SF, MF, and NF was understandable, since from a spectral standpoint, SF is a transitional class between NF (i.e., agriculture/pasture) and MF. The initial stages of regeneration are spectrally similar to agriculture/pasture. Conversely, final stages of regrowth are more related to MF. The values reported in the two previous studies are much higher than those from this study, and that could be associated with the spatial resolution of the datasets that were used: ~1-km SPOT 4 VEGETATION and NOAA AVHRR vs. the 30-m spatial resolution Landsat TM and ETM+ data. The spatial arrangement of SF in the BLA, dominated by small previously abandoned areas, especially in Rondônia (e.g., [52]), creates an additional difficulty when trying to discriminate these from the surrounding land cover classes with coarse spatial resolution data. Kimes et al., 1999 [53] used 20-m spatial resolution SPOT High Resolution Visible (HRV) data between 1986 and 1994 to

discriminate MF, NF and SF in Rondônia and reported an accuracy of 95% in the SF class with misclassification happening mostly with the MF class. Metzger, 2002 [54] used Landsat TM data over the Bragantina region (Pará state) in 1996 to map various land cover classes and reported omission and commission errors in SF age classes ranging from 8–11% and 7–10%, respectively, with misclassification occurring mainly from confusion with MF. Vieira et al., 2003 [55] used Landsat 7 ETM+ data to map different stages of SF in São Francisco do Pará (northeastern Pará state) and reported omission and commission errors between 17–25% and 22–40% respectively, due mainly to misclassification among SF stages. Kuplich, 2006 [56] used Landsat TM and Synthetic Aperture Radar (SAR) data to discriminate various land cover classes in Manaus (included the BDFFP) and reported omission and commission errors in the SF age classes of 10–99% and 32–80% respectively, but those errors decreased when the SF age classes were aggregated into a unique class (78% and 25%, respectively). Although errors are still high in some studies (e.g., [18,20,56]), others have been able to discriminate SF from MF and NF with a high degree of accuracy (e.g., [53–55]).

Prates-Clark et al., 2009 [38] validated the 1995 and the 2001 land cover maps over Manaus and Santarém respectively using field-based reference data. For the validation of the 1995 land cover map over Manaus, field data collected in 1995 were used [22,42] and the authors reported an overall accuracy of ~99%, with omission errors in the MF, NF and SF classes of 0.3%, 3.5% and 1.9%, respectively, and commission errors of 0.5%, 7.0% and 9.5%, respectively. The major source of error in the SF class (higher commission error) was misclassification as MF class. At Santarém, field data collected in 2002 [57] were used to validate the 2001 land cover map. The overall accuracy was 88%, the omission errors were 2.2%, 1.1% and 4.2% for the MF, NF and SF classes respectively, and the commission errors were 0.6%, 28.8% and 71.9% respectively. The particularly high commission error in the SF class was essentially the consequence of misclassification as MF. The NF class displayed a high commission error that was related to misclassification as both MF and SF. Higher errors indicated in Prates-Clark et al., 2009 [38] for the Santarém site were also depicted in our accuracy assessment of the 2010 land cover map, although of much lower magnitude.

A major limitation of the accuracy assessment carried out in this study it that only one date (classification) of each time-series per study area was validated with very high spatial resolution data. Nevertheless, a 10-fold cross-validation procedure was implemented for those dates classified with the random forests algorithm, i.e., the 21 Landsat TM images of the Machadinho d'Oeste time-series (1984–2011) and the post 2003 Santarém Landsat TM images (2005–2010). In Machadinho d'Oeste, the average overall accuracy was ~98%, the omission errors of the MF, NF and SF classes ranged from 0.0–4.4%, 0.0–2.1% and 0.7–7.6% respectively, and the commission errors were between 0.0–4.2%, 0.0–0.3% and 1.7–7.1% respectively. At Santarém, the 2005–2010 time-series of classified Landsat TM data resulted in an average overall accuracy of 95%, the omission errors of the MF, NF and SF classes ranged between 1.8–6.7%, 0.0–0.9% and 8.1–16.7%, respectively, and the commission errors were between 3.7–7.5%, 0.0–0.6% and 4.1–13.2% respectively. Both the omission and commission errors of the NF class were very low, thus indicating that the major source of error of the SF class was misclassification as MF.

6.2. Differences of land cover change patterns among sites

Extensive deforestation across the BLA started in the 1970s and was concentrated along the southern and eastern rims of the region [58]. The main deforestation drivers were conversion of forest to cropland and/or cattle ranching, which were carried out both by small farmers and large landholders [59]. According to the data provided by INPE, 2013 [13], the states where the three sites are located displayed different deforestation rates (Figure 10). Deforestation rates were always higher in Pará, followed by Rondônia and Amazonas. However, since 2003–2005, deforestation rates have been decreasing, with Rondônia experiencing a rapid decrease leading to values closer to those observed in the Amazonas state. Several factors contributed to this decline, namely the improvement of market-driven environmental governance [60]. Pará experienced the highest number of settled families by agrarian reform in the nine states composing the BLA (~31,000 settled families per year in the period 2003–2006), and the number has been increasing since the 1960s [61]. Although the vast majority of deforestation in the BLA can be tracked to large landholders occupying the land for cattle ranching, the implementation of planned settlements is not negligible at all, especially in Rondônia, which is known by its small farmers' radial, fishbone and watershed deforestation patterns (e.g., [58]). For example, up to the mid-1990s, the number of settled families in Rondônia was only second to Pará, with 1,423 settled families per year against 1,462 settled families per year in Pará [61].

The three study areas have different socio-economic drivers leading to deforestation, different land uses and subsequent regeneration. The Manaus site is mostly included in the Manaus municipality, although 24% is also included in the Rio Preto da Eva municipality. This study area basically encompasses several environmental protection areas. As such, it seems that these conservation areas have been critical at preserving most of the forest in the region. The conservation units (CUs) limits accessed from the Brazilian Ministry of the Environment (MMA, available at http://mapas.mma.gov.br/i3geo/datadownload.htm) were used to estimate the ratio of relative incidence of deforestation (NF and SF) outside and inside the CUs in the last date of the time-

series for each site. The three sites had a significant proportion of the area included in the CUs: 43% in Manaus, 37% in Santarém and 23% in Machadinho d'Oeste. In Manaus, the relative incidence of deforestation outside the CUs was twice that inside CUs and in Santarém and Machadinho d'Oeste that value increased to five times. In fact, several studies have concluded that protected areas in the Amazon region are indeed effective at reducing deforestation rates (e.g., [60,62,63]). Figure 2A illustrates the lower amount of land cover change in Manaus, with approximately 20% of MF lost between 1973 and 2011, when compared with Santarém (more than 30% of MF lost between 1984 and 2010) and Machadinho d'Oeste (~40% of MF lost between 1984 and 2011). Deforestation in the Manaus study area started in the 1970s, with the construction of the BR-174 highway connecting Manaus and Boa Vista. Early deforestation is traceable along this highway, but many areas were abandoned as a result of poor soil fertility; indeed, many small farmers moved to Roraima, which was considered to have younger fertile soils [64]. The area of forest in 2011 was approximately 69% and 81% of the territory of the Manaus and Rio Preto da Eva municipalities respectively [13]. Figure 2A shows that the proportion of MF in the Manaus site decreased from approximately 90% in 1973 to 71% in 2011, which is similar to the value reported by INPE, 2013 [13] for the Manaus municipality.

The Santarém site is located south of the city of Santarém, in the Belterra municipality, along the BR-163 highway, and including part of the Tapajós National Forest. This highway links Santarém and Cuiabá in the south. According to Scatena et al., 1996 [65] and Brondizio and Moran, 2012 [66], this region has a tradition of agriculture and agroforestry since the early 1900s, which has further escalated as a consequence of the construction of the BR-163 in the 1970s. In fact, of the three sites Santarém had the lowest proportion of MF at the beginning of the time-series (~80%) and the highest proportion of SF (~20%), thus suggesting that deforestation was commonplace in the area in the early 1980s. According to INPE, 2013 [13], approximately 68% of the forest area in the Belterra municipality was remaining in 2011, whereas we report a proportion of MF in 2010 of 46%. A vast majority of the Belterra municipality is included in the Tapajós National Forest, mainly the part that is west of the BR-163 highway. So, it is not surprising that the proportion of remaining forest in this municipality is higher than the proportion of remaining forest in the study area that includes the fraction east of the BR-163 highway that is not a protected area.

The Machadinho d'Oeste site (Rondônia state) is included mostly in the Machadinho d'Oeste municipality, although 20% overlaps the Vale do Anari municipality. Of the three sites, this was the one with the highest degree of land cover change, with a reduction of ~60% in the MF cover from 1984 (91%) to 2011 (32%). Deforestation in this region started in the 1980s with the implementation of settlements by INCRA, of which Machadinho d'Oeste and Vale do Anari are just two examples [33]. According to Batistella and Moran, 2005 [33], the proportion of MF in 1988 and 1998 in Machadinho d'Oeste was around 80% and 66% respectively, while our estimates were 82% in 1987 (1988 was not mapped) and 55% in 1998. INPE, 2013 [13] reports an area of MF in the Machadinho d'Oeste municipality in 2011 of approximately 61%, and our study estimated the percentage of MF in the Machadinho d'Oeste study area as 32%. Google Earth data from 2013 shows that the areas of the Machadinho d'Oeste municipality to the east and north of the study area have much less deforestation and so this is a possible explanation for the higher proportion of MF detected by Batistella and Moran, 2005 [33] in 1998 and INPE, 2013 [13] up to 2011. Also, Batistella and Moran,

2005 [33] identified the land cover types occurring in the deforested areas of Machadinho d'Oeste. In 1988, approximately 2% was covered by SF, the value increasing to approximately 13% in 1998; our study mapped SF in 1987 and 1998 and the proportion was ~13% and ~29% respectively. Again, when considering the entire Machadinho d'Oeste municipality, there are extensive areas to the north and east of the study area that remained intact, with these covered by MF. Nevertheless, there is a coincident temporal trend that is the increase of SF from 1988 to 1998.

6.3. Land abandonment and the emergence of secondary forest

Land abandonment in the BLA is a consequence of a range of factors that have changed across the last decades since the inception of large scale deforestation in the region. Namely these are; reduced crop productivity as a consequence of poor soil fertility; lack of financial incentives, migratory patterns; non-traditional land uses and market fluctuations [67]. At Manaus, in 2011, land abandonment resulted in SF that was comprised mainly of areas with advanced ASF (50% ≥16 years), although large areas of intermediate ASF (37% between 6–15 years) were also present (Figure 3A and Figure 6). This suggests that the conservation areas were not only effective at preserving MF but also SF areas. As mentioned before, deforested areas in this study area were abandoned mainly because of poor soil fertility, with settlers moving to the nearby state of Roraima [64]. It is clear from Figure 3A that the higher proportion of advanced ASF in the study area is in the northern half of the site. These areas undergoing regeneration in 2011 had mostly short PALU (64%) as a consequence of intensification of land use and requirement for new land, and 65% was deforested only once. In comparison, the Santarém site, which included parts of the Tapajós National Forest, had regrowth dominated by intermediate ASF in 2010. Those areas experienced essentially short PALU (88% ≤2 years) and low FC (52% deforested once). According to Brondizio and Moran, 2012 [66], land abandonment in the region, and associated regeneration, was high until 1999. At this point, large-scale soybean cultivation started, mainly because of the decay of primary and secondary roads, lack of social services, and limited access to water. On the other hand, land abandonment in Machadinho d'Oeste resulted in SF that was mainly in the initial (46%) and intermediate (41%) ASF classes, with these areas experiencing mostly (75%) short PALU and low FC (57%). This was a planned settlement and most of the deforested area is under agriculture and or pasture use, suggesting that a vast majority of the area undergoing SF could be indeed forest fallow that might be under cattle ranching use or being subjected to subsistence agriculture in a crop/fallow cycle [66].

One important aspect worth mentioning is the large proportion of SF in the Santarém site that resulted from the 1997 wildfires. This area of regeneration was not the consequence of land abandonment but a natural process following fire-induced disturbance. This wildfire was just one among many others that occurred in the BLA as a consequence of the 1997–1998 *El Niño* Southern Oscillation (ENSO) event (e.g., [68]). In fact, the proportion of NF in 1997 (prior to the wildfire) was around 6% and in 1998 had increased to approximately 15%, mainly as a consequence of that large wildfire, perfectly visible in Figure 3B as the orange irregular zone of intermediate ASF in the centre of the study area. As a consequence of the natural regeneration process, the SF proportion increased to approximately 44% in 2000 (~32% in 1997). Therefore, approximately 20% of the area of SF

in 2010 was estimated as resulting from natural regeneration following that wildfire event.

6.4. Implications of prior land use for biomass accumulation and biodiversity restoration in secondary forests

Regenerating forests in the three selected sites have been classified in terms of ASF, PALU prior to abandonment and FC. Prior LULCC practices have a fundamental influence on the vegetation regenerating following land abandonment in the Amazon [10,22,69–75]. Vegetation structure, species composition and dominance are just some of the parameters of the regenerating vegetation that have been studied to identify the impact of prior land use practices (e.g., [76]). As such, these have an impact on the capability of these forests to accumulate biomass (and to act as a C sink at a greater or lesser extent) and to restore biodiversity (e.g., [77]).

Uhl et al., 1988 [71] studied areas undergoing secondary succession following pasture abandonment in Pará (Paragominas municipality) and identified major patterns related to prior land use intensity; light use sites, characterized by a lower PALU (0–4 years) and with no reclearance or just one slash-and-burn episode, had high species richness dominated by pioneer species (e.g., *Cecropia* sp. and *Solanum* sp.), and a ~13 m closed canopy with an AGB accumulation rate of 10 Mg ha^{-1} yr^{-1} (8-year old sites); comparatively, moderate use sites, having higher PALU (6–12 years) and with 1 to 5 clearing episodes, had lower species richness dominated by a 7–8 m *Vismia* sp. partially developed canopy and an AGB accumulation rate of 5 Mg ha^{-1} yr^{-1}; heavy use sites, subjected to large-scale mechanized operations, had poor species richness (*Solanum* sp. and *Cecropia* sp.), mainly composed by scattered trees and an AGB accumulation rate of only 0.6 Mg ha^{-1} yr^{-1}. Mesquita et al., 2001 [75] studied regeneration pathways in a region north of Manaus that is included in our Manaus site (BDFFP) and identified major differences related to prior land use. Basically, areas that were deforested and abandoned immediately were dominated by *Cecropia* spp. trees, whereas those that were abandoned after some years of pasture use were dominated by *Vismia* spp. trees. Lucas et al., 2002 [22], also identified two distinct regeneration pathways in a region north of Manaus and in the same way as Mesquita et al. [75]; the authors identified *Vismia* spp. dominated regeneration associated to sites with more intensive land use and those dominated by *Cecropia* spp. to less intensive land use. According to Chazdon et al., 2007 [74], in terms of AGB accumulation, tropical secondary forests are characterized by a rapid growth rates in the first years after land (agriculture or pasture) abandonment. Furthermore, Zarin et al., 2001 [78] showed that there is a strong correlation between the AGB of secondary forests and the number of years following abandonment (i.e., ASF), soil texture and climate data in the Amazon region. On this pretext, the secondary forests in Manaus have the potential to accumulate more AGB, followed by Santarém and Machadinho d'Oeste. In 2011 (the last date of the time-series), Machadinho d'Oeste had approximately 45% of the area of SF with less than five years of age, and around 40% with less than three years of age. Machadinho d'Oeste was initially a settlement project implemented by the Brazilian government and today most of its people are dependent on subsistence agriculture [32]. Therefore, it is possible that some of the area mapped as SF was indeed forest fallow, part of a crop/fallow system of subsistence agriculture, which have been acknowledged by several studies (e.g., [79,80]).

Concluding Remarks

An accurate assessment of large-scale land use and land cover change dynamics over remote areas (e.g., Amazon) can only be carried out with an adequate monitoring system. Remote sensing data, in this case annual or quasi-annual time-series of high spatial resolution optical data (Landsat program), has proven its ability to accurately discriminate MF, SF and NF over three sites in the BLA experiencing several decades of deforestation. Lower deforestation rates and a greater proportion of intermediate and advanced ASF classes were characteristic of the Manaus site. On the other hand, Machadinho d'Oeste had the highest deforestation rates and lower regrowth rates, with most of its regeneration occurring in areas abandoned over the past 5 years. LULCC at Santarém displayed an intermediate behavior, with intermediate deforestation rates and the highest regrowth rates, which lead to regeneration dominated by areas of intermediate ASF. Conservation units were effective at reducing deforestation at all three sites. The temporal evolution of the spatial arrangement of the various parcels of land identified as one of the three classes could provide new insights about the fragmentation patterns in the region (Ewers et al. 2013 [81]).

Although several generalizations about the type and composition of secondary forest occurring in each of the three sites can be made, a correct assessment can only be made after an *in situ* assessment. Therefore, the research presented here will provide the opportunity to assess the influence of previous LULCC dynamics on the biomass accumulation and biodiversity restoration and to further investigate hypothesis related to differences among the three sites. A field campaign is foreseen for 2014 to collect forest inventory data that will be used to study the influence of prior LULCC on the capability of regrowth areas to accumulate AGB and to restore biodiversity. In the end, a framework related to the influence of prior LULCC on the biomass accumulation and biodiversity will be built and could be used to inform land management policies in the region.

Acknowledgments

The United States Geological Survey (USGS) and Brazil's National Institute for Space Research (INPE) are particularly thanked for providing free access to the Landsat sensor data used in this study. Nathan Thomas (Department of Geography and Earth Sciences, Aberystwyth University, Aberystwyth, United Kingdom) is particularly acknowledged for reviewing earlier versions of the manuscript.

Author Contributions

Conceived and designed the experiments: JC RL. Performed the experiments: JC JJ RL CG. Analyzed the data: JC JJ RL. Contributed reagents/materials/analysis tools: JC JJ RL CG. Wrote the paper: JC JJ RL.

References

1. Falkowski P, Scholes RJ, Boyle E, Canadell J, Canfield D, et al. (2000) The global carbon cycle: A test of our knowledge of earth as a system. Science 290: 291–296.

2. Raupach M (2011) Pinning down the land carbon sink. Nature Climate Change 1: 148–149.

3. Heimann M, Reichstein M (2008) Terrestrial ecosystem carbon dynamics and climate feedbacks. Nature 451: 289–292.

4. Pan Y, Birdsey R, Fang J, Houghton R, Kauppi P, et al. (2011) A Large and Persistent Carbon Sink in the World's Forests. Science 333: 988–993.

5. Gibson L, Lee T, Koh L, Brook B, Gardner T, et al. (2011) Primary forests are irreplaceable for sustaining tropical biodiversity. Nature 478: 378–381.

6. Achard F, Eva H, Stibig H, Mayaux P, Gallego J, et al. (2002) Determination of deforestation rates of the world's humid tropical forests. Science 297: 999–1002.

7. Hooper D, Adair E, Cardinale B, Byrnes J, Hungate B, et al. (2012) A global synthesis reveals biodiversity loss as a major driver of ecosystem change. Nature 486: 105–108.

8. Lederer M (2011) From CDM to REDD+ − What do we know for setting up effective and legitimate carbon governance? Ecological Economics 70: 1900–1907.

9. Olson D, Dinerstein E, Wikramanayake E, Burgess N, Powell G, et al. (2001) Terrestrial ecoregions of the worlds: A new map of life on Earth. Bioscience 51: 933–938.

10. Chazdon RL (2003) Tropical forest recovery: legacies of human impact and natural disturbances. Perspectives in Plant Ecology Evolution and Systematics 6: 51–71.

11. Cardinale B, Duffy J, Gonzalez A, Hooper D, Perrings C, et al. (2012) Biodiversity loss and its impact on humanity. Nature 486: 59–67.

12. Goulding M, Barthem R, Ferreira EJG (2003) The Smithsonian atlas of the Amazon. Washington, D.C.: Smithsonian Books. 253 p.

13. INPE (2013) Projeto PRODES: Monitoramento da Floresta Amazônica Brasileira por Satélite. Instituto Nacional de Pesquisas Espaciais. Available: http://www.obt.inpe.br/prodes/index.php. Accessed 11 March 2014.

14. Davidson EA, de Araujo AC, Artaxo P, Balch JK, Brown IF, et al. (2012) The Amazon basin in transition. Nature 481: 321–328.

15. Stone TA, Schlesinger P, Houghton RA, Woodwell GM (1994) A map of the vegetation of South America based on satellite imagery. Photogrammetric Engineering and Remote Sensing 60: 541–551.

16. Schroeder PE, Winjum JK (1995) Assessing Brazil carbon budget: II. Biotic fluxes and net carbon balance. Forest Ecology and Management 75: 87–99.

17. Fearnside P (1996) Amazonian deforestation and global warming: Carbon stocks in vegetation replacing Brazil's Amazon forest. Forest Ecology and Management 80: 21–34.

18. Lucas RM, Honzak M, Curran PJ, Foody GM, Milne R, et al. (2000) Mapping the regional extent of tropical forest regeneration stages in the Brazilian Legal Amazon using NOAA AVHRR data. International Journal of Remote Sensing 21: 2855–2881.

19. Cardille J, Foley J (2003) Agricultural land-use change in Brazilian Amazonia between 1980 and 1995: Evidence from integrated satellite and census data. Remote Sensing of Environment 87: 551–562.

20. Carreiras JMB, Pereira JMC, Campagnolo ML, Shimabukuro YE (2006) Assessing the extent of agriculture/pasture and secondary succession forest in the Brazilian Legal Amazon using SPOT VEGETATION data. Remote Sensing of Environment 101: 283–298.

21. Neeff T, Lucas RM, dos Santos JR, Brondizio ES, Freitas CC (2006) Area and age of secondary forests in Brazilian Amazonia 1978–2002: An empirical estimate. Ecosystems 9: 609–623.

22. Lucas RM, Honzak M, Amaral ID, Curran PJ, Foody GM (2002) Forest regeneration on abandoned clearances in central Amazonia. International Journal of Remote Sensing 23: 965–988.

23. Fearnside PM (2000) Global warming and tropical land-use change: Greenhouse gas emissions from biomass burning, decomposition and soils in forest conversion, shifting cultivation and secondary vegetation. Climatic Change 46: 115–158.

24. Houghton RA (2005) Aboveground forest biomass and the global carbon balance. Global Change Biology 11: 945–958.

25. Silva JMN, Carreiras JMB, Rosa I, Pereira JMC (2011) Greenhouse gas emissions from shifting cultivation in the tropics, including uncertainty and sensitivity analysis. Journal of Geophysical Research-Atmospheres 116. doi:10.1029/2011JD016056.

26. Laurance S, Laurance W, Andrade A, Fearnside P, Harms K, et al. (2010) Influence of soils and topography on Amazonian tree diversity: a landscape-scale study. Journal of Vegetation Science 21: 96–106.

27. Saatchi S, Houghton R, Alvala R, Soares J, Yu Y (2007) Distribution of aboveground live biomass in the Amazon basin. Global Change Biology 13: 816–837.

28. Laurance WF, Camargo JLC, Luizao RCC, Laurance SG, Pimm SL, et al. (2011) The fate of Amazonian forest fragments: A 32-year investigation. Biological Conservation 144: 56–67.

29. van Gardingen PR, Valle D, Thompson I (2006) Evaluation of yield regulation options for primary forest in Tapajos National Forest, Brazil. Forest Ecology and Management 231: 184–195.

30. Bacha CJC, Rodriguez LCE (2007) Profitability and social impacts of reduced impact logging in the Tapajos National Forest, Brazil - A case study. Ecological Economics 63: 70–77.

31. Espírito-Santo FDB, Shimabukuro YE, Aragão LEOC, Machado ELM (2005) Analysis of the floristic and phytosociologic composition of Tapajós national forest with geographic support of satellite images. Acta Amazonica 35: 155–173.

32. Miranda EE (2013) Sustentabilidade Agrícola na Amazonia - 23 anos de monitoramento da agricultura em Machadinho d'Oeste (RO). Embrapa Monitoramento por Satélite. Available: http://www.machadinho.cnpm.embrapa.br/. Accessed 11 March 2014.

33. Batistella M, Moran EF (2005) Dimensões humanas do uso e cobertura das terras na Amazônia: uma contribuição do LBA. Acta Amazonica 35: 239–247.

34. Silver WL, Neff J, McGroddy M, Veldkamp E, Keller M, et al. (2000) Effects of soil texture on belowground carbon and nutrient storage in a lowland Amazonian forest ecosystem. Ecosystems 3: 193–209.

35. Bierregaard RO (2001) Lessons from Amazonia: the ecology and conservation of a fragmented forest. New Haven, USA: Yale University Press. 478 p.

36. Laurance WF, Fearnside PM, Laurance SG, Delamonica P, Lovejoy TE, et al. (1999) Relationship between soils and Amazon forest biomass: A landscape-scale study. Forest Ecology and Management 118: 127–138.

37. Keller M, Varner R, Dias JD, Silva H, Crill P, et al. (2005) Soil-atmosphere exchange of nitrous oxide, nitric oxide, methane, and carbon dioxide in logged and undisturbed forest in the Tapajos National Forest, Brazil. Earth Interactions 9: 1–28.

38. Prates-Clark C, Lucas R, dos Santos J (2009) Implications of land-use history for forest regeneration in the Brazilian Amazon. Canadian Journal of Remote Sensing 35: 534–553.

39. Chander G, Markham B, Helder D (2009) Summary of current radiometric calibration coefficients for Landsat MSS, TM, ETM+, and EO-1 ALI sensors. Remote Sensing of Environment 113: 893–903.

40. Lillesand TM, Kiefer RW (2000) Remote sensing and image interpretation. New York, USA: John Wiley & Sons. 724 p.

41. Navulur K (2007) Multispectral image analysis using the object-oriented paradigm. Boca Raton, USA: CRC Press/Taylor & Francis. 165 p.

42. Lucas R, Xiao X, Hagen S, Frolking S (2002) Evaluating TERRA-1 MODIS data for discrimination of tropical secondary forest regeneration stages in the Brazilian Legal Amazon. Geophysical Research Letters 29: 42-1-42-4.

43. Hastie T, Tibshirani R, Friedman JH (2009) The elements of statistical learning: data mining, inference, and prediction. New York (USA): Springer. 745 p.

44. Roberts D, Numata I, Holmes K, Batista G, Krug T, et al. (2002) Large area mapping of land-cover change in Rondonia using multitemporal spectral mixture analysis and decision tree classifiers. Journal of Geophysical Research-Atmospheres 107: LBA 40-1-LBA 40-18.

45. Puyravaud J (2003) Standardizing the calculation of the annual rate of deforestation. Forest Ecology and Management 177: 593–596.

46. Pereira MC, Setzer AW (1993) Spectral characteristics of fire scars in Landsat-5 TM images of Amazonia. International Journal of Remote Sensing 14: 2061–2078.

47. Cochrane M, Souza C (1998) Linear mixture model classification of burned forests in the Eastern Amazon. International Journal of Remote Sensing 19: 3433–3440.

48. Foody GM (2009) Sample size determination for image classification accuracy assessment and comparison. International Journal of Remote Sensing 30: 5273–5291.

49. Gallego FJ (2004) Remote sensing and land cover area estimation. International Journal of Remote Sensing 25: 3019–3047.

50. Walsh TA, Burk TE (1993) Calibration of satellite classifications of land area. Remote Sensing of Environment 46: 281–290.

51. Tenenbein A (1972) A double sampling scheme for estimating from misclassified multinomial data with applications to sampling inspection. Technometrics 14: 187–202.

52. Ferraz SFD, Vettorazzi CA, Theobald DM, Ballester MVR (2005) Landscape dynamics of Amazonian deforestation between 1984 and 2002 in central Rondonia, Brazil: assessment and future scenarios. Forest Ecology and Management 204: 67–83.

53. Kimes DS, Nelson RF, Salas WA, Skole DL (1999) Mapping secondary tropical forest and forest age from SPOT HRV data. International Journal of Remote Sensing 20: 3625–3640.

54. Metzger JP (2002) Landscape dynamics and equilibrium in areas of slash-and-burn agriculture with short and long fallow period (Bragantina region, NE Brazilian Amazon). Landscape Ecology 17: 419–431.

55. Vieira ICG, de Almeida AS, Davidson EA, Stone TA, de Carvalho CJR, et al. (2003) Classifying successional forests using Landsat spectral properties and ecological characteristics in eastern Amazonia. Remote Sensing of Environment 87: 470–481.

56. Kuplich TM (2006) Classifying regenerating forest stages in Amazonia using remotely sensed images and a neural network. Forest Ecology and Management 234: 1–9.

57. Prates-Clark CC (2004) Remote sensing of tropical regenerating forests in the Brazilian Amazon. PhD dissertation. Aberystwyth, Wales, UK: University of Wales.

58. Fearnside PM (2005) Deforestation in Brazilian Amazonia: History, rates, and consequences. Conservation Biology 19: 680–688.

59. Kirby KR, Laurance WF, Albernaz AK, Schroth G, Fearnside PM, et al. (2006) The future of deforestation in the Brazilian Amazon. Futures 38: 432–453.

60. Nepstad DC, McGrath DG, Soares B (2011) Systemic Conservation, REDD, and the Future of the Amazon Basin. Conservation Biology 25: 1113–1116.

61. Pacheco P (2009) Agrarian Reform in the Brazilian Amazon: Its Implications for Land Distribution and Deforestation. World Development 37: 1337–1347.

62. Rosa IMD, Purves D, Souza C Jr, Ewers RM (2013) Predictive Modelling of Contagious Deforestation in the Brazilian Amazon. PLoS ONE 8(10): e77231. doi:10.1371/journal.pone.0077231.

63. Nolte C, Agrawal A, Silvius KM, Soares-Filho BS (2013) Governance regime and location influence avoided deforestation success of protected areas in the Brazilian Amazon. Proceedings of the National Academy of Sciences of the United States of America 110: 4956–4961.

64. Fearnside PM, Graca P (2006) BR-319: Brazil's manaus-porto velho highway and the potential impact of linking the arc of deforestation to central Amazonia. Environmental Management 38: 705–716.

65. Scatena FN, Walker RT, Homma AKO, deConto AJ, Ferreira CAP, et al. (1996) Cropping and fallowing sequences of small farms in the "terra firme" landscape of the Brazilian Amazon: A case study from Santarem, Para. Ecological Economics 18: 29–40.

66. Brondizio ES, Moran EF (2012) Level-dependent deforestation trajectories in the Brazilian Amazon from 1970 to 2001. Population and Environment 34: 69–85.

67. Perz SG, Skole DL (2003) Social determinants of secondary forests in the Brazilian Amazon. Social Science Research 32: 25–60.

68. Nepstad DC, Verissimo A, Alencar A, Nobre C, Lima E, et al. (1999) Large-scale impoverishment of Amazonian forests by logging and fire. Nature 398: 505–508.

69. Uhl C (1987) Factors controlling succession following slash-and-burn agriculture in Amazonia. Journal of Ecology 75: 377–407.

70. Buschbacher R, Uhl C, Serrao EAS (1988) Abandoned pastures in eastern Amazonia. II. Nutrient stocks in the soil and vegetation. Journal of Ecology 76: 682–699.

71. Uhl C, Buschbacher R, Serrao EAS (1988) Abandoned pastures in eastern Amazonia. I. Patterns of plant succession. Journal of Ecology 76: 663–681.

72. Saldarriaga JG, West DC, Tharp ML, Uhl C (1988) Long-term chronosequence of forest succession in the upper Rio Negro of Colombia and Venezuela. Journal of Ecology 76: 938–958.

73. Uhl C, Jordan CF (1984) Succession and nutrient dynamics following forest cutting and burning in Amazonia. Ecology 65: 1476–1490.

74. Chazdon RL, Letcher SG, van Breugel M, Martinez-Ramos M, Bongers F, et al. (2007) Rates of change in tree communities of secondary Neotropical forests following major disturbances. Philosophical Transactions of the Royal Society B-Biological Sciences 362: 273–289.

75. Mesquita R, Ickes K, Ganade G, Williamson G (2001) Alternative successional pathways in the Amazon Basin. Journal of Ecology 89: 528–537.

76. Uhl C, Clark K, Clark H, Murphy P (1981) Early plant succession after cutting and burning in the upper Rio Negro region of the Amazon basin. Journal of Ecology 69: 631–649.

77. Chazdon RL (2008) Beyond deforestation: Restoring forests and ecosystem services on degraded lands. Science 320: 1458–1460.

78. Zarin DJ, Ducey MJ, Tucker JM, Salas WA (2001) Potential biomass accumulation in Amazonian regrowth forests. Ecosystems 4: 658–668.

79. Brown S, Lugo A (1990) Tropical Secondary forests. Journal of Tropical Ecology 6: 1–32.

80. Lu DS, Moran E, Batistella M (2003) Linear mixture model applied to Amazonian vegetation classification. Remote Sensing of Environment 87: 456–469.

81. Ewers RM, Didham RK, Pearse WD, Lefebvre V, Rosa IMD, et al. (2013) Using landscape history to predict biodiversity patterns in fragmented landscapes. Ecology Letters 16: 1221–1233.

20

Tree Spatial Structure, Host Composition and Resource Availability Influence Mirid Density or Black Pod Prevalence in Cacao Agroforests in Cameroon

Cynthia Gidoin[1,2]*, Régis Babin[2,3], Leïla Bagny Beilhe[2,4], Christian Cilas[2], Gerben Martijn ten Hoopen[2,4], Marie Ange Ngo Bieng[5]

1 Supagro, UMR System, Montpellier, France, 2 CIRAD, UPR Bioagresseurs, Montpellier, France, 3 icipe, Plant Health Division, Coffee Pest Project, Nairobi, Kenya, 4 IRAD, Yaoundé, Cameroon, 5 CIRAD, UMR System, Montpellier, France

Abstract

Combining crop plants with other plant species in agro-ecosystems is one way to enhance ecological pest and disease regulation mechanisms. Resource availability and microclimatic variation mechanisms affect processes related to pest and pathogen life cycles. These mechanisms are supported both by empirical research and by epidemiological models, yet their relative importance in a real complex agro-ecosystem is still not known. Our aim was thus to assess the independent effects and the relative importance of different variables related to resource availability and microclimatic variation that explain pest and disease occurrence at the plot scale in real complex agro-ecosystems. The study was conducted in cacao (*Theobroma cacao*) agroforests in Cameroon, where cocoa production is mainly impacted by the mirid bug, *Sahlbergella singularis*, and black pod disease, caused by *Phytophthora megakarya*. Vegetation composition and spatial structure, resource availability and pest and disease occurrence were characterized in 20 real agroforest plots. Hierarchical partitioning was used to identify the causal variables that explain mirid density and black pod prevalence. The results of this study show that cacao agroforests can be differentiated on the basis of vegetation composition and spatial structure. This original approach revealed that mirid density decreased when a minimum number of randomly distributed forest trees were present compared with the aggregated distribution of forest trees, or when forest tree density was low. Moreover, a decrease in mirid density was also related to decreased availability of sensitive tissue, independently of the effect of forest tree structure. Contrary to expectations, black pod prevalence decreased with increasing cacao tree abundance. By revealing the effects of vegetation composition and spatial structure on mirids and black pod, this study opens new perspectives for the joint agro-ecological management of cacao pests and diseases at the plot scale, through the optimization of the spatial structure and composition of the vegetation.

Editor: Morag McDonald, Bangor University, United Kingdom

Funding: The authors thank the West African Scientific Partnership Platform on diversity of production systems and agroecological management of pests and diseases of the French Agricultural Research Centre for International Development (CIRAD) and the research project "Search for trade-offs between production and other ecosystem services provided by tropical agroforestry systems" (SAFSE) for funding. The funders had no role in study design, data collection and analysis, decision to publish, or preparation of the manuscript.

Competing Interests: The authors have declared that no competing interests exist.

* Email: gidoin@supagro.inra.fr

Introduction

Maintaining production and improving ecosystem services of agro-ecosystems while simultaneously reducing dependence on external inputs such as pesticides and fertilizers is a major challenge in agriculture today [1]. To achieve this goal, two concepts based on the "intensification in the use of the natural functionalities that ecosystems offer" are ecological intensification and agro-ecology [2]. A central idea behind these concepts is that associated biodiversity could provide a large range of ecosystem services that may have the same effect as chemical inputs, including pesticides, for pest regulation [3]. With regard to pest and disease regulation services, combining crop plants with other plant species can impact pest and disease occurrence through divers and often complex mechanisms [4]–[5]. However, the

regulation of pests and diseases depends much more on specific characteristics of plant associations than on species richness *per se*, yet these characteristics are largely unknown.

Variability in resource availability is one of the main hypotheses proposed to explain the impact of plant diversity on pest and disease occurrence in agro-ecosystems [5]–[6]. Two mechanisms are involved in this hypothesis: (i) the "resource dilution" mechanism, which consists of a reduction in pest and disease occurrence through a decrease in host abundance [7] and (ii), the "alternate resource introduction" mechanism, which consists of an increase in pest and disease occurrence through the introduction of alternative hosts [8]. In general, "resource dilution" holds true for specialist pests and diseases that have only one or a few host plants [9], while the "alternate resource" mechanism generally holds true for more generalist pests and diseases.

Another mechanism involved in pest and disease regulation in multispecies agro-ecosystems is linked to variations in microclimatic conditions [6]. For example, shade trees associated with a crop reduce light availability and wind speed, buffer temperature, and increase relative humidity [10]. These microclimatic variations, which can be grouped under the term shading [11], are determined by several factors, including the spatial structure of shade trees. Tree spatial structure can be described by the characteristics of its vertical and horizontal structure including the number of strata, the density of individual trees in a stratum, and the horizontal distribution of shade trees [4]. These characteristics determine the mean and variance of sunlight transmitted to the understory and can directly affect processes related to the life cycles of pests and pathogens of the crop [6]–[8]–[12].

Under natural conditions, the microclimatic and resource alteration mechanisms are interlinked and interact. For example, microclimatic variations affect the vegetative growth of host plants and hence the availability of sensitive tissue resources for pests and diseases [13]–[14]. However, microclimatic and resource alteration mechanisms are rarely studied together and, as a result, their independent effects and relative importance in pest and disease occurrence at the plot scale are still not known. Although these mechanisms have been confirmed by both empirical research on plant pests and diseases [15] and by epidemiological models [16], the present study is one of the first to try to assess the relative importance of these different mechanisms in a real complex agro-ecosystem. Furthermore, depending on their biology, pests and diseases are differently impacted by each of these mechanisms [17]. In a previous work in cacao agroforest plots in Costa Rica, we studied the independent and joint effects of variables related to the two mechanisms on the impact of one cacao fungal disease (frosty pod rot caused by *Moniliophthora roreri*) [18]. The results of this study revealed the significant influence of the spatial structure of shade trees in the regulation of cacao disease.

In this paper, we focus on cacao (*Theobroma cacao* L.) agroforests in Cameroon. An agroforest is defined as an agro-ecosystem that is structurally close to a natural forest ecosystem with diverse vegetation composition and complex spatial structure [19]. Cameroon is currently the fifth cocoa producer of the world [20], but cocoa production is seriously affected by the mirid bug *Sahlbergella singularis* Hagl. [21] and black pod disease, caused by *Phytophthora megakarya* Brasier & Griffin [22]. Cocoa mirids are not specific to the crop and can be found on several plant species belonging to the Malvaceae family (e.g. *Ceiba pentandra*, *Cola acuminata*, *Cola lateritia* and *Sterculia rhinopetala*) [23], some of which are often associated with cacao trees in agroforests [21]. Black pod disease (BP) caused by *Phytophthora megakarya* can be considered as a specialized pathogen of cacao trees since it does not damage other plant species present in cacao based agroforests and has not been isolated from diseased plant material outside cacao based agroforests [24]–[25]. *Phytophthora megakarya* zoospores are dispersed by rain splash between 5–6 meters [26] and may spread further when free flowing water is available. Adult mirids are able to fly 1–2 km on average [27] but the larvae cannot fly.

The balance of shading effects on pest and disease regulation can be complex [4]–[8], as is the case in the regulation of mirids and black pod disease. Firstly, heavy shade reduces mirid density, probably by limiting cacao vegetative growth, i.e. the availability of young shoots (also named flushes), which are a primary food resource for mirids [21]. At the same time, the introduction of shade trees belonging to the Malvaceae family, known to include alternative host plants for mirids, could reduce the positive impact of shading on mirid density [28]. In the case of black pod, shading

is considered to provide microclimatic conditions favorable to the spread and development of the disease [29]. However, the introduction of shade trees may result in resource dilution through a decrease in cacao tree abundance and hence in the availability of potentially sensitive tissue (mainly cacao fruits, also known as cacao pods), which could then reduce the negative impact of shading on the regulation of black pod. Therefore, assessing the respective effects of host composition, the availability of sensitive tissue, and shade tree spatial structure is particularly important as it could improve the joint integrated management of mirids and black pod disease in cacao agroforests.

The question we addressed in this study was "Which part of the variation in mirid density and black pod prevalence among agroforests is explained by host composition, the spatial structure of the vegetation, and resource availability?" To answer this question, mirid density, black pod prevalence and key variables related to spatial structure and composition of the vegetation and the amount of available sensitive tissue were characterized in 20 cacao agroforest plots in the region Centre in Cameroon. We examined the independent contributions of key variables explaining mirid density and disease prevalence using a hierarchical partitioning protocol. We then discuss the implications of our results for the joint agro-ecological management of mirids and black pod disease in cacao agroforests through the optimization of the spatial structure and composition of the vegetation.

Materials and Methods

Study site and sampling plots

The Lékié department in Cameroon was chosen because it represents an intermediate situation on the north–south gradient characterizing natural conditions in the cocoa growing region of central Cameroon [30]. In this department, cacao agroforests are located in a mosaic of cacao based agroforestry plantations owned by different farmers and of forest zones with substantial human activity. Individual farmers generally own one or two cacao plantations with an average size of 2.69 ha (SD = 0.09, n = 687) [30]. Twenty 2,500 m^2 sampling units (50×50 m, see Figure 1) were selected in 20 cacao agroforest plantations belonging to different farmers', based on the criterion that the unit contained more than 10 shade trees associated with the cacao trees. Each sampling unit (hereafter referred to as 'plot') was representative of the cacao agroforest plantation as a whole. The 20 plots were located in a total area of 10×10 km spread over three localities: Nkolobang (9 plots), Zima (7 plots) and Mbakomo (4 plots). The rainfall pattern is bimodal with between 1,500 and 1,600 mm of rainfall per year [30]. Data were collected between September 2011 and November 2012.

Characterization of vegetation structure

In September 2011, vegetation composition and spatial structure were characterized in each plot. The x,y cartesian coordinates of each plant (2 m tall or over) were recorded in each plot using a theodolite (Leica Builder 409, Leica Geosystems, Heerbrugg, Switzerland). Plants were identified to either species or family level, and in a few cases, were not identified. The trees associated with the cacao trees, whether identified or not, were classified in two categories: forest trees or fruit trees. This classification is based on the management of trees in agroforests and is that used in the literature, it states that the fruit trees are mainly introduced by farmers whereas the forest trees are mainly left standing in agroforests when the cacao plantation is established [19]–[31].

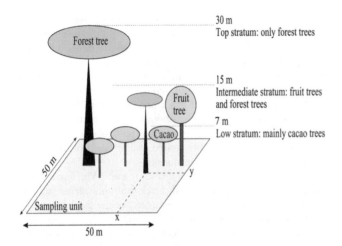

Figure 1. Schematic representation of the vertical structure of cacao agroforest in the Lékié department in Cameroon.

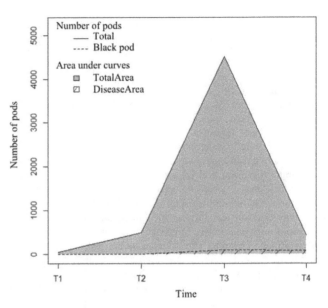

Figure 3. Number of total and damaged pods on the 80 cacao trees sampled in one plot. Number of pods from T1 (May 2012) to T4 (November 2012) on the 80 cacao trees sampled in the plot studied and mapped in Figure 2.

Pest and disease resources were described using variables related to host composition. The cacao tree was the main host for mirids and black pod in the agroforests we studied. For that reason, the relative abundance of cacao trees (*Abca*, i.e. their abundance with respect to all the vegetation mapped in the plot) was assumed to be a valid proxy to study the resource dilution mechanism. For

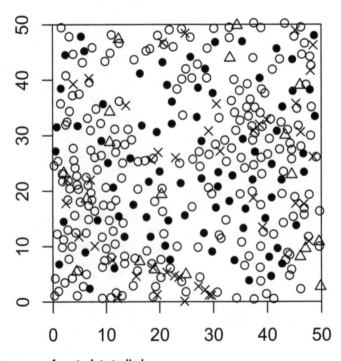

Figure 2. A sample map of a cacao agroforest plot studied.

Table 1. List of variables used to describe the 20 cacao agroforests.

Categories	Variables	Code	Unit or modalities	Min	Max	Mean	Transf.
Mirid pest	Mirid density 2011/12	Dmir	number of ind./ha	0	842	117	log
Black pod disease	Black pod prevalence	BPP	%	0	5	1	sqrt
Host composition "resource Hypothesis"	Cacao abundance	Abca	%	74	94	84	-
	Alternative host	Alter	Presence				
			Absence				
Sensitive tissue "resource Hypothesis"	Number of pods 2011/12	Prod1	number of pod/ha	78	22178	8497	sqrt
	Number of pods 2012	Prod2	number of pod x day	80.10^4	31.10^5	21.10^5	sqrt
	Flush presence	Flush	-	0.2	0.5	0.3	sqrt
Spatial structure "microclimate Hypothesis"	Total plant density	Dtot	number of trees/ha	580	1660	1129	-
	Density of associated shade trees	Dsha	number of shade trees/ha	40	168	97	log
	% intermediate trees	%Inter	%	18	92	51	sqrt
	Spatial structure of forest trees	HSFo	Low density				
			Aggregated				
			Random				
	Spatial structure of fruit trees	HSFu	Low density				
			Random				
			Regular				

Variables for density and number of pods are presented at the hectare scale but are used at the plot scale (1/4 ha) in statistical analyses.

Table 2. Information on plots concerned by the variable forest tree horizontal structure (*HSFo*).

HSFo modalities	Low density	Aggregated	Random
Number of plots	5	7	8
Density of forest trees	<10 ind./plot	>10 ind./plot	>10 ind./plot
L(r) curves *outside confidence envelope	No calculation of *L(r)* due to low forest tree densities	*L(r)*>0	*L(r)*=0
Log(*Dmir*) ANOVA: F=3.9* Tukey HSD[1]	3.5 a	2.7 ab	2.1 b

[1]means with different letters are significantly different, P<0.05.

mirids, alternative hosts to cacao trees were present in the agroforests concerned [23] so the presence/absence of alternative mirid hosts (*Alter*) was also taken into account. Concerning black pod, in our study, the cacao tree was considered to be the only host.

Microclimatic conditions were described using variables related to spatial structure of the vegetation. The vertical and horizontal structure of the cacao trees and associated shade trees (forest trees and fruit trees) affect mean and variance of the light transmitted to the pest and disease present in the understory. The vertical spatial structure was characterized using four strata: the top stratum (at a distance of approximately from 15 to 30 m from the ground); the intermediate stratum (approximately 7 to 15 m from the ground), the low stratum with vegetation in the same stratum as cacao trees (approximately 3 to 7 m from the ground) and the lowest stratum with vegetation lower than cacao trees (approximately 2 to 3 m from the ground) (Figure 1). This classification is justified by field observations to distinguish the amount of shade received by the

cacao trees (the top and intermediate canopy strata provided shade for the cacao trees) and the self-shading produced by cacao trees and the vegetation in the same strata. Each plant was recorded as belonging to one of these strata. The total cover (*Dtot*) produced by the cacao trees (self-shading) and associated shade trees was estimated by the total tree density from the cacao to the top canopy strata (the lowest stratum was not included in this calculation). The shade tree cover (*Dsha*) was estimated by the sum of tree densities in the top and intermediate canopy levels. The proportion of shade tree cover provided by the intermediate stratum was estimated by the proportion of shade trees in the intermediate stratum compared with the total number of shade trees (%*Inter*).

The horizontal spatial structure, which is rarely taken into account in agro-ecological studies, was characterized using the *L*(r) function (Besag in Ripley [32]). Although this function is often used in forest research [14] it has rarely been used for agro-ecosystems. The *L*(r) function is based on the calculation of the

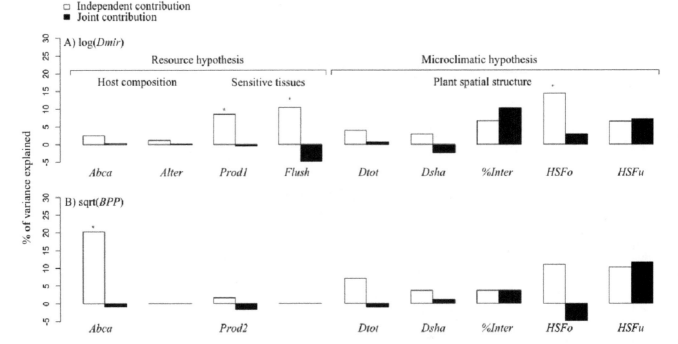

Figure 4. Independent and joint contribution (% of variance explained) of explanatory variables on mirid density and black pod prevalence. Results of the hierarchical partitioning analyses of A) mirid density (*Dmir*) and B) black pod prevalence (*BPP*). Significant independent contributions of explanatory variables are indicated by *(Z-score value>1.65, determined by randomization tests with 1,000 iterations). See Table 1 for the definition of the variables.

Table 3. Table of correlations.

Categories	Codes	Modalities	Dmir	BPP
Host composition	Abca		−0.02	−0.36
	Alter		0.00	-
		Presence	0.04	-
		Absence	−0.05	-
Sensitive tissues	Flush		0.25	-
	Prod1		0.34	-
	Prod2		-	−0.01
Spatial structure	Dtot		0.39	0.39
	Dsha		0.10	−0.02
	%Inter		0.33	0.20
	HSFo		0.13	0.21
		Low density	0.64	0.35
		Aggregated	0.03	−0.27
		Random	−0.43	0.02
	HSFu		0.11	0.17
		Low density	−0.44	−0.24
		Random	0.02	0.04
		Regular	0.48	0.24

Pearson coefficients and R^2 ANOVA values respectively, between continuous or categorical explanatory variables and residuals of models for mirid density (*Dmir*) or black pod prevalence (*BPP*) after controlling for the effects of the other hypotheses. The mean of residual values are indicated for modalities of categorical variables. See Table 1 for code and definition of variables.

expected number of neighbors within a distance below or equal to r of any point of the point pattern, i.e. in our study, of tree Cartesian positions in a plot. The $L(r)$ function makes it possible to distinguish three types of tree distribution patterns: regular ($L(r) < 0$), random ($L(r) \approx 0$) or aggregated ($L(r) > 0$). The statistical significance of the regular or aggregated pattern compared to a random pattern was determined by comparing the observed $L(r)$ value to a 95% confidence envelope based on simulated $L(r)$ values of a complete spatial randomness (CSR) of tree horizontal structure, simulated by 1,000 Monte Carlo simulations. We applied the spatial structure analysis separately to forest trees and fruit trees. Indeed, the $L(r)$ function is applied to a set of individuals subjected to the same process of spatial distribution [33]. We assumed that this was the case for the two types of shade trees in the stands: forest trees and fruit trees. To mitigate edge effects (taken into account in our analyses by the isotropic edge correction as proposed in Ripley [32]), the outcome of the $L(r)$ function should be interpreted for r values less than half or a quarter of the width of the plot [33]. Thus, $L(r)$ was calculated for r values less than 12 m, i.e. less than 1/4 of 50 m. For statistical reasons, the $L(r)$ function cannot be used on patterns at low stand densities [34]. Therefore, the $L(r)$ function was calculated only in plots with at least 10 forest trees or 10 fruit trees per plot. The 12 values of $L(r)$ calculated for forest trees (r from 1 m to 12 m) and the 12 values of $L(r)$ calculated for fruit trees were used in two separate Principal Component Analyses (PCA). With respect with the Kaiser rule, we retained the PCA axes that explained at least 80% of the inertia [35]. Subsequently, the observation coordinates on these axes were used in two different hierarchical cluster analyses based on the Euclidean distance and Ward's criterion [36]. This resulted in clusters of plots with a similar horizontal spatial structure (regular, random, or aggregated) for forest trees (*HSFo*) and for fruit trees (*HSFu*). Plots not included in the

typologies due to low densities of forest or fruit trees were attributed to a modality denoted "Low density" in the *HSFo* and *HSFu* categorical variables.

Pest and disease assessment

In each plot, 80 adult cacao trees were selected for observations of the presence of sensitive tissues and for pest and disease monitoring. The trees were selected so as to be homogeneously distributed in the plot (Figure 2).

Plots were not treated with insecticide by farmers for the two years of the study. Mirid density was evaluated in September 2011 and August 2012, i.e. during the usual period of mirid outbreaks in the region [21]. For sampling, the 80 cacao trees were sprayed with an insecticide composed of 20 g/L of lambda cyhalothrin and 20 g/L of imidaclopride using a motorized mist blower (Stihl 420, Stihl AG, Waiblingen, Germany) equipped with a formulation pump and a 0.8 restrictor, and fitted with the baffle plate provided by Stihl adjusted to 25 mL/ha. Treatments were carried out at 6:00 AM, when mirids are not yet very active to limit the risk of winged adults escaping. Seven hours after the treatment, dead insects were collected on 4×4 m plastic sheets positioned under the cacao trees before spraying. The collected insects were preserved in glass hemolysis tubes containing 70% alcohol. The French Agricultural Research Centre for International Development (CIRAD) and the Agricultural research institute for the development (IRAD) have granted permission. In the laboratory, the individual mirids belonging to *Sahlbergella singularis* collected under the 80 selected cacao trees in each plot were counted. For each year, mirid density per plot (*Dmir*) was estimated from the mean number of mirids per tree and the density of cacao trees per plot.

Data on disease damage were collected at four dates: T1: May 2012, T2: July–August 2012, T3: October 2012, T4: November

2012. At each date, healthy and diseased pods (infected by black pod) were counted on the 80 selected cacao trees. Only pods at least 10 cm in length were considered. For each plot, the mean of the total number of pods (healthy and diseased pods) and the mean of the number of diseased pods per tree were calculated for the 80 selected cacao trees and multiplied by the number of cacao trees present in the plot. Based on these values, for each plot, we plotted the curves of the total number of pods from T1 to T4 and the number of diseased pods over the same period (Figure 3). Next, black pod prevalence (BPP) was calculated using the formula: $BPP = \frac{DiseaseArea \times 100}{TotalArea}$, where TotalArea is the area under the total number of pods curve and DiseaseArea is the area under the number of diseased pods curve [18]. Fungicide treatments by farmers were homogeneous between plots with a recommended rate of one 50 g bag of Ridomil Gold plus 66 WP (Active ingredients: metalaxyl-M 6% and copper(1)oxide 60%) applied with a side lever knapsack sprayer containing 15 L water, four times between T1 to T4. Recorded differences between farmers' practices did not have a significant effect on black pod prevalence (data not shown) and were therefore not taken into account in our analysis.

Assessment of available sensitive tissue

We also describe pest and disease resources using variables that account for the amount of sensitive tissue available. Since black pod only attacks cacao tree pods and mirids preferentially attack the young shoots and pods, the attractiveness of a plot varied as a function of the amount of these sensitive tissues it contained. The amount of tissues sensitive to mirids was evaluated by the proportion of cacao trees with young shoots (called flushes) and the number of pods. For each plot, the flush presence variable (Flush) was the mean proportion of the 80 selected cacao trees with flushes during the study period, i.e. from T0 = September 2011 to T4 = November 2012. The number of pods available to mirids was estimated as the total number of pods, healthy and diseased, at the plot scale at T0 and T3 (the two periods when mirid outbreaks occur). The variable for pod production was denoted Prod1 and is expressed as the number of pods.

The amount of tissues sensitive to *P. megakarya* was evaluated by the number of pods. For each plot, the number of pods was assessed using the value of the area under the curve of the total number of pods from T1 to T4 (Figure 3). The number of pods available for black pod was denoted Prod2 and is expressed as fruit x day.

In all, 9 explanatory variables related to the host composition (Abca, Alter), shade tree spatial structure (Dtot, Dsha, %Inter, HSFo, HSFu), sensitive tissue amount (Flush, Prod1 or Prod2) were calculated to explain mirid density (Dmir) or black pod prevalence (BPP) at the plot scale (see Table 1 for the list of variables).

Statistical models

Statistical models were developed to explain Ys variables, i.e. mirid density and black pod prevalence, by explanatory variables Xs for host composition, sensitive tissue availability and tree spatial structure. The approach we used comprised three main steps:

(1) Ys and Xs variables were log or square-root transformed to respect the normal distribution assumption (Table 1). Variation in mirid density between 2011 and 2012 was compared by ANOVA. No significant effect was found ($F = 2.10$, $P = 0.16$). Consequently, the data for 2011 and 2012 were pooled for subsequent analysis.

(2) Correlations between the Xs explanatory variables were tested using the Spearman rank correlation test for correlations between continuous variables and the Kruskal-Wallis test with the Dunn correction for pairwise comparisons for correlations between continuous and categorical variables. Some explanatory variables were not independent, with Spearman's correlation coefficient values ranging from -0.54 to 0.50. Therefore, a hierarchical partitioning protocol [37] was used to examine the independent effect of each explanatory variable on mirid and black pod variables. Two hierarchical partitioning protocols were used: (i) one to explain mirid density (Dmir) with the variables Abca, Alter, Prod1, Flush, Dtot, Dsha, %Inter, HSFo and HSFu and (ii) one to explain black pod prevalence (BPP) with the variables Abca, Prod2, Dtot, Dsha, %Inter, HSFo and HSFu. Hierarchical partitioning is a method that considers all possible models in a multiple regression setting to identify the most likely causal factors (explanatory approach). The independent effect (denoted I_i) of an explanatory variable X_i on Y is then obtained by averaging all the increases in fit generated by including the variable X_i in all the models in which X_i appears. R-squared, which expresses the Y variance explained by a model, was used as goodness-of-fit measure. The joint effect J_i of the X_i variable (effect caused jointly with the other Xs variables) is obtained by subtracting I_i from R_i, with R_i the R-squared of the model explaining Y with only X_i. For each X_i variable, the independent and joint contributions are expressed as a percentage of the total explained variance R_i, with $R_i = I_i + J_i$. The significance of independent effects is evaluated by Mac Nally's Z-scores using a randomization test with 1,000 iterations [38]. There is no measurement of the significance of joint effects. However, it is important to not only look at the independent "explanatory" power (I) of variables but also the ratio between I and J. Because, even if X_i has an apparently high independent effect on Y (great I_i), if X_i appeared to have complex inter-relationships with other explanatory variables (great J_i), then manipulating X_i alone may not have the desired outcomes [37].

(3) Hierarchical partitioning does not provide information on the direction of the relationship (positive or negative) between the Y variable and each explanatory variable. Therefore, to test the effect of microclimatic hypothesis variables on pest density or disease prevalence after controlling for the effect of resource hypothesis variables: (i) we built all the possible linear models that explain Y with the combination of variables of the resource hypothesis, i.e. the variables describing host composition and the sensitive tissue availability, (ii) the best model, with the smallest AIC (Akaike Information Criterion [39]) was chosen, (iii) we used the residuals of this model and we tested their relationship (positive or negative) with each variable of the microclimatic hypothesis, i.e. Dtot, Dsha, %Inter, HSFo and HSFu. To analyze the relationship, a Pearson correlation test (for continuous explanatory variables) or an ANOVA (for categorical explanatory variables) was performed. The same procedure was applied to study the effect of resource hypothesis variables after controlling for the effect of the microclimatic hypothesis [40]–[41].

Statistical analyses were performed with the packages "stats" for statistical hypothesis testing (e.g. ANOVA, Spearman rank correlation test, Kruskal-Wallis test), "ads" for the point pattern analyses (L(r) function), "FactoMineR" for the Principal Component Analyses and "hier.part" for the Hierarchical Partitioning protocol of the R software 2.15.0 [42]–[43].

Results

Vegetation composition and spatial structure

The relative abundance of cacao tree (*Abca*) ranged from 74% to 94% with a mean of 84% with respect to all plants 2 m-tall or over. In 11 plots, at least one of the alternative mirid hosts (*Alter*), i.e. *Ceiba pentandra*, *Cola acuminata*, *Cola lateritia* or *Sterculia rhinopetala,* was present (Table 1).

The total density of plants 2 m-tall or over (*Dtot*) ranged from 580 to 1,660 with a mean of 1,129 individuals per hectare (Table 1). Shade tree density (*Dsha*), i.e. only associated trees that were taller than the cacao trees, ranged from 40 to 168 with a mean of 97 shade trees per hectare (Table 1). The percentage of shade trees belonging to the intermediate stratum (*%Inter*) ranged from 18 to 92 with a mean of 51%. Concerning the horizontal structure of the forest trees (*HSFo*), five plots had a forest tree density <10 individuals per plot (40 trees per hectare, Table 2) and were classified as "low density". In the 15 other plots, forest tree density ranged from 48 to 156 trees per hectare. Seven of these plots displayed a significant aggregated structure of forest trees, with positive values of the *L(r)* function > the 95% confidence envelope for at least one *r* value (Table 2). In the eight other plots, forest trees displayed a structure that did not significantly differ from a random distribution. Concerning horizontal structure of the fruit trees (*HSFu*), eight plots had a fruit tree density of less than 10 individuals per plot (40 trees per hectare) and were classified as "low density". In the 12 other plots, fruit tree density ranged from 40 to 104 trees per hectare. Three plots had significant negative values of the *L(r)* function, and a trend towards a regular distribution was observed in four plots. In the five remaining plots, fruit trees displayed a structure that did not significantly differ from a random distribution.

Factors affecting mirid density

Mirid density ranged from 0 to 842 *Sahlbergella singularis* individuals per hectare with a mean of 117 mirids per hectare (Table 1). The variables sensitive tissue availability and shade tree spatial structure partially explained mirid density at the plot scale (Figure 4A).

The presence of flushes (*Flush*) and the number of pods (*Prod1*) had significant independent effects on mirid density (*Dmir*). They explained respectively 10.4% and 8.5% of mirid density variance independently of the other variables (Figure 4A). Flush presence and the number of pods were significantly positively correlated with mirid density (Table 3), after controlling for the effects of variables for tree spatial structure.

The horizontal structure of forest trees (*HSFo*) had a significant independent effect on mirid density (*Dmir*) and explained 14.5% of mirid density variance independently of the other variables (Figure 4A). Mean mirid density at the plot scale decreased when *HSFo* went from "Low density" to "Aggregated", and then from "Aggregated" to "Random" structure, after controlling for the effects of host composition and the availability of sensitive tissue (Table 2 and Table 3).

None of the host composition variables had a significant independent effect on mirid density (Figure 4A).

The joint effects of *Flush*, *Prod1* and *HSFo* with the other explanatory variables on mirid density were lower than their independent effects (Figure 4A).

Factors affecting black pod prevalence

Black pod prevalence (*BPP*) ranged from 0 to 5% infected pods of the total number of pods in 2012 at the plot scale with a mean of 1% (Table 1).

Among the host composition variables, cacao tree abundance (*Abca*) had a significant independent effect on black pod prevalence and explained 20.3% of the variance independently of the other variables (Figure 4B). Cacao tree abundance was negatively correlated with black pod prevalence (Table 3), after controlling for the effects of variables for tree spatial structure. The joint effect of *Abca* and the other explanatory variables on black pod prevalence was lower than its independent effect (Figure 4).

Neither sensitive tissue availability, i.e. the number of pods, nor the variables related to tree spatial structure had a significant independent effect on black pod prevalence (Figure 4B).

Discussion

The aim of this study was to assess the independent effects and the relative importance of different variables related to resource availability and microclimatic variation that explain pest and disease occurrence in real complex agro-ecosystems, at the plot scale.

Our study showed that some characteristics of vegetation structure related to the different study hypotheses influence pest and disease occurrence in the farmers' agroforests studied here.

Concerning mirids

Our hypothesis was that tree spatial structure influences mirid density through microclimatic variations. We assumed that the low density of shade trees or the microclimatic heterogeneity generated by the aggregation of shade trees could increase mirid density at the plot scale due to the higher density of mirids on cacao trees in direct sunlight than on cacao trees in the shade [21]. We also expected that when there were relatively more shade trees in the top stratum than in the intermediate stratum, mirid density would be reduced due to the decrease in mean transmitted light [21–12]. Concerning plant spatial structure variables, our results showed that only the variable horizontal structure of forest trees impacted mirid density independently of other variables. Mirid density decreased when a minimum number of randomly distributed forest trees were present compared with an aggregated distribution or when the density of forest trees was low. Babin et al. [21] found that mirid pockets, which are areas highly infested by mirids, were generally located in the sunniest zones of plots. In cacao agroforests, large forest trees, unlike fruit trees, tend to homogenize the distribution of the light resource and consequently limit the development of mirid pockets [21]. Our results confirm that the presence of forest shade trees is correlated with a decrease in mirid density. Several architectural characteristics distinguish forest trees from fruit trees. The most obvious difference is their height as only forest trees reached the top canopy level. However, this difference is probably not the only one involved, since otherwise the variable vertical structure (*%Inter*) would have been significant rather than the variable forest tree horizontal structure (*HSFo*). Thus, the difference in architecture of forest trees which generally have a larger and more porous canopy than fruit trees (a trend revealed by our data, data not shown) could thus also explain why forest trees are most likely to provide more uniform shading. Moreover, the reduction in mirid density is not only observed when forest trees are left standing in cacao agroforests, but this reduction appears to be more effective when forest trees are randomly distributed rather than aggregated. The interaction between the spatial organization of individuals and the distribution of the light resource has only rarely been studied in complex plant ecosystems. Using simulations, Martens et al. [12] showed that the variance in understory light is dependent on the spatial patterns of trees, with an increase in the variance of understory light when the horizontal structure of canopy trees progresses from random to aggregate. Thus, forest tree aggregation is likely to introduce heterogeneity in light distribution, with an alternation of shaded (under aggregates) and sunny (outside aggregates) environments, with sunny environments favoring the development of mirid pockets. Regular patterns of forest trees were not observed at our study site whereas they were observed for fruit trees. This is probably due to differences in the management of forest trees, which are left over from the previous forest ecosystem, compared with fruit trees, which are mainly planted by farmers [19]–[44]. In a previous study, we showed that compared with other horizontal structures or low density, regular forest tree structure reduced frosty pod rot (caused by *Moniliophthora roreri*) prevalence on cacao pods in an agroforest in the Talamanca region in Costa Rica [18]. The results we obtained concerning mirid density and the above result concerning frosty pod rot prevalence underline the potential of promoting forest tree horizontal structure for the agricultural management of cacao pests and diseases. It is interesting to note that in our study, the classical shade tree density or total tree density variables had no significant effect on pest density, whereas the variable tree horizontal structure, which is rarely taken into account, did. Our results thus open new perspectives for agro-ecological research on the regulation of pest and disease by shading through the tree spatial structure.

We also assumed that host composition and sensitive tissue availability influence mirid density through variations in resources. Agroforests are often described as complex agro-ecosystem with significant planned and spontaneous plant diversity [4]–[19]. We studied real agroforests and one of our observations was that cacao agroforests in the Lékié department varied little with respect to the abundance of cacao trees: cacao tree relative abundance ranged from 74% to 94%. This range may not be large enough to empirically test the resource dilution mechanism with the host abundance as proxy to describe the resource amount. Plus, the amount of available sensitive tissue is rarely taken into account in studies on the dilution mechanism. However, our study showed that the amount of sensitive tissue is probably a more valid proxy for describing the resources available for pests and diseases than host composition. Indeed, our results showed that, unlike cacao tree abundance, pod and flush production both had significant independent effects on mirid density. This effect is in agreement with the dilution mechanism since a decrease in the resource (here the amount of sensitive tissue) leads to a decrease in the pest density. In a previous study, we showed that the amount of sensitive tissue was more important than host composition in explaining variations in cacao tree disease [18]. Therefore, besides host abundance, it is important to include the availability of sensitive tissues to identify the effect of variations in resource availability on the pest and disease in real multispecies agro-ecosystems. This variable is particularly important since it makes it possible to determine whether microclimatic variation has a direct or indirect effect on the pest and disease, through variations in plant growth.

A major aim of this study was also to disentangle the effects of the mechanisms involved in the "microclimatic" and the "resource" hypotheses on mirid density in cacao agroforests. Our results showed that the variables forest tree spatial structure and the amount of sensitive tissue both had independent effects on mirid density, thus indicating that hypotheses concerning both the "microclimatic" and "resource" mechanisms may be involved in mirid infestation. Crucial life history traits of cocoa mirids could enlighten us: they are photophobic insects [23], yet at the plot level, they are often located in the sunniest environments [21]. To explain this contradiction, Babin et al. [21] suggested that higher resource availability related to flush intensity, which is greater in cacao trees exposed to sunlight, may attract and allow the development of larger mirid populations. From this assumption, we could conclude that, for cocoa mirids, "microclimatic" mechanisms may act indirectly through "resource" mechanisms. But our results also showed that the effects of the structure of shade trees and the amount of available resources were independent. Moreover, there was no significant difference in the proportion of cacao trees with flushes and in the number of pods between plots with different forest tree horizontal structure (unpublished data). Our study thus did not reveal a relationship between "resource" and "microclimatic" mechanisms. Consequently it is difficult to validate or invalidate the hypothesis that mirids are present in the sunniest areas because of the larger amount of sensitive tissues. A further study that considers the different variables at the cacao tree scale, (and not only at a plot scale as was the case in our study), would help answer this question. However, in terms of tradeoff between pest regulation and pod production services, our results are particularly interesting because we showed that mirid density could be reduced by optimizing the design of tree associations without affecting average pod production at the plot scale.

Concerning black pod

The low percentages of black pod prevalence (*BPP*) in our study plots, which did not reach more than 5% of the production, partly due to homogenized treatment with fungicides, preclude any definitive conclusions. Indeed, without treatment, yield losses as high as 50% are commonly observed and when conditions are favorable, yield losses can reach 80% [45]–[46]. We also presume that the year our observations were conducted was a year with relatively low losses and also that the four point observations of prevalence in the year may also underestimate real losses accumulated over entire year. Nevertheless, despite the fungicide treatment in the plots and the low prevalence rates observed, the host composition effect is probably a particularly strong effect. For that reason, we discuss this effect below, as we believe that it may be even more pronounced in cacao plots without fungicide treatments.

One of our hypotheses was that host composition and sensitive tissue availability influences black pod prevalence through resource variation. In our study plots, black pod prevalence was only explained by cacao tree abundance, independently of other variables, and black pod prevalence increased with a decrease in cacao tree abundance. Based on the "resource dilution" mechanism the opposite result would be expected, in other words a decrease in host tree abundance should have led to a decrease in disease prevalence [47]. Other hypotheses could be proposed to explain this result. Firstly, the relation that exists between the age of cacao agroforests and vegetation structure linked to the farmer's choice of management: On the one hand, old cacao trees that die are not systematically replaced with other cacao trees, or if they are replaced, the fact the cacao tree seedlings were small meant they were not taken into account in our mapping (which only included plants over 2 m in height). Moreover, cacao tree seedlings are sometimes planted together with Musaceae to provide shade during seedling growth [48] and also to provide an income for farmers from the sale of Musaceae fruits. This type of management can lead to a decrease in cacao tree density but also in cacao relative abundance linked to the age of the plantation. On the other hand, an increase in the age of the cacao plot may also lead to an increase in the amount of primary inoculum of *P. megakarya* in the plot, which has accumulated over the years. The negative relationship between cacao tree abundance and black pod prevalence could thus be partly explained by these two effects of the age of the plantation. Secondly, we could also expect that the increase in cacao tree abundance, which was negatively correlated with plant species richness in our plots (unpublished data), would lead to a decrease in fauna associated with non-cacao trees, like ants, at the plot scale. For example, ants of the genus *Crematogaster* are very common in cacao agroforests in our study region, and move in large numbers on cacao trees, where they tend honeydew hemiptera and prey on various arthropods [49]. These ants nest in dead wood or in hollow branches and trunks, so that they benefit from the presence of big forest trees [50]. These ants, like other insects, are assumed to play a determining role in the propagation of black pod by transporting soil particles that contain *Phytophthora* spores [51]–[52]. The potential decrease in the population of ants nesting in big shade trees, due to the increase in the relative abundance of cacao trees, could thus explain our results.

Another of our hypotheses was that the spatial structure of shade trees influences black pod prevalence by causing microclimatic variations. Concerning black pod prevalence, we assumed that the low density of self shading ($Dtot$) and of shade trees ($Dsha$) could reduce black pod prevalence at the plot scale due to aeration of the plot and a concomitant decrease in relative humidity [22]. We also expected that a decrease in the percentage of shade trees in the intermediate stratum (rather than in the top stratum) would decrease black pod prevalence due to more even rain interception and better plot aeration in the cacao tree stratum [22]. However, our results showed that, in our study, none of the variables related to plant spatial structure had an independent effect on disease prevalence. Mfegue [53] found no relationship between shade classes and black pod incidence either. In a previous study in Costa Rican cacao agroforests, we showed that only a regular spatial structure of forest trees reduced frosty pod rot prevalence (caused by *Moniliophthora roreri*) on cacao pods in comparison with plots with a low density of forest trees [18]. Like *M. roreri*, *P. megakarya* is primarily a cacao pod pathogen and the germination of these two pathogens is favored by high humidity, and hence by the presence of shade trees [52]. However, they have two different main modes of dispersal, *M. roreri* by wind and *P. megakarya* by rain. Consequently, shade reduces the spore dispersal of *M. roreri* while its impact on the spore dispersal of *P. megakarya* is not yet known. To control the development of frosty pod rot, it is generally advised to provide moderate and uniform shade at the plot scale, while to control black pod development, it is generally advised to reduce shade. In the Costa Rican study, in the case of *M. roreri*, we showed that although lower disease prevalence was not automatically achieved by conserving forest trees, this reduction was effective only if forest trees were regularly distributed throughout the plot. In the plots in the study area in Cameroon, the lack of impact of the variable tree spatial structure on black pod prevalence is certainly related to the generally high density of shade trees, more dense than the shade tree pattern that could reduce the development of the disease (Table S1).

This study also aimed to disentangle the effects of the mechanisms involved in the "microclimatic" and the "resource" hypotheses on the prevalence of black pod disease in cacao agroforests. One possible explanation for the surprising negative relationship between cacao tree abundance and black pod prevalence could be an increase in the relative abundance of shade trees due to the decrease in the abundance of the cacao trees. This increase in the relative abundance of shade trees could create a shadier environment, which supposedly increases black pod development. However, this explanation can be excluded, since if this were the case, one of the tree spatial structure variables, such as shade tree density or total tree density, would be

significant in the hierarchical partitioning analysis. Due to the very low joint effect of the variable cacao tree abundance in the hierarchical partitioning analysis, to find an explanation for this result, we need to look for a variable that was not taken into account in our analysis and is correlated with cacao tree abundance, as proposed in the previous paragraph with regard to the age of the plot and the effect of inoculum accumulation.

Further studies, including studies involving classical variables known to influence the disease (e.g. inoculum potential) and plant structure variables, are needed to elucidate the natural mechanisms involved in the regulation of black pod. However, our results already show that optimizing host composition could be an effective way to decrease black pod prevalence.

Joint agro-ecological management implications for mirids and black pod disease and future outlook

One of the recommendations for the control of mirids is to maintain a uniform shade level in cacao tree plantations, while one of the recommendations for the control of black pod is to avoid excessive shade [21]–[22]. "Thus, to be effective, shade management strategies have to find a balance between shade conditions unfavourable for both mirids and black pod [21]". The present study provides new information on how tree structure characteristics affect the development of the pest and the disease. This is important, since knowledge of such characteristics will help improve pest and disease management strategies. Forest trees are expected to be better shade trees to reduce the occurrence of both mirids and black pod [21]–[54]. In our study, we confirmed that forest trees are better than fruit trees because they appear to provide better shade conditions for mirid control. What is more, our results suggest that reduced pest density is not automatically achieved by preserving forest trees. Indeed, we observed that aggregated forest trees are less efficient than a more randomised tree pattern probably because the latter homogenizes the light under the canopy. Our results also showed that variation in the spatial structure of forest trees did not significantly affect black pod prevalence at the plot scale. Consequently, in our study area, it appears that an agro-ecological approach to mirid and black pod management could be the preservation of natural resources, in this case native forest trees, but under farmer management. However, the preservation of forest trees in cacao agroforests could also favor the presence of a certain fauna, for example ants, that may be implicated in the dissemination of black pod. Thus a more detailed understanding is needed of the causes of the relationship between black pod prevalence and cacao tree abundance, along with studies that involve classical variables known to influence the disease and plant structure variables, are important.

Based on previous observations, we might expect similar results with respect to the impact of the spatial structure of canopy trees on biodiversity in general [55] and on other pests and diseases in particular. Indeed, we also assume that the importance of the spatial structure of shade trees for mirid or frosty pod rot [18] regulation also applies in coffee plantations, where leaf rust caused by *Hemileia vastatrix* is less severe under moderate shade [8]. However, the horizontal spatial structure of shade trees is a structural characteristic that has not received adequate attention to date, and the description of tree patterns and their impact on ecosystem functioning, particularly on pest and disease regulation services, need to be studied in more detail.

Many studies on the relationships between plant diversity and pest or disease occurrence have focused on the resource dilution mechanism. However, the main resource for the pests and diseases is often also the main financial resource for farmers, as is the case of the cacao tree and its fruits here. This study, which takes into account microclimatic and resource mechanisms, showed that the effect of the spatial structure of forest trees is independent of the amount of sensitive tissue (leaves and pods) and therefore provides interesting information on the tradeoff between pest regulation and pod production services.

Acknowledgments

The authors thank the cacao farmers of Nkolobang, Zima and Mbakomo for welcoming us in their plots. We thank Nga Marcus and IRAD-Cameroon technicians Mbarga Joseph Bienvenu, Owona Benoit and Petchayo Tigang Sandrine for their help with data collection. We also thank Daphne Goodfellow for reviewing the English. Finally, we acknowledge the two reviewers for their detailed and useful comments which enabled us to significantly improve the quality of this manuscript.

Author Contributions

Conceived and designed the experiments: CG MANB RB LBB GMTH CC. Performed the experiments: CG MANB RB LBB GMTH. Analyzed the data: CG. Contributed reagents/materials/analysis tools: CG. Contributed to the writing of the manuscript: CG MANB RB LBB GMTH CC.

References

1. Tilman D, Cassman KG, Matson PA, Naylor R, Polasky S (2002) Agricultural sustainability and intensive production practices. Nature 418: 671–677.
2. Chevassus-au-Louis B, Griffon M (2008) La nouvelle modernité: une agriculture productive à haute valeur écologique. In: Déméter 2008: économie et stratégie agricole. Paris: Club Déméter. 7–48.
3. Sabatier R, Wiegand K, Meyer K (2013) Production and Robustness of a Cacao agroecosystem: Effects of Two Contrasting Types of Management Strategies. PLoS ONE 8(12): e80352.
4. Malézieux E, Crozat Y, Dupraz C, Laurans M, Makowski D, et al. (2009) Mixing plant species in cropping systems: concepts, tools and models. A review. Agron Sustain Dev 29: 43–62.
5. Ratnadass A, Fernandes P, Avelino J, Habib R (2012) Plant species diversity for sustainable management of crop pests and diseases in agroecosystems: a review. Agro Sustain Dev 32: 273–303.
6. Avelino J, ten Hoopen GM, DeClerck F (2011) Ecological mechanisms for pest and disease control in coffee and cacao agroecosystems of the Neotropics. In: Rapidel B, DeClerck F, Le Coq J-F, Beer J, editors. Ecosystem Services from Agriculture and Agroforestry Measurement and Payment. London: Earthscan. 91–117.
7. Burdon JJ (1987) Diseases and plant population biology. Cambridge: Cambridge University Press. 222 p.
8. Schroth G, Krauss U, Gasparotto L, Aguilar JAD, Vohland K (2000) Pests and diseases in agroforestry systems of the humid tropics. Agroforest Syst 50: 199–241.
9. Mitchell CE, Tilman D, Groth JV (2002) Effects of grassland plant species diversity, abundance, and composition on foliar fungal disease. Ecology 83: 1713–1726.
10. Beer J, Muschler R, Kass D, Somarriba E (1997) Shade management in coffee and cacao plantations. Agroforest Syst 38: 139–164.
11. Somarriba E, Domínguez L, Harvey C (2005) ¿Cómo evaluar y mejorar el dosel de sombra en cacaotales? Agroforestería en las Américas 41: 120–128.
12. Martens SN, Breshears DD, Meyer CW (2000) Spatial distributions of understory light along the grassland/forest continuum: effects of cover, height, and spatial pattern of tree canopies. Ecol Modell 126: 79–93.
13. Calonnec A, Burie JB, Langlais M, Guyader S, Saint-Jean S, et al. (2013) Impacts of plant growth and architecture on pathogen processes and their consequences for epidemic behaviour. Eur J Plant Pathol 135: 479–497.
14. Ngo Bieng MA, Perot T, Coligny F, Goreaud F (2013) Spatial pattern of trees influences species productivity in a mature oak–pine mixed forest. Eur J For Res 132: 841–850.
15. Burdon JJ, Chilvers GA (1982) Host density as a factor in plant-disease ecology. Annual Review of Phytopathology 20: 143–166.
16. Anderson RM, May RM (1981) The population dynamics of microparasites and their invertebrate hosts. Philosophical Transactions of the Royal Society of London. Series B, Biological Sciences. 451–524.
17. Garrett KA, Mundt CC (1999) Epidemiology in mixed host populations. Phytopathology 89: 984–990.
18. Gidoin C, Avelino J, Deheuvels O, Cilas C, Ngo Bieng MA (2014) Shade tree spatial structure and pod production explain frosty pod rot intensity in cacao agroforests, Costa Rica. Phytopathology 104: 275–281.
19. Schroth G, Harvey CA, Vincent G (2004) Complex agroforests: Their structure, diversity, and potential role in landscape conservation. In: Schroth G, Da Fonseca G, Harvey C, Gascon C, Vasconcelos H, Izac A-M editors. Agroforestry and Biodiversity Conservation in Tropical Landscapes. Washington: Island Press. 227–260.
20. FAO (2013) Production. Part of FAOSTAT – FAO Database for food and agriculture. Rome: Food and Agriculture Organization of the United Nations (FAO). Available: http://faostat3.fao.org/faostat-gateway/go/to/home/E. Accessed 2013 Sept 1.
21. Babin R, ten Hoopen GM, Cilas C, Enjalric F, Yede, et al. (2010) Impact of shade on the spatial distribution of *Sahlbergella singularis* in traditional cocoa agroforests. Agric For Entomol 12: 69–79.
22. Cilas C, Despréaux D (2004) Improvement of Cocoa Tree Resistance to *Phytophthora* Diseases. Montpellier: CIRAD. 177 p.
23. Entwistle PF (1972) Pests of cocoa. London: Longman Group Limited. 779 p.
24. Holmes KA, Evans HC, Wayne S, Smith J (2003) *Irvingia*, a forest host of the cocoa black-pod pathogen, *Phytophthora megakarya*, in Cameroon. Plant Pathol 52: 486–490.
25. Opoku IY, Akrofi AY, Appiah AA (2002) Shade trees are alternative hosts of the cocoa pathogen *Phytophthora megakarya*. Crop Prot 21: 629–634.
26. Ten Hoopen GM, Sounigo O, Babin R, Yede, Dikwe G, et al. (2010) Spatial and temporal analysis of a Phytophthora megakarya epidemic in a plantation in the centre region of Cameroon. Proceedings of the 16th International Cocoa Research Conference. 683–687.
27. Leston D (1973) The Flight Behaviour of cocoa-capsids (Hemiptera: *Miridae*). Entomologia Experimentalis et Applicata 16: 91–100.
28. Mpé JM, Vos J, Neuenschwander P (2001) Integrated management of cocoa mirids in Cameroon. In: Vos J, Neuenschwander P, editors. West Africa Regional Cocoa IPM Workshop, Cotonou, Benin, 13–15 November, 2001. Proceedings CPL Press. 39–43.
29. Evans HC (1998) Disease and sustainability in the cocoa agroecosystem. In: Proceedings of the First International Workshop on Sustainable Cocoa Growing. Panama City, 30 March to 2 April 1998. p 12.
30. Jagoret P, Michel-Dounias I, Malézieux E (2011) Long-term dynamics of cocoa agroforests: a case study in central Cameroon. Agroforestry systems 81: 267–278.
31. Sonwa DJ, Nkongmeneck BA, Weise S, Tchatat M, Adesina A, et al. (2007) Diversity of plants in cocoa agroforests in the humid forest zone of Southern Cameroon. Biodivers Conserv 16: 2385–2400.
32. Ripley BD (1977) Modeling Spatial Patterns. J R Stat Soc Series B Stat Methodol 39: 172–212.
33. Goreaud F (2000) Apports de l'analyse de la structure spatiale en forêt tempérée à l'étude et la modélisation des peuplements complexes. Thèse de doctorat en Sciences forestières. ENGREF. 362 p.
34. Condit R (2000) Spatial Patterns in the Distribution of Tropical Tree Species. Science 288: 1414–1418.
35. Kaiser HF (1991) Coefficient alpha for a principal component and the Kaiser-Guttman rule. Psychol Rep 68: 855–858.
36. Lebart L (1994) Complementary use of correspondence analysis and cluster analysis. In: Greenacre M, Blasius, editors. Correspondence analysis in the social sciences. Academic Press. 162–178.
37. Mac Nally R (2000) Regression and model-building in conservation biology, biogeography and ecology: The distinction between and reconciliation of « predictive » and « explanatory » models. Biodivers Conserv 9: 655–671.
38. Mac Nally R (2002) Multiple regression and inference in ecology and conservation biology: further comments on identifying important predictor variables. Biodivers Conserv 11: 1397–1401.
39. Akaike H (1974) A new look at the statistical model identification. IEEE Trans Automat Contr 19: 716–723.
40. Gaba S, Chauvel B, Dessaint F, Bretagnolle V, Petit S (2010) Weed species richness in winter wheat increases with landscape heterogeneity. Agric Ecosyst Environ 138: 318–323.
41. Leprieur F, Beauchard O, Blanchet S, Oberdorff T, Brosse S (2008) Fish invasions in the world's river systems: when natural processes are blurred by human activities. PLoS Biol 6: e28.
42. R Development Core Team (2013) R: A Language and Environment for Statistical Computing. Vienna, Austria: R Foundation for Statistical Computing.
43. Walsh C, Mac Nally R (2009) Package "hier.part". Documentation for R (R. project for statistical computing. Available: http://cran.r-project.org/web/packages/hier.part/hier.part.pdf/.

44. Ngo Bieng MA, Gidoin C, Avelino J, Cilas C, Deheuvels O, et al. (2013) Diversity and spatial clustering of shade trees affect cacao yield and pathogen pressure in Costa Rican agroforests. Basic Appl Ecol 14: 329–336.

45. Berry D, Cilas C (1994) Etude génétique de la réaction à la pourriture brune des cabosses chez des cacaoyers (*Theobroma cacao* L.) issus d'un plan de croisement diallèle. Agronomie 14: 599–609.

46. Despréaux D, Cambrony D, Clément D, Partiot M (1988) Etude de la pourriture brune des cabosses du cacaoyer au Cameroun: définition de nouvelles méthodes de lutte. In: Proc. 10th Int. Cocoa Res. Conf., Santo Domingo, Dominican Republic. Cocoa Producers' Alliance (COPAL), UK. 407–412.

47. Keesing F, Holt RD, Ostfeld RS (2006) Effects of species diversity on disease risk. Ecol Lett 9: 485–498.

48. Jagoret P (2011) Analyse et évaluation de systèmes agroforestiers complexes sur le long terme: Application aux systèmes de culture à base de cacaoyer au Centre Cameroun. Thèse de doctorat. Montpellier Supagro. 288 p.

49. Tadu Z, Djiéto-Lordon C, Yede, Messop Youbi E, Aléné CD, et al. (2014) Ant mosaics in cocoa agroforestry systems of Southern Cameroon: influence of shade on the occurrence and spatial distribution of dominant ants. Agroforest Syst, doi:10.1007/s10457-014-9676-7.

50. Richard FJ, Fabre A, Dejean A (2001) Predatory behavior in dominant arboreal ant species: the case of *Crematogaster* sp. (Hymenoptera: Formicidae). J Insect Behav 14: 271–282.

51. Konam JK, Guest DI (2004) Role of beetles (Coleoptera: Scolytidae and Nitidulidae) in the spread of Phytophthora palmivora pod rot of cocoa in Papua New Guinea. Australas Plant Pathol 33: 55–59.

52. Evans HC (1998) Disease and sustainability in the cocoa agroecosystem. In Proceedings of the First International Workshop on Sustainable Cocoa Growing, Panama City.

53. Mfegue CV (2012) Origine et mécanismes de dispersion des populations de *Phytophthora megakarya*, pathogène du cacaoyer au Cameroun. Thèse de doctorat Montpellier SupAgro. 186 p.

54. Akrofi AY, Appiah AA, Opoku IY (2003) Management of *Phytophthora* pod rot disease on cocoa farms in Ghana. Crop Prot 22: 469–477.

55. Bartels SF, Chen HYH (2010) Is understory plant species diversity driven by resource quantity or resource heterogeneity? Ecology 91: 1931–1938.

Permissions

List of Contributors

Xuejiao Bai
College of Forestry, Shenyang Agricultural University, Shenyang, Liaoning Province, China
State Key Laboratory of Forest and Soil Ecology, Institute of Applied Ecology, Chinese Academy of Sciences, Shenyang, Liaoning Province, China

Ji Ye, Xugao Wang, Zuoqiang Yuan and Zhanqing Hao
State Key Laboratory of Forest and Soil Ecology, Institute of Applied Ecology, Chinese Academy of Sciences, Shenyang, Liaoning Province, China

Tania Brenes-Arguedas
Centro Cientifico Tropical, San Jose, Costa Rica

Shuai Shi, Dingliang Xing and Fei Lin
University of Chinese Academy of Sciences, Beijing, China
State Key Laboratory of Forest and Soil Ecology, Institute of Applied Ecology, Chinese Academy of Sciences, Shenyang, Liaoning Province, China

Theresa M. Nogeire and Jennifer M. Duggan
School of Environmental and Forest Sciences, University of Washington, Seattle, Washington United States of America

Frank W. Davis
Bren School of Environmental Science and Management, University of California Santa Barbara, Santa Barbara, California, United States of America

Kevin R. Crooks
Department of Fish, Wildlife, and Conservation Biology, Colorado State University, Fort Collins, Colorado, United States of America

Erin E. Boydston
U.S. Geological Survey, Western Ecological Research Center, Thousand Oaks, California, United States of America

Ruth Ann Atchley and Paul Atchley
Department of Psychology, University of Kansas, Lawrence, Kansas, United States of America

David L. Strayer
Department of Psychology, University of Utah, Salt Lake City, Utah, United States of America

Yaoxin Guo and Gaihe Yang
College of Agronomy, Northwest A&F University, Yangling, Shaanxi, China

Gang Li
College of Life Science, Northwest A&F University, Yangling, Shaanxi, China

Youning Hu, Di Kang and Dexiang Wang
College of Forestry, Northwest A&F University, Yangling, Shaanxi, China

Junwei Luan and Shirong Liu
The Research Institute of Forest Ecology, Environment and Protection, Chinese Academy of Forestry, Key Laboratory of Forest Ecology and Environment, China's State Forestry Administration, Beijing, PR China

Jingxin Wang
West Virginia University, Division of Forestry and Natural Resources, Morgantown, West Virginia, United States of America

Xueling Zhu
Baotianman Natural Reserve Administration, Tuandong, Chengguan Town, Neixiang County, Henan Province, PR China

Aurora Saucedo-García
Posgrado en Ciencias Biológicas, Instituto de Ecología, Universidad Nacional Autónoma de México, Distrito Federal, México

Departamento de Ecología Funcional, Instituto de Ecología, Universidad Nacional Autónoma de México, Distrito Federal, México

Ana Luisa Anaya
Departamento de Ecología Funcional, Instituto de Ecología, Universidad Nacional Auto´noma de México, Distrito Federal, México

Francisco J. Espinosa-García
Laboratorio de Ecología Química, Centro de Investigaciones en Ecosistemas, Universidad Nacional Auto´noma de México, Morelia, Michoacán, México

María C. González
Departamento de Botánica, Instituto de Biología, Universidad Nacional Autónoma de México, Distrito Federal, México

Jacqueline T. Davis and Kerrie Mengersen
School of Mathematical Sciences, Queensland University of Technology, Brisbane, Australia

Nicola K. Abram
Durrell Institute for Conservation and Ecology, School of Anthropology and Conservation, University of Kent, Canterbury, Kent, United Kingdom
HUTAN –Kinabatangan Orang-utan Conservation Programme, Sandakan, Sabah, Malaysia
Borneo Futures Project, People and Nature Consulting International, Jakarta, Indonesia

Marc Ancrenaz
HUTAN –Kinabatangan Orang-utan Conservation Programme, Sandakan, Sabah, Malaysia
Borneo Futures Project, People and Nature Consulting International, Jakarta, Indonesia

Jessie A. Wells
ARC Centre of Excellence for Environmental Decisions, University of Queensland, Brisbane, Australia

Erik Meijaard
Borneo Futures Project, People and Nature Consulting International, Jakarta, Indonesia
ARC Centre of Excellence for Environmental Decisions, University of Queensland, Brisbane, Australia
School of Archaeology & Anthropology, Australian National University, Canberra, Australia

Dongsheng Zhao, Shaohong Wu and Yunhe Yin
Institute of Geographical Sciences and Natural Resources Research, Chinese Academy of Sciences, Anwai, Beijing, China

Marion Schrumpf and Ernst-Detlef Schulze
Max Planck Institute for Biogeochemistry, Hans-Knöll-Straße 10, 07745 Jena, Germany

Klaus Kaiser
Soil Sciences, Martin Luther University Halle-Wittenberg, 06120 Halle (Saale), Germany

Hervé Bertin Daghela Bisseleua
MDG Centre West and Central Africa, Dakar, Senegal
Laboratory of Entomology, IRAD, Yaoundé, Cameroon
Georg-August-University Goettingen, Department of Crop Science, Entomological Section, Goettingen, Germany
Centre for Environmental Research and Conservation, The Earth Institute of Columbia University, New York, New York, United States of America

Daniel Fotio, Yede and Alain Didier Missoup
Laboratory of Entomology, IRAD, Yaoundé, Cameroon

Stefan Vidal
Georg-August-University Goettingen, Department of Crop Science, Entomological Section, Goettingen, Germany

Tobias Plieninger
Department of Geosciences and Natural Resource Management, University of Copenhagen, Frederiksberg, Denmark

Cang Hui
Centre for Invasion Biology, Department of Mathematical Sciences, Stellenbosch University, Matieland, South Africa
Mathematical and Physical Biosciences, African Institute for Mathematical Sciences, Cape Town, South Africa

Mirijam Gaertner
Centre for Invasion Biology, Department of Mathematical Sciences, Stellenbosch University, Matieland, South Africa

Lynn Huntsinger
Department of Environmental Science, Policy, and Management, University of California, Berkeley, California, United States of America

Valerie E. Peters
Odum School of Ecology, University of Georgia, Athens, Georgia, United States of America

Chunyu Zhang, Yanbo Wei and Xiuhai Zhao
Key Laboratory for Forest Resources and Ecosystem Processes of Beijing, Beijing Forestry University, Beijing, China

Klaus von Gadow
Faculty of Forestry and Forest Ecology, Georg-August-University Göttingen, Göttingen, Germany

Gail Campbell-Smith
Durrell Institute of Conservation and Ecology, University of Kent, Canterbury, Kent, United Kingdom

Miran Campbell-Smith
Orangutan Information Centre, Human-Orangutan Conflict and
Mitigation Programme, Medan, North Sumatra, Indonesia

Ian Singleton
Sumatran Orangutan Conservation Programme, PanEco Foundation, Medan, North Sumatra, Indonesia

Matthew Linkie
Durrell Institute of Conservation and Ecology, University of Kent, Canterbury, Kent, United Kingdom

Fauna & Flora International, Cambridge, United Kingdom

Sonia M. Hernandez, Valerie E. Peters and C. Ron Carroll
Odum School of Ecology, University of Georgia, Athens, Georgia, United States of America

Brady J. Mattsson and Robert J. Cooper
Daniel B. Warnell School of Forestry and Natural Resources, University of Georgia, Athens, Georgia, United States of America

Michael Kessler
Systematic Botany, University of Zurich, Zurich, Switzerland

Dietrich Hertel and Christoph Leuschner
Plant Ecology, University of Göttingen, Göttingen, Germany

Hermann F. Jungkunst
Geoecology/Physical Geography, Institute for Environmental Sciences, University of Koblenz-Landau, Landau, Germany

Jürgen Kluge
Systematic Botany, University of Zurich, Zurich, Switzerland
Faculty of Geography, University of Marburg, Marburg, Germany

Stefan Abrahamczyk
Systematic Botany, University of Zurich, Zurich, Switzerland
Systematic Botany and Mycology, Department of Biology, University of Munich, Munich, Germany

Merijn Bos
Agroecology, University of Göttingen, Göttingen, Germany
Louis Bolk Institute, LA Driebergen, The Netherlands

Damayanti Buchori
Department of Plant Protection, Faculty of Agriculture, IPB, Bogor Agricultural University, Kampus Darmaga, Bogor, Indonesia

Gerhard Gerold
Landscape Ecology, Institute of Geography, University of Göttingen, Göttingen, Germany

S. Robbert Gradstein
Muséum National d'Histoire Naturelle, Departement Systematique et Evolution (UMS 602), C.P. 39, Paris, France

Shahabuddin Saleh
Department of Agrotechnology, Faculty of Agriculture, University of Tadulako, Palu, Indonesia

Stefan Köhler
Landscape Ecology and Land Evaluation, Faculty for Agricultural and Environmental Sciences, University of Rostock, Rostock, Germany

Gerald Moser
Plant Ecology, University of Göttingen, Göttingen, Germany
Department of Plant Ecology, University of Giessen, Giessen, Germany

Ramadhanil Pitopang
Department of Biology, Faculty of Mathematics and Natural Sciences, Tadulako University, Palu, Indonesia

Christian H. Schulze
Department of Animal Biodiversity, Faculty of Life Sciences, University of Vienna, Vienna, Austria

Simone G. Sporn
FEMS Central Office, CL Delft, The Netherlands

Ingolf Steffan-Dewenter
Department of Animal Ecology and Tropical Biology, Biocenter, University of Würzburg, Würzburg, Germany

Sri S. Tjitrosoedirdjo
Department of Biology, Faculty of Mathematics and Natural Sciences, IPB, Bogor Agricultural University, Kampus Darmaga, Bogor, Indonesia

Teja Tscharntke
Agroecology, University of Göttingen, Göttingen, Germany

Jeanne L. Nel, David C. Le Maitre, Lara van Niekerk, Laurie Barwell, Belinda Reyers, Greg G. Forsyth, Andre K. Theron and Patrick J. O'Farrell
Natural Resources and the Environment, Council for Scientific and Industrial Research (CSIR), Stellenbosch, Western Cape, South Africa

Deon C. Nel
World Wide Fund for Nature (WWF), Cape Town, Western Cape, South Africa

Jean-Marc Mwenge Kahinda, Francois A. Engelbrecht, Evison Kapangaziwiri
Natural Resources and the Environment, Council for Scientific and Industrial Research (CSIR), Pretoria, Gauteng, South Africa

Brian W. van Wilgen
Natural Resources and the Environment, Council for Scientific and Industrial Research (CSIR), Stellenbosch, Western Cape, South Africa
Centre for Invasion Biology, Department of Botany and Zoology, Stellenbosch University, Stellenbosch, Western Cape, South Africa

Sally Archibald
School of Animal, Plant and Environmental Sciences, University of the Witwatersrand, Johannesburg, Gauteng, South Africa
Natural Resources and the Environment, Council for Scientific and Industrial Research (CSIR), Pretoria, Gauteng, South Africa

Ran Meng and Philip E. Dennison
Department of Geography, University of Utah, Salt Lake City, Utah, United States of America

Carla M. D'Antonio
Department of Ecology, Evolution and Marine Biology, University of California Santa Barbara, Santa Barbara, California, United States of America

Max A. Moritz
Department of Environmental Science, Policy, and Management, University of California, Berkeley, California, United States of America

Edith Bai, Weiwei Dai, Ping Jiang and Shijie Han
State Key Laboratory of Forest and Soil Ecology, Institute of Applied Ecology, Chinese Academy of Sciences, Shenyang, China

Wei Li, Shanlong Li, Jianfei Sun and Bo Peng
State Key Laboratory of Forest and Soil Ecology, Institute of Applied Ecology, Chinese Academy of Sciences, Shenyang, China
College of Resources and Environment, University of Chinese Academy of Sciences, Beijing, China

Tiefeng Piao
Institute of Soil and Water Conservation, Northwest A&F University, Yangling, China
Division of Forest Ecology, Department of Forest Conservation, Korea Forest Research Institute, Seoul, Republic of Korea

Jung Hwa Chun, Hee Moon Yang and Kwangil Cheon
Division of Forest Ecology, Department of Forest Conservation, Korea Forest Research Institute, Seoul, Republic of Korea

João M. B. Carreiras
Tropical Research Institute (IICT), Lisboa, Portugal

National Centre for Earth Observation (NCEO), Centre for Terrestrial Carbon Dynamics (CTCD), University of Sheffield, Sheffield, South Yorkshire, United Kingdom

Joshua Jones
Department of Geography and Earth Sciences, Aberystwyth University, Aberystwyth, Ceredigion, United Kingdom

Richard M. Lucas
Department of Geography and Earth Sciences, Aberystwyth University, Aberystwyth, Ceredigion, United Kingdom
Centre for Ecosystem Science, School of Biological, Earth and Environmental Sciences, University of New South Wales, Sydney, New South Wales, Australia

Cristina Gabriel
Tropical Research Institute (IICT), Lisboa, Portugal

Alberto Bernués
Department of Animal and Aquacultural Sciences, Norwegian University or Life Sciences, Ås, Norway
Departamento de Tecnología en Producción Animal, Centro de Investigación y Tecnología Agroalimentaria de Aragón, Zaragoza, Spain

Tamara Rodríguez-Ortega
Departamento de Tecnología en Producción Animal, Centro de Investigación y Tecnologı´a Agroalimentaria de Aragón, Zaragoza, Spain

Raimon Ripoll-Bosch
Departamento de Tecnología en Producción Animal, Centro de Investigación y Tecnología Agroalimentaria de Aragón, Zaragoza, Spain
Animal Production Systems Group, Wageningen University, Wageningen, The Netherlands

Frode Alfnes
School of Economics and Business, Norwegian University or Life Sciences, Ås, Norway

Gregory J. Pec, Gary C. Carlton
Department of Biological Sciences, California State Polytechnic University, Pomona, California, United States of America

Cynthia Gidoin
Supagro, UMR System, Montpellier, France
CIRAD, UPR Bioagresseurs, Montpellier, France

Régis Babin
CIRAD, UPR Bioagresseurs, Montpellier, France

icipe, Plant Health Division, Coffee Pest Project, Nairobi, Kenya

Leïla Bagny Beilhe and Gerben Martijn ten Hoopen
CIRAD, UPR Bioagresseurs, Montpellier, France

IRAD, Yaoundé, Cameroon

Christian Cilas
CIRAD, UPR Bioagresseurs, Montpellier, France

Marie Ange Ngo Bieng
CIRAD, UMR System, Montpellier, France

Index

Printed in the USA
CPSIA information can be obtained
at www.ICGtesting.com
JSHW051441221024
72173JS00006B/1546

9 781632 397959